Climate Change Litigation Cases in Context

ELGAR CASES IN CONTEXT

Elgar's Cases in Context series provides a space for the analysis of notable cases in dynamic areas of law and practice. Volumes in the series feature commentary on key and illustrative cases in their respective fields highlighting examples of both foundational principles and departures from precedent, raising important questions for further research. The series will foreground law in action and invite critical reflection on the impact of case law not only in the respective field of study, but also in a wider legal, regulatory, and social context. The series will be of interest to legal scholars, higher level students and in some instances, those working in legal practice.

For a full list of Edward Elgar published titles, including the titles in this series, visit our website at www.e-elgar.com.

Climate Change Litigation Cases in Context

Edited by

Camille Cameron KC

Dalhousie University, Canada

Patricia Galvão Ferreira

Dalhousie University, Canada

Riley Weyman

Pape Salter Teillet LLP, Canada

ELGAR CASES IN CONTEXT

Edward Elgar PUBLISHING

Cheltenham, UK • Northampton, MA, USA

Published by

Edward Elgar Publishing Limited
The Lypiatts
15 Lansdown Road
Cheltenham
Glos GL50 2JA
UK

Edward Elgar Publishing, Inc.
William Pratt House
9 Dewey Court
Northampton
Massachusetts 01060
USA

Authorised representative in the EU for GPSR queries only: Easy Access System Europe – Mustamäe tee 50, 10621 Tallinn, Estonia, gpsr.requests@easproject.com

A catalogue record for this book is available from the British Library

Library of Congress Control Number: 2026931792

This book is available electronically in the **Elgar**online
Law subject collection
https://doi.org/10.4337/9781035343188

MIX
Paper | Supporting responsible forestry
FSC® C013604

ISBN 978 1 0353 4317 1 (cased)
ISBN 978 1 0353 4318 8 (eBook)
ISBN 978 1 0353 9701 3 (ePub)

Printed and bound by CPI Group (UK) Ltd, Croydon, CR0 4YY

Contents

List of contributors

Funmilola (Funmi) Abioye (LLD) is a legal academic and researcher trained in Nigeria and South Africa. She is an Assistant Professor of Law at the College of Law, University of Saskatchewan and a member of the International Law Association (ILA) committee on international law and migration. Her research focuses on the rule of law at play, particularly in developing countries. Her research interests stem from this and include the legal aspects of international migration, good governance and legitimacy of legal systems, and environmental governance and climate litigation.

Isabela Soares Bicalho is a master's student in Public International Law at the University of São Paulo, under the supervision of Professor Alberto do Amaral Junior. She is a law graduate from the Federal University of Uberlândia (UFU), with academic mobility at the Lusophone University of Humanities and Technologies (ULHT). She specialises in environmental law and climate law.

Camille Cameron KC is a Full Professor in the Schulich School of Law, Dalhousie University. She was a civil litigation lawyer in Canada for ten years before becoming an academic. She has held teaching positions in Hong Kong, Australia, and Canada and has been a consultant for international development organisations on various judicial and legal reform projects. Her research interests include the administration of and access to civil justice, class actions, climate change litigation, and third-party litigation funding.

Matheus Henrique Cano holds a bachelor's degree in Law from the University of São Paulo (USP) and is currently pursuing a bachelor's degree in Philosophy. His work focuses on the intersections between legal frameworks, environmental governance, and ethical approaches to sustainability. He is a lawyer specialised in environmental law, climate change, and sustainability.

Nathalie Chalifour is a Full Professor at the University of Ottawa's Centre for Environmental Law and Global Sustainability and a leading expert on the rapidly evolving legal framework for climate change in Canada. She has published widely on jurisdictional authority to enact climate laws and youth-led Charter litigation, including *Climate Change and the Canadian Constitution – Federalism and Charter Rights in a Warming World* (2025, University of Toronto Press). Professor Chalifour has represented interveners in climate-related litigation, and her work has been cited by the Supreme Court of Canada and the Federal Court of Appeal. She is frequently featured in the media and is a member of the Royal Society of Canada's College of New Scholars. She has received awards for *Excellence in Teaching*, *Excellence in Research*, and *Public Engagement* from the Law Faculty. She earned her JSM and PhD in Law from Stanford University, where she was both a Fulbright Scholar and a Stanford Fellow.

Olaf Dilling is an attorney and partner at re|Rechtsanwälte, working primarily in the areas of administrative law, environmental law, and public infrastructure law. He has a doctorate in Product and Waste Law from the University of Bremen. Before becoming a practitioner in law, he worked as a researcher at various institutions such as Helmholtz Centre for Environmental Research and the German Advisory Council for the Environment (SRU). He published several books and articles, particularly on transnational law, comparative environmental law, and common resources.

Erin Dobbelsteyn is a PhD candidate in environmental law at the University of Ottawa. She holds a BSc in Neuroscience from McGill University, a JD from Dalhousie's Schulich School of Law, and an LLM in Global Sustainability and Environmental Law from the University of Ottawa. Erin is engaged in research and community initiatives related to international environmental law, environmental rights, climate justice, and public and planetary health. Most recently, she represented Friends of the Earth Canada as an intervener in *Mathur v Ontario*, the first human rights-based climate change lawsuit in Canada to have a hearing on the merits. Erin is the Deputy Director of the North American region of the Global Network for Human Rights and the Environment, the vice-chair of the Board of East Coast Environmental Law, and a part-time Assistant Professor at the Schulich School of Law.

Patricia Galvão Ferreira is an Associate Professor with the Marine and Environmental Law Institute at Schulich School of Law, Dalhousie University, and a leading scholar on transnational climate justice. Her work critically examines how international law and national law can address the disproportionate burdens and costs of climate change on vulnerable nations and communities, with a focus on principles like Common but Differentiated Responsibilities (CBDR), climate finance, and climate litigation. Her work is driven by a passion for bridging legal theory with on-the-ground justice. Professor Galvão Ferreira's research agenda also explores critical intersections of ocean and climate, food systems and climate, and renewable energy law. She is a co-editor of Brill Nijhoff Publishers' *Ocean Yearbook* and currently serves as the chairperson of the Brazil advisory board for the Center for Justice and International Law (CEJIL).

Helene Krasnoff is a senior attorney at the Environmental Law Institute (ELI) and the director of its Climate Judiciary Project, a judicial education initiative providing evidence-based information about climate science and how it is arising in the law. She has litigated and supervised dozens of high-profile cases in all levels of both the federal and state courts on matters that involved complex issues of constitutional, statutory, procedural, and administrative law.

Parul Kumar is a doctoral candidate at the Faculty of Law and Criminology in KU Leuven, Belgium. She holds a bachelor's degree in law (BA, LLB (Hons.)) from the National Law School of India University, Bangalore (India) and a master's degree in law (LLM) from the University of Hamburg, Ghent University, and the Warsaw School of Economics. She has professional experience as a dispute resolution lawyer as well as a public policy expert in think tanks focusing on climate policy in India and Germany.

Gabriel Mantelli holds a PhD in Philosophy and Legal Theory from the University of São Paulo (USP), where he also earned his bachelor's degree in Law, and a master's degree in Law and Development from the Getulio Vargas Foundation (FGV). He is a Brazilian legal researcher in climate justice and critical international law, and his work focuses on climate colonialism and the intersection of human rights and environmental law.

Rebekkah Markey-Towler is a PhD candidate at the Melbourne Law School and a Research Fellow with Melbourne Climate Futures at the University of Melbourne. Prior to working at the University of Melbourne, she was the Executive Associate to a judge on the Federal Court of Australia. Her research is focused on regulating issues at the intersection of corporate, financial, and climate change law, as well as climate change litigation.

Jarryd Page is an attorney at the Environmental Law Institute (ELI). Page's work focuses on climate litigation, including on fundamental rights and the intersection between science and law. He regularly presents on the scope and trends of climate litigation and educates judges about the science of climate change and how climate is arising in the law through ELI's Climate Judiciary Project.

Benjamin Ralston (LLM) is an Assistant Professor at the University of Saskatchewan College of Law and Associate Faculty with the University of Saskatchewan School of Environment and Sustainability. He has practised law since 2011 and taught it since 2015. Professor Ralston's practice areas and research interests overlap and encompass environmental law, administrative law, human rights, and Indigenous rights, with particular interest in the intersections between, and interdisciplinary approaches to researching, these topics.

Johannes Reich is Full Professor of Public Law, Environmental Law, and Energy Law at the Faculty of Law of the University of Zurich. His research centres on international and domestic Climate Change Law, Swiss and Comparative Constitutional Law, and the foundations of Administrative Law. He read law at University of Zurich (lic. iur.) and Yale Law School (LLM) and was awarded a doctorate in law by the University of Basel (Dr iur.; PhD equivalent).

Vernon Rive is an Associate Professor at Auckland Law School. Prior to commencing a full-time academic career, Vernon practised environmental, planning, and public law in New Zealand. His teaching and research focus on public, environmental, and climate change law, with particular interests in international and domestic climate litigation, climate mobility, and fossil fuel subsidy reform.

Romina Luz Schaller is a postdoctoral researcher in the Department of Environmental and Planning Law at the Helmholtz Centre for Environmental Research-UFZ (Germany). She also coordinates the Network of Competence on Future Challenges of Environmental Law (KomUR). Her work focuses on international and European environmental law, particularly in climate change mitigation (e.g. forest protection through REDD+) as well as on ecosystem services and sustainable development. Her current research explores the legal aspects for the regulation of carbon dioxide removal technologies (CDR) and the emerging challenges linked to their governance.

Marie-Louise Siemons holds a law degree from the Pontifical Catholic University of São Paulo (PUC-SP). She was a PIBIC-CNPq research fellow, during which she conducted a study on mining in Indigenous lands in Brazil. She coordinated the Young Ambassadors Programme Brazil (YAP-BR) and was a member of the Paulo Nogueira Neto Environmental Law Clinic at the

University of São Paulo. Her current research focuses on socio-environmental law, climate litigation, Indigenous rights, and human rights.

Riley Weyman is a senior associate with Pape Salter Teillet LLP in Toronto and specialises in Indigenous rights law, with an emphasis on issues at the intersection of Indigenous rights, governance, and the environment. Riley received his JD and a Certificate in Environmental Law from the Schulich School of Law at Dalhousie University. Riley has co-authored several journal articles and a book chapter on climate change litigation. His other research interests include Indigenous laws and legal issues; class-actions; and regulation and assessment of energy, mining, and infrastructure projects.

Mingzhe Zhu is a senior lecturer in Just Transition at the University of Glasgow. He is a legal theorist and historian who explores the interaction between state violence, capitalism, and socio-environmental disasters in the past, present, and future.

Foreword

Climate change is a perplexing challenge – not only because of its scale and impact, but because it unsettles the boundaries of every discipline and our concepts of law, governance, and justice.

Amitav Ghosh argues that the failure to address climate change is, at least partly, a crisis of imagination. Literature, he suggests, has struggled to make climate change thinkable and felt. I suspect that the same is true of law. If literature has struggled to imagine climate change, law has struggled to situate it within its normative frameworks. For years, legal writing on climate litigation was cartography – "objective" maps of facts, frameworks, and judicial pronouncements – while leaving the terrain of meaning and context unexplored. These accounts strip cases of their depth, reducing them to Euclidean geometry, when, in truth, they are layered narratives of struggle, innovation, and institutional response.

Every climate case is an onion, with endless layers of significance. Beneath the surface lie lived realities – the motivations of plaintiffs, the dilemmas of judges, the collisions of institutions. Cases cannot be seen as sterile texts; they are stories of rights asserted, duties contested, and futures imagined. To abstract them of their contexts is to rob them of their true contributory value to jurisprudential, political, and social development. Scholarship bears the burden of unearthing these narratives so that they resonate with judges, lawyers, and students – not as isolated rulings, but as narratives that evoke empathy, inspiration, and meaningful inquiry.

The earliest cases were battles for recognition – forcing juridical orders to acknowledge climate change as a legal subject. That frontier has shifted. Today, courts grapple with questions of obligation, accountability, equity, access to justice, and competing interests. The recent advisory opinion of the International Court of Justice on state responsibilities is emblematic of this evolution and marks a new epoch. With the proliferation of climate litigation, law's promises are being tested against our planetary realities.

This book is a gateway. It does more than catalogue cases; it is an interrogation that seeks to situate key decisions within their sociological context, revealing the values and politics that enrich them. It offers comparative depth, drawing connections across legal traditions. It is a step towards equipping judges, practitioners, and scholars to appreciate not only the "what" of a case, but the "why" – and to grapple with deeper questions of law, justice, and governance.

As someone who approached the courts in Pakistan seeking the recognition of climate obligations, I have seen firsthand how litigation can transform discourse. Following other cases, I have also witnessed how these lawsuits expose broader tensions – between politics and law, between immediate relief and systemic change. Plaintiffs often seek urgent remedies; lawyers craft justiciable frameworks; judges wrestle with balancing meaningful relief against fears of judicial overreach and institutional backlash.

The cases in this volume are more than precedents; they are the yardsticks in conversations about justice in the age of climate crisis. Climate litigation is not a panacea. It can inspire symbolic victories, yet risks practical regression if judgements fail to translate into enforceable actions. May this book inspire readers to engage with the layered realities that shape our collective future – and to imagine creative pathways toward progress.

Asghar Leghari
Partner
PrimeLegalCo.
Lahore, Pakistan
Climate Fellow – Jackson School of Global Affairs (2024)

Acknowledgements

We have had the privilege of working with a talented group of contributing authors on this project. We thank them for their expertise, their willingness to participate, their excellent contributions, and their patience as we pestered them with inquiries during the editorial process.

We have been assisted in many ways by Claire Nicholson, Katherine Silins, and Emily Huang. Claire, JD '24 was with us when the project began. She helped to keep us on track and organised, and was instrumental in getting this project across the finish line. Katherine, JD '26 and Emily, JD '26 provided valuable editorial assistance throughout the course of the project.

A close analysis of a collection of climate litigation cases like those considered in this volume reveals many things, including the role of courts in modern climate governance. This and other themes are explored in this volume, but there is one we think worthy of mention and acknowledgement here. It is the vital role that dedicated individuals, environmental justice advocates, charities, clinics, NGOs, pro bono lawyers, and other supporters play in bringing cases like these to courts. The opportunity for courts to clarify the relevant laws as they relate to climate change, to offer possible solutions, and perhaps to point the way forward for future litigants would not be possible but for these invaluable contributions.

Camille Cameron
Patricia Galvão Ferreira
Riley Weyman

List of abbreviations

AAPCL	Atmospheric Pollution Prevention and Control Law (China)
ACCC	Australian Competition and Consumer Commission
ADC	Declaratory Action of Constitutionality (Brazil)
ADI	Direct Action of Unconstitutionality (Brazil)
ADO	Direct Action of Unconstitutionality by Omission (Brazil)
ADPF	Allegation of Violation of Fundamental Principle (Brazil)
ASIC	Australian Securities and Investments Commission
BGB	*Bundesblattgesetz* (German Civil Code)
BNDES	Brazilian Development Bank
CDR	carbon dioxide removal
CMA	Senate Environmental Committee (Brazil)
CNA	National Confederation of Agriculture (Brazil)
CNC	National Confederation of Commerce (Brazil)
CNI	National Confederation of Industry (Brazil)
CNS	National Confederation of Services (Brazil)
CNT	National Confederation of Transportation (Brazil)
CoE	Council of Europe
COPW	Centre for Oil Pollution Watch
CPL	Civil Procedure Law (China)
CSDDD	Corporate Sustainability Due Diligence Directive
CSRD	Corporate Sustainability Reporting Directive
CTCA	Cap and Trade Cancellation Act (Ontario, Canada)
DEQ	Department of Environmental Quality (Montana)
DETEC	Federal Department of the Environment, Transport, Energy and Communication (Switzerland)
DPSP	Directive Principle of State Policy (India)
ECHR	European Convention on Human Rights

ECtHR	European Court of Human Rights
EEDCL	ecological and environmental damage compensation litigation
EIA	environmental impact assessment
ELD	Environmental Liability Directive
EPL	Environmental Protection Law (China)
ERA	Environmental Rights Amendments (Montana)
ESG	Environmental, social, and governance
ETS	emissions trading scheme
EU	European Union
FEPA	Federal Environmental Protection Agency (Nigeria)
FON	Friends of Nature Institute
GHG	greenhouse gas
GW	gigawatts
ICCPR	International Covenant on Civil and Political Rights
ICESCR	International Covenant on Economic, Social and Culture Rights
INESC	Institute of Systems and Computer Engineering
IPCC	Intergovernmental Panel on Climate Change
kWh	kilowatt-hours
LNG	liquified natural gas
LRF	Fiscal Responsibility Law (Brazil)
LULUCF	Land Use, Land-Use Change and Forestry
MDEQ	Montana Department of Environmental Quality
MEIC	Montana Environmental information Center
MEPA	Montana Environmental Policy Act
MMP	mixed-member proportional
MNRE	Ministry of New and Renewable Energy (India)
MOEFCC	Ministry of Environment, Forest and Climate Change (India)
NAE	National Administration of Energy (China)
NDC	Nationally Determined Contribution
NDRC	National Development Reform Commission (China)
NGO	non-governmental organization
NNPC	Nigerian National Petroleum Corporation

OECD	Organization for Economic Co-operation and Development
PIL	public interest litigation
PNMA	Brazilian National Environmental Policy
PNMC	National Policy on Climate Change (Brazil)
PSB	Brazil Socialist Party
PSOL	Socialism and Freedom Party (Brazil)
REDE	Sustainability Network Party (Brazil)
REL	Renewable Energy Law (China)
RMA	Resource Management Act 1991 (New Zealand)
RWE	RWE AG, German multinational energy company
SAN	Senior Adocates of Nigeria
SASAC	State-owned Assets Supervision and Administration Commission (China)
SNMC	National Climate Change System (Brazil)
SOE	state-owned enterprise
SPC	Supreme People's Court of China
STF	Brazilian Supreme Court
tWh	terawatt-hours
UN	United Nations
UNDRIP	United Nations Declaration on the Rights of Indigenous Peoples
UNFCCC	United Nations Framework Convention on Climate Change
WEF	World Economic Forum
ZPO	*Zivilprozessordnung* (German Code of Civil Procedure)

1. The climate governance tightrope: novel strategies and the role of courts in climate change

Patricia Galvão Ferreira, Camille Cameron, and Riley Weyman

1. INTRODUCTION

1.1 The Aims of This Volume

The aim of the Edward Elgar *Cases in Context* series, to which this volume contributes, is to provide a space to analyze notable cases in thematic areas of law that depart from precedent and raise important questions for further investigation. In this volume, authors analyze eleven climate cases, some concluded and some in progress, from domestic courts in the Global South and Global North. The cases are those in which parties advocated for novel approaches and in which courts either took up such invitations or made groundbreaking decisions on their own initiative. They offer examples of decisions that have been variously described as 'novel', 'groundbreaking', 'landmark', 'pivotal', or 'a first'. They also invite questions and analysis about what the real impacts of these 'groundbreaking' decisions will be.

Each of the chapters in this volume analyzes the claims and defenses; the identities and roles of the parties and other relevant actors; access to justice challenges; procedural options and strategic choices; outcomes; and the national and transnational impacts of these decisions. By prioritizing jurisdictional diversity and close contextual analysis, this volume contributes to the broader scholarly conversation by examining how Global South and North courts are navigating their roles in climate governance in landmark cases *in practice*. This includes how courts are addressing procedural, legal, and political challenges across a wide range of settings and with a diversity of claims and actors.

Courts have emerged as key sites of climate governance contestation and accountability. Recent data identifies 2,967 cases and notes that while the number of cases is increasing, the rate of growth is slowing.[1] Most of these cases originate in the Global North, with about 260 (or 9%), coming from the Global South.[2] Estimates of the number of climate lawsuits, underway or completed, will vary depending on how 'climate change litigation' is defined.[3] This volume adopts a deliberately broad definition, recognizing that choices about whether, when, and how to pursue litigation that affects climate outcomes vary considerably across jurisdictions, especially between Global South and Global North countries. Rather than applying a pre-determined definition of climate change litigation, contributors to this volume were asked to select climate-related cases they deemed as representative and groundbreaking in the jurisdictions about which they were writing. As Bouwer and others note in their discussion of climate litigation in Africa, the meaning of climate justice and the strategies for pursuing it differ considerably across jurisdictions.[4] Adopting this broad approach 'creates the space to read litigation in which climate issues are not front and centre or the main priority, as nonetheless worth studying'.[5]

The eleven cases considered in this volume reflect the reality that across jurisdictions, a diverse array of litigants are turning to courts to assert, defend, or challenge climate-related duties. Their motivations include compelling government action, securing protection from climate harms, and enforcing existing rights. The claimants in these eleven cases include youth, Indigenous groups, seniors, and climate justice NGOs or not-for-profit organizations. Eight of the defendants are governments, and three are corporate actors. Even where NGOs and not-for-profit organizations are not named claimants, they have been instrumental in other ways in most of the eleven cases, whether by organizing the litigation, acting as legal counsel, or offering financial or other

1 Joana Setzer and Catherine Higham, 'Global Trends in Climate Change Litigation: 2025 Snapshot' (Grantham Research Institute LSE 2025) 3–4. Note that while the overall rate of growth is slowing, there is strong growth in climate litigation in the Global South, 'particularly in the higher-emitting emerging economies of Brazil, South Africa and India' (ibid 4).

2 ibid 4.

3 Kim Bouwer and others, 'Africa, Climate Justice and the Role of the Courts' in Kim Bouwer and others (eds), *Climate Litigation and Justice in Africa* (Bristol University Press 2024) 6–7; Kate McKenzie and others, 'Climate Change Litigation: One Definition to Rule Them All …?' in Francesco Sindico and others (eds), *Research Handbook on Climate Change Litigation* (Edward Elgar 2024) 2–11; Setzer and Higham (n 1).

4 Bouwer and others (n 3) 6–7; McKenzie and others (n 3).

5 Bouwer and others (n 3) 7.

supports. Intervenors and *Amicus Curiae* (Friends of the Court) also figure in many of the eleven cases, including with acknowledgments of their contributions by judges in their decisions.

The importance of domestic courts in climate litigation has been reinforced by recent advisory opinions from the International Tribunal for the Law of the Sea (ITLOS),[6] the Inter-American Court of Human Rights,[7] and the International Court of Justice (ICJ).[8] These opinions affirm that States have legal obligations under international law to address climate change and are likely to generate a new wave of national and international climate litigation.[9]

In this overview chapter, we propose three questions as a useful framework for better understanding the role and limitations of judicial engagement in climate governance.

(1) Should courts review climate disputes?
This threshold question raises issues of justiciability, institutional legitimacy, and gatekeeping, including how courts can distinguish between legal disputes and political questions, and how they can manage standing and access without exceeding their institutional capacity.

(2) Can courts determine climate-related obligations?
Once courts determine that they can hear a climate case, can they identify and apply legal standards to assess the existence and scope of climate-related obligations? This raises questions about the adequacy of existing doctrines—whether constitutional, administrative, tort-based, or rights-based—to address climate harms. In some cases, courts rely on well-established legal principles. In others, they stretch or adapt existing

---

6 ITLOS Advisory Opinion, 'Request for an Advisory Opinion Submitted by the Commission of Small Island States on Climate Change and International Law' (21 May 2024) ITLOS Case No 31.

7 *Climate Emergency and Human Rights*, Advisory Opinion AO-32/25, Inter-American Court of Human Rights Series A No 32 (29 May 2025).

8 *Obligations of States in Respect of Climate Change* (Advisory Opinion) 2025 <https://www.icj-cij.org/sites/default/files/case-related/187/187-20250723-adv-01-00-en.pdf> accessed 29 August 2025.

9 See Lynsi Burton, 'Unpacking What the ICJ's Opinion Means for Climate and Environmental Action' (*Stockholm Environment Institute*, 28 July 2025) <www.sei.org/perspectives/icj-opinion-climate-environmental-action/> accessed 29 August 2025; Joe Udel and Floris Tan, 'New Standards in Government Framework Litigation: Legal Implications of the ICJ Advisory Opinion on Climate Change' (*Sabin Centre*, 5 August 2025) <https://blogs.law.columbia.edu/climatechange/2025/08/05/new-standards-in-government-framework-litigation-legal-implications-of-the-icj-advisory-opinion-on-climate-change/> accessed 29 August 2025.

frameworks to fit novel claims in response to evolving scientific and normative understandings. A challenge is whether and how courts can recognize or expand legal obligations in the climate context without overstepping their institutional boundaries vis-à-vis the legislative and executive branches of government.

(3) Can courts offer meaningful redress?
Even if courts can determine legal obligations, can they offer meaningful redress in climate-related disputes? This question relates to the scope and limitations of judicial remedies in the context of the global and polycentric nature of climate change.

We use these three questions as a lens through which to consider the cases in this volume and to illuminate how courts are operating within the legal, institutional, and normative constraints of climate governance.

2. SHOULD COURTS HEAR CLIMATE DISPUTES?

Central to this question is how courts address justiciability and standing. While sometimes discussed together, these are distinct inquiries. As the Supreme Court of Nigeria clarified in *Centre for Oil Pollution Watch v Nigerian National Petroleum Corporation* (*COPW v NNPC*, Chapter 5), standing asks whether the claimant is the proper party to bring the case, whereas justiciability asks whether the issue itself is appropriate for judicial resolution. In that case, the main issue before the Supreme Court was whether the NGO Centre for Oil Pollution Watch had standing to challenge environmental harms caused by an oil spill. Traditionally, Nigerian courts have required a claimant to demonstrate the breach of a direct and personal legal interest to satisfy locus standi. This approach has often excluded NGOs and public interest claimants. In *COPW v NNPC*, the Court departed from this narrow standard, recognizing the pressing nature of environmental and climate issues and broadening the conception of 'sufficient interest' in this context. As Funmi Abioye notes in Chapter 5, this marked a significant jurisprudential shift in Nigerian environmental law.[10]

The extent to which standing has been a barrier to claimants in climate litigation varies across jurisdictions.[11] It was an issue in *Smith v Fonterra* in

10 Funmi Abioye, *Centre for Oil Pollution Watch (COPW) v Nigeria National Petroleum Corporation (NNPC)*: Locus Standi and the Emerging Trends in Climate Litigation in Nigeria, ch 5.

11 Randal S Abate, 'Standing' in Margaretha Wewerinke-Singh and Sarah Mead (eds), *The Cambridge Handbook on Climate Litigation* (CUP 2025).

the context of the public nuisance claim advanced by Mr. Smith.[12] As Vernon Rive explains in Chapter 4, the Court of Appeal had 'side-stepped the issue of whether a standing requirement for public nuisance claims remained appropriate in the contemporary New Zealand context'.[13] On further appeal to the Supreme Court, however, that Court rejected the traditional justifications for the standing requirement in public nuisance cases. A striking feature of that decision is that the Supreme Court identified 'the implications of ubiquitous harms such as pollution (including from GHGs)' as a reason for a reconsideration of the traditional rules of standing.[14] The Supreme Court's willingness to consider this departure is informed by its openness to the potential capacity of tort law to address climate change harms. Rive explains that this would be a major departure from previous decisions.[15]

The question of who may speak on behalf of the public or vulnerable groups is increasingly central. In *Miszdi Yikh v Canada*,[16] for example, and in *Smith v Fonterra*, claimants are seeking to advance claims on behalf of Indigenous communities, raising questions about collective rights and representational standing. By contrast, in *Verein KlimaSeniorinnen v Switzerland* (*KlimaSeniorinnen v Switzerland*), national courts found that the elderly female claimants failed to show individualized injury, while at the European Court of Human Rights, standing was not contested, suggesting a broader willingness to recognize associations and representative actions in transnational fora.[17] In jurisdictions where civil society organizations are already accepted as legitimate public interest litigants—such as India, the United States, Brazil, Canada, and China—standing has not posed a major hurdle. The evolving nature of climate harms, which are diffuse, long-term, and often impact collectives rather than individuals, will continue to press courts to adapt their standing doctrines to reflect contemporary environmental realities.

Justiciability is another threshold issue in climate litigation. Courts must assess whether the legal questions raised fall within their institutional competence or whether they are precluded by the separation of powers. In many

---

12 *Smith v Fonterra Cooperative Group Ltd* [2024] NZSC 5, [2024] 1 NZLR 134 (*Smith v Fonterra (Supreme Court)*).

13 Vernon Rive, 'Common Law at the Coal Face: *Smith v Fonterra and others'*, ch 4, p 90.

14 ibid; *Smith v Fonterra* (n 12) [151].

15 Rive (n 13), p 88.

16 *La Rose v Canada* 2023 FCA 241; *Misdzi Yikh v Canada* 2023 FCA 241.

17 David Estrin and Patricia Galvão Ferreira, 'Advancing Climate Justice: The New IBA Model Statute for Proceedings Challenging Government Failure to Act on Climate Change' in Proc WCEL 2nd World Environmental Law Congress <https://ssrn.com/abstract=3559045> accessed 29 August 2025.

cases, including some considered in this volume, governments have argued that climate-related claims implicate matters of political discretion, such as national energy policy, budgetary allocations, or international commitments, that should remain outside the courts' purview.[18] Courts are increasingly recognizing, however, that not all exercises of discretion are immune from judicial review. How the case is framed can facilitate judicial scrutiny. Where claimants assert that government action or inaction violates constitutional rights, statutory obligations, or established duties of care, courts have been more willing to intervene.[19] The issue is not whether the matter involves political dimensions; climate policy does. The issue is whether it also involves legal obligations that courts are both empowered and duty-bound to interpret and enforce.

The growing legal architecture of climate governance, including framework statutes, sectoral targets, constitutional environmental provisions, and international commitments, has created a body of norms that courts can interpret and enforce. The justiciability barrier has not disappeared, but it is evolving. Courts are recognizing that where governments have committed to specific legal obligations, judicial oversight is not only permissible but necessary to preserve the integrity of legal systems and prevent commitments from becoming legally meaningless.[20]

In countries where courts have a well-established role in adjudicating environmental disputes, for example, India and Brazil, climate-related claims are often considered justiciable with little controversy. In jurisdictions where courts have historically shown greater deference in environmental matters, however, particularly where complex balancing of economic, environmental, and social priorities is involved, climate litigation is forcing a re-examination of the judicial role. Courts in these contexts are being asked to define the contours of their authority in climate-related claims.[21]

Mathur v Ontario is the first Canadian case in which claims cleared the justiciability hurdle at the preliminary motion to strike stage and at the hearing on the merits.[22] Notwithstanding that the claim was dismissed at the merits

---

18 See eg *Mathur v His Majesty the King in Right of Ontario*, 2023 ONSC 2316 (*Mathur v Ontario*), 2024 ONCA 762 (*Mathur v Ontario*) (ch 2), *Smith v Fonterra* (n 12) (ch 4), *Juliana v United States* 947 F.3d 1159 (9th Cir 2020).

19 See eg *Held v Montana* 2024 MT 312, 419 Mont 403 (2024), *Mathur v Ontario* (n 18).

20 See eg *Mathur v Ontario* (n 18); STF, ADPF 708 DF, Relator: Min Roberto Barroso, 04.07.2022 (Brazil) (Climate Fund Brazil); *Misdzi Yikh v Canada* (n 16); *La Rose v Canada* (n 16).

21 See eg discussion of *Mathur v Ontario* (ch 2).

22 ibid.

hearing (that dismissal was overturned on appeal),[23] the ruling on justiciability survived. That was a first and can fairly be described as groundbreaking. Similarly, in *Smith v Fonterra*, the conclusion of the Court of Appeal that climate change is a polycentric issue not amenable to judicial resolution was rejected on further appeal to the Supreme Court of New Zealand. As Rive states, this was 'a major departure from the approach taken in other courts on the issue of whether complex questions of polycentric climate policy are suitable for determination' by judges.[24] In *PSB v Brazil* (*Climate Fund*), the government's argument that courts should not interfere in polycentric matters best left to the legislative and executive branches was rejected. Not only did the Court reject it, but it also ruled that the Court had the responsibility to ensure that the Executive fulfilled its valid legislative obligation to disburse the earmarked climate funds for climate projects.[25]

International courts are also addressing the question of whether courts should hear climate disputes. In its 2025 Advisory Opinion, the ICJ addressed concerns that it might exceed its judicial function by opining on States' climate obligations. The Court dismissed this critique, recalling its own precedent:

> The Court states the existing law and does not legislate. This is so even if, in stating and applying the law, the Court necessarily has to specify its scope and sometimes note its general trend.[26]
>
> (Legality of the Threat or Use of Nuclear Weapons, Advisory Opinion, I.C.J. Reports 1996 (I), p. 237, para. 18.)

This reinforces the view that interpreting legal obligations, including in complex or politically charged areas like climate change, is a judicial function. Courts are not required to resolve every policy question, but they are obligated to state what the law requires.[27]

3. CAN COURTS DETERMINE LEGAL OBLIGATIONS?

Once courts have accepted that climate cases fall within their jurisdiction, the next challenge is to determine the legal obligations at stake. In doing so, courts

23 *Mathur v Ontario* (n 18).
24 Rive (n 13), ch 4, p 102.
25 Climate Fund (n 21), ch 7.
26 ICJ Advisory Opinion 2025 (n 8).
27 See Setzer and Higham (n 1) 16 and the authorities cited there for considerations of the impacts that this and other international cases might have on future domestic climate litigation, including court interpretations of states' obligations under the Paris Agreement.

are navigating varied legal landscapes, where existing legislative or constitutional mandates may provide clear standards or where courts are asked to innovate within legal frameworks that were not designed with the complexities of climate change in mind. These varied legal landscapes include general legal principles, such as reasonableness, causation, due diligence, and proportionality; rights-based claims, including human rights and constitutional environmental rights; plural legal systems, notably Indigenous legal traditions interacting with common or civil law; federalism; and evolving obligations under international climate law.

A number of the cases considered in this volume offer examples of courts evaluating claims grounded in existing legal commitments. In Brazil, for example, as Gabriel Mantelli, Matheus Henrique Cano, Isabela Soares Bicalho, and Marie-Louise Siemons explain in Chapter 7, courts have interpreted climate obligations not as novel duties but as mandates already embedded in legislation. In this view, the judicial role is to ensure the validity and implementation of legislative objectives in the face of executive inaction.

Mingzhe Zhu identifies a similar pattern in China. The cases he discusses in Chapter 10, *FON* (*Friends of Nature Environmental Institute of Chaoyang District, Beijing (FON) v State Grid Gansu Electric Power Corporation*, and *FON v State Grid Ningxia Electric Power Corporation*) concern litigation around curtailment rates in energy grids in two provinces. Courts were asked in those cases to ensure that renewable energy laws and national climate policy goals were properly implemented by state-owned grid operators. In these cases, courts were not crafting new obligations but were enforcing compliance with existing regulatory mandates.

This approach is also visible in the *Australian Investment and Securities Commission (ASIC) v Mercer* where, as Rebekkah Markey-Towler discusses in Chapter 11, regulators and courts applied existing corporate and financial laws to find climate-related misrepresentations. In *ASIC v Mercer*, a superannuation fund was challenged for misleading disclosures regarding climate risk management. This case is another example of how courts are being asked to interpret long-standing duties, such as those related to truthfulness, due diligence, and fiduciary responsibility, in a climate context and are adapting traditional doctrines to new risk environments.

Zhu and Markey-Towler note, however, that these cases also offer examples of the limitations of existing legal principles in some settings in addressing the complexities of climate change. Markey-Towler explains how traditional corporate and securities law principles are inadequate to address current climate change challenges and complexities raised in 'greenwashing' cases.[28] Zhu

---

[28] Markey-Towler, ch 11.

notes that climate governance in China lacks public participation, transparency, and accountability and presents challenges that extend far beyond the scope of any single lawsuit.[29]

In other jurisdictions, courts are still struggling to articulate enforceable obligations. In *Mathur v Ontario*, analyzed by Erin Dobbelsteyn and Nathalie Chalifour in Chapter 2, the Court was willing to hear the case but hesitant to find that the provincial government's weakening of its climate target violated constitutional rights. While the plaintiffs argued that Ontario's actions threatened their rights to life, security, and equality under the *Canadian Charter of Rights and Freedoms* (Charter), the Ontario Superior Court ultimately found that the plaintiffs did not meet the existing legal tests to substantiate their claims. This decision has been successfully appealed and will return to the Ontario Superior Court, where the Applicants will be seeking remedies for the Ontario government's violation of their *Charter* rights. This case highlights the difficulty of translating diffuse and long-term climate harms into justiciable legal standards, especially in jurisdictions like Canada where environmental rights are not explicitly constitutionally entrenched.

By contrast, courts in jurisdictions with an explicit constitutional right to a healthy environment have been more willing to interpret these rights as encompassing the right to a stable climate, thereby recognizing corresponding obligations. In *Held v Montana*, discussed by Jarryd Page and Helene Krasnoff in Chapter 8, Montana's constitutional right to a 'clean and healthful environment' served as the basis for a successful youth-led challenge. The Court found that a statutory restriction preventing state agencies from considering climate change in environmental reviews violated this constitutional guarantee. The Court did not need to craft a new legal standard; it interpreted existing constitutional text in the light of the best available scientific knowledge.

Elsewhere, courts are being asked to fuse various legal frameworks to determine the scope of climate-related obligations. In *Misdzi Yikh v Canada*, discussed in Chapter 9, Benjamin Ralston explores how the plaintiffs, members of the Wet'suwet'en Nation, framed their claims in relation to Charter rights but grounded their arguments in Wet'suwet'en law and relationships to land. The plaintiffs argued that climate change and its impacts can impair their freedom to choose where they move to and live on their territories, and that this harm is exacerbated by the 'kinship-based and place-dependent' nature of their culture.[30] In so pleading, they have placed familiar constitutional arguments in the broader context of Wet'suwet'en law and society. This case has survived

---

29 Minghe Zhu, ch 10 [31].
30 Ralston, ch 9 [at fn 108].

a motion to strike by the government and is scheduled to proceed to an eight-week trial beginning on October 26, 2026.[31]

Similarly, in *Smith v Fonterra*, the New Zealand Supreme Court has allowed a tort claim against major GHG emitters to proceed, noting that tikanga Māori must inform the development of tort law.[32] This case is striking not only for the fact that the pleading of a new climate change tort has been accepted by the Supreme Court and will now be an issue at the merits hearing, but also because it offers the possibility of a new climate change tort that focuses less on 'individualized relational proximity' and more on 'ecocentric conceptions of harm and proximity'.[33] The case stands out for its openness to doctrinal evolution in tort and its recognition that climate harms may require a redefinition of harm and proximity that departs from individualized, anthropocentric frames.

As noted earlier in the context of *ASIC v Mercer* and the two cases brought in China by the FON,[34] not all court decisions in climate cases that find legal obligations based on the application of existing legal principles offer progressive or expansive interpretations. As Parul Kumar notes in her analysis of *MK Ranjitsinh v Union of India* in Chapter 12, the Indian Supreme Court weighed the protection of the endangered Great Indian Bustard against the development of renewable energy infrastructure promoted as part of India's climate commitments. While the Court recognized a constitutional right to be free from the adverse effects of climate change, and while this decision can fairly be described as groundbreaking, the Court ultimately declined to grant the relief sought by conservationists, prioritizing energy transition goals over biodiversity protection. This case illustrates what Kumar describes as an anthropocentric bias in the Court's reasoning and raises broader concerns about whether courts are always reliable allies in climate justice. It serves as a reminder that judicial engagement does not guarantee progressive outcomes and that courts

---

31 See 'La Rose v. Her Majesty the Queen' (*Climate Change Litigation Database*) <https://climatecasechart.com/non-us-case/la-rose-v-her-majesty-the-queen/#:~:text=In%20December%202023%2C%20the%20Federal,beginning%20on%20October%2026%2C%202026> accessed 29 August 2025.

32 Tikanga is defined as 'correct procedure, custom, habit, lore, method, manner, rule, way, code, meaning, plan, practice, convention, protocol – the customary system of values and practices that have developed over time and are deeply embedded in the social context'. See 'tikanga' (*Te Aka Māori Dictionary*) <https://maoridictionary.co.nz/search?keywords=tikanga#:~:text=1.%20(noun)%20correct%20procedure%2C%20custom%2C%20habit%2C%20lore%2C,convention%2C%20protocol%20%2D%20the%20customary%20system%20of> accessed 29 August 2025.

33 Rive (n 13), ch 4, p 103.

34 See pp. 8–9 of this chapter.

may also reinforce dominant priorities, such as economic growth or climate mitigation, at the expense of marginalized interests or ecological integrity.

Regional and international human rights frameworks have helped articulate obligations where domestic standards remain underdeveloped. In *KlimaSeniorinnen v Switzerland*, analyzed by Johannes Reich in Chapter 6, the European Court of Human Rights held that Switzerland violated the right to private and family life by failing to take sufficient climate mitigation measures. The Court's ruling set out a framework of positive obligations that will bind all Council of Europe member states and may reshape the way national courts interpret their own human rights obligations in the climate context. Similarly, as Funmi Abioye explains in Chapter 5, the Nigerian litigants in *COPW v NNPC* relied on Article 24 of the African Charter on Human and Peoples' Rights to assert environmental rights, and the Supreme Court's recognition of locus standi for an NGO reflects an openness to public interest climate litigation grounded in international law.

The cases in this volume show that courts are moving, whether cautiously, boldly, or controversially, to identify legal standards applicable to climate harms. Where climate obligations are already clearly stated in law, through constitutional rights, legislation, or regulatory frameworks, courts have been more willing to act. Where obligations are less explicit, or the legal doctrines are ill-suited to collective and long-term risks, courts have shown more reluctance. Even in these cases, however, litigants are forcing national courts to consider how existing legal traditions, including private law, Indigenous law, and international human rights, might be used to construct new standards fit for current climate challenges.

The recent ITLOS, IACtHR, and the ICJ cases underline the role of national courts in the climate governance landscape.[35] The ICJ in particular grounded its conclusions in customary international environmental principles, including the duty to prevent transboundary harm, the obligation to cooperate, and the obligations articulated in the UNFCCC and the Paris Agreement.[36] In doing so, the ICJ has offered a model of how national courts can define legal obligations while remaining within the bounds of their institutional role.

4. CAN COURTS OFFER MEANINGFUL REDRESS?

Even where courts determine that legal obligations exist and have been breached, the question of meaningful redress arises. The harms at stake are

---

35 ITLOS Advisory Opinion (n 6); IACtHR Advisory Opinion (n 7); ICJ Advisory Opinion (n 8).

36 ICJ Advisory Opinion (n 8).

grave and are typically the result of cumulative, diffuse, and polycentric causes. No single court order can remedy these harms fully or, arguably, significantly. While courts can clarify obligations, affirm rights, and compel compliance with existing legal frameworks, the actual mitigation of climate harm depends on coordinated regulatory, political, and economic action across jurisdictions. The remedies courts provide, especially in domestic cases, may be legally and symbolically powerful, but they cannot fully address the scale and complexity of the climate crisis. Courts' institutional legitimacy may be strained if they appear to usurp the role of legislatures or executives, but offering no redress or adopting a narrow conception of judicial remedies risks rendering courts irrelevant.

The question of appropriate remedies goes to the heart of the judiciary's institutional role in legal orders structured by the separation of powers. The cases in this collection show that domestic courts are treading cautiously, especially when asked to order structural remedies such as wide-reaching injunctions or expansive declarations that may be seen to intrude on legislative or executive authority.

The challenges of redressability vary significantly across jurisdictions, depending on the underlying legal framework and the judiciary's constitutional mandate. In systems where courts are explicitly empowered to review both negative and positive obligations of the State, such as under the constitution of Brazil or in the European human rights system, courts are accustomed to issuing injunctions that require legislative or policy action to uphold fundamental rights. In *PSB v Brazil*, the Supreme Court held that the Executive's failure to give effect to the national Climate Fund violated constitutional rights. It ordered the government to restructure and resume the Fund's operations, basing its decision on the constitutional right to an ecologically balanced environment.[37] Similarly, in *Klimaseniorinnen Schweiz v Switzerland*, the European Court of Human Rights recognized that States have positive obligations to adopt adequate regulatory frameworks to protect against the harms of climate change. In cases like these, the judiciary operates within a legal structure that affirms both the right and the duty of the State to act, enabling courts to craft remedies that go beyond declaratory relief, even if the remedies remain open-ended.

However, in jurisdictions where courts have more limited authority to enforce positive obligations, or where the separation of powers is interpreted more rigidly, judicial remedies tend to be more contested and more restrained. These variations underscore that redressability is not only a question of legal

37 Climate Fund Brazil (n 21).

causation or remedy design, but also of institutional competence and authority within each legal system.[38]

In cases grounded in constitutional or human rights frameworks, plaintiffs often ask for declaratory relief, alone or with other remedies. In such cases, they are trying to extend the scope of such remedies to the climate context. The cases include plaintiffs asking courts to affirm that a stable climate is part of the right to life (*Mathur v Ontario*, *Misdzi Yikh v Canada*), privacy and family life (*Klimaseniorinnen v Switzerland*), or the broader right to a healthy environment (*Held v Montana*, *PSB v Brazil*, *MK Ranjitsinh v India*). While such declarations may lack direct coercive effect, they can perform other useful functions by clarifying legal duties; anchoring public discourse in evidence, the best available science, and rights-based principles; and providing authoritative rebuttals to climate denialism and regulatory or enforcement delays.

The cases in this volume, among others, illustrate that due to the systemic and long-term nature of the climate crisis, many courts are considering or relying on forms of dialogic remedies or remedial supervision, particularly in strategic cases against governments.[39] Rather than issuing overly prescriptive orders that might overstep institutional boundaries, these courts are choosing to affirm legal obligations while leaving a degree of discretion to governments as to how they comply. The courts then supervise implementation to ensure that compliance is meaningful, or they delegate such supervision to maintain the separation of their judicial role from political branches. These approaches fall along a continuum of judicial engagement. In *Neubauer v Germany*,[40] the Federal Constitutional Court explicitly ordered the legislature to revise the national climate law by a fixed deadline. This was a strong form of judicial oversight anchored in constitutional rights.[41] Similarly, in another landmark climate case, *Leghari v Federation of Pakistan*,[42] the Lahore High Court created a Climate Change Commission with an ongoing mandate to oversee

38 See eg *Juliana v United States* (n 18); *Mathur v Ontario* (ch 2); *La Rose v Canada* (n 16); *Misdzi Yikh v Canada* (ch 9).

39 The debate on remedies for human rights violations and the role of courts is extensive and can inform the discussions on the role of courts in climate governance. See Rosalind Dixon and Po Jen Yap, 'Responsive judicial remedies' (2025) 14:2 GlobCon 323; Kent Roach, *Remedies for Human Rights Violations: A Two-Track Approach to Supra-national and National Law* (CUP 2021); David Landau, 'Aggressive Weak-Form Remedies' (2014) 5:1 Constitutional Court Rev 244; Kent Roach, 'Dialogic remedies' (2014) 17:3 ICON 860.

40 1 BvR 2656/18 (Germany).

41 Louis J Kotzé, 'Neubauer et al. versus Germany: Planetary Climate Litigation for the Anthropocene?' (2021) 22:8 German L J 1423.

42 *Leghari v Pakistan* (2015) WP No 25501/2015.

implementation and provide periodic reports to the court.[43] This allowed the judiciary to retain a supervisory role but at some distance. *MK Ranjitshinh v Union of India* arguably reflects a more deferential approach. The Supreme Court proposed the creation of an expert advisory committee to guide the government's ecological obligations, without retaining jurisdiction or requiring formal reporting.

Klimaseniorinnen v Switzerland presents a distinct model. The ECtHR affirmed a justiciable right to climate protection under Article 8 but declined to identify specific remedial steps, deferring supervision to the Committee of Ministers of the Council of Europe, the political organ tasked with ensuring compliance under the ECHR system.

Although differing in institutional design and intensity, these cases share a common remedial logic. They aim to institutionalize accountability mechanisms that enable courts to support the implementation of open-ended orders while respecting the separation of powers.

Not all courts are willing to embrace or devise mechanisms for a supervisory role. As some decisions in this collection demonstrate, courts may affirm duties or allow claims to proceed but still refrain from retaining jurisdiction or establishing oversight structures to ensure compliance. In *PSB v Brazil*, while the Supreme Court recognized a constitutional right to climate protection and issued an order prohibiting the government from withholding financial contributions to the Climate Fund, the majority declined a proposal by one of the justices to institute periodic judicial review of implementation, reflecting concern about institutional overreach. This judicial hesitation is also visible in the Canadian context. In *Misdzi Yikh v Canada* (and *La Rose v Canada*,[44] heard on appeal with *Misdzi Yikh*), lower courts ruled that the claims were not justiciable, noting the systemic nature of the relief sought and the risk that any effective remedy would involve judicial supervision of climate policy implementation. These decisions were reversed on appeal and will proceed to hearings on the merits.

These examples illustrate that even where courts are open to recognizing climate-related rights and issuing remedies, they may still decline to construct or adopt mechanisms for ongoing oversight. This reveals a tension between

---

43 See for a discussion of this case Emily Baritt and Boitumelo Sediti, 'The Symbolic Value of Leghari v Federation of Pakistan: Climate Change Adjudication in the Global South' (2019) 30:2 KLJ 203; Jacqueline Peel and Hari M Osofsky, 'A Rights Turn in Climate Litigation?' (2017) 7:1 TEL 37; 'Leghari v. Federation of Pakistan' (*Climate Change Litigation Database*) <https://climatecasechart.com/non-us-case/ashgar-leghari-v-federation-of-pakistan/>.

44 *La Rose v Canada* (n 16).

judicial engagement and judicial restraint in the face of complex governance challenges.

The issue of justiciability often becomes entangled with the question of remedies in judicial reasoning, particularly where courts perceive that adjudicating a case could require them to engage in ongoing policy supervision. Professor Kent Roach addresses this issue in his argument for 'remedial modesty'.[45] His view is that the risk of a case being found non-justiciable can be reduced if litigants tailor their requested remedies to emphasize outcomes rather than means. By avoiding demands that require courts to dictate or monitor policy implementation, litigants may improve their chances of clearing justiciability hurdles. Professor Roach acknowledges that courts are more comfortable affirming rights or duties where the remedy leaves space for executive or legislative discretion.[46]

While this view aligns with a more restrained model of judicial engagement, it does not necessarily foreclose meaningful climate adjudication. Rather, it reflects a strategic approach to navigating the institutional constraints that many courts have emphasized when declining supervisory roles. As such, it offers a framework for understanding how litigation design and the content of pleadings and claims might respond to and accommodate the remedial caution seen in several of the cases discussed previously. Overly expansive remedial requests risk triggering institutional concerns that obscure the question of redressability with the threshold question of the legitimacy of judicial review, or the question of whether a legal duty exists. As the debate over climate remedies continues to evolve, preserving this distinction allows for a more principled and flexible understanding of how courts can engage with complex climate harms and give redress, without requiring them to overstep their constitutional or institutional roles.

The recent ICJ Advisory Opinion on climate change clarifies that international climate-related obligations, while not necessarily obligations of result, are obligations of conduct that require States to take all appropriate measures to prevent significant harm and to ensure that domestic actions are capable of

---

[45] Kent Roach, 'Judicial Remedies for Climate Change: Less May be More' (*Oxford Business Law Blog*, 15 December 2021) <https://blogs.law.ox.ac.uk/business-law-blog/blog/2021/12/judicial-remedies-climate-change-less-may-be-more > accessed 29 August 2025.

[46] Roach also notes that 'Judges may have an incentive to overstate remedial demands when deciding, as they frequently do, that climate change is not justiciable or that standing or causation has not been established'. See Kent Roach, 'Judicial Remedies for Climate Change' (2021) 17:1 Journal of Law and Equality 105, 116.

meeting international and national climate objectives.[47] While courts may be cautious not to prescribe specific outcomes, they can and do assess whether States' conduct is plausibly directed toward fulfilling the underlying legal purpose, such as achieving emissions reductions in line with the Paris Agreement or protecting human rights threatened by climate inaction.[48] This reinforces the idea that even when remedies take a declaratory or open-ended form, the obligations they affirm can and should be understood as legally binding and substantively effective. It still leaves open, however, the question of what forms of supervisory roles courts should retain or delegate to hold States to account without overstepping their institutional role.

Another dimension of the remedies debate concerns the transformative potential of declaratory relief. Even where courts stop short of issuing coercive orders, a judicial declaration that a government has breached its legal duty to adopt effective climate policies or to protect vulnerable populations from foreseeable harm can carry significant weight. As discussed prior, many courts affirm substantive obligations of conduct through declaratory relief, even when they refrain from issuing coercive or supervisory orders. Such declarations provide an authoritative legal record, affirming that climate change is real, that it is caused by humans, and that it is inflicting harm. Even without binding enforcement mechanisms, declaratory relief can empower civil society, bolster public demands for accountability, and create legal baselines. We see such dynamics in cases like *Held v Montana*, *Mathur v Ontario*, and *Klimaseniorinnen v Switzerland*, where courts have affirmed the existence of State obligations grounded in constitutional, statutory, or human rights law, even as they exercised caution regarding remedies. These judgments demonstrate that declaratory relief, when grounded in substantive obligations of conduct, can be a powerful instrument for shaping climate discourse and reinforcing legal standards.

Even in cases that have not resulted in enforceable outcomes, such as *Lliuya v RWE*, courts have signaled that climate liability for carbon majors is legally conceivable. As Olaf Dilling and Romina Schaller explain in Chapter 3, the Higher Regional Court of Hamm's ruling that a major emitter could, in principle, be held liable under German property law for contributing to climate-related risks abroad, even if specific harm and causation must still be proved,

---

[47] ICJ Advisory Opinion (n 8).

[48] Sandrine Maljean-Dubois, 'Climate Litigation: The Impact of the Paris Agreement in National Courts' (2022) 324 The Taiwan Law Review 211; Brian J Preston, 'The Influence of the Paris Agreement on Climate Litigation: Legal Obligations and Norms (Part 1)' (2021) 33:1 JEL 1; Laura Wegener, 'Can the Paris Agreement Help Climate Change Litigation and Vice Versa?' (2020) 9:1 TEL 17.

goes to the issue of private accountability. Such a decision can lay groundwork for future action.[49]

While courts face real constraints in offering effective climate remedies, they are not powerless. Through creative, proportionate, and principled approaches, courts are helping to shape the legal landscape of climate accountability. Their remedies may not be immediate or sweeping, but they are contributing to a framework that enables broader political, legal, and societal change.

5. CONCLUSION

The close analysis of eleven 'groundbreaking' cases considered in this volume offers insights into two key issues: the roles courts are playing in the complex climate governance landscape, and the roles they should be playing. As the cases also indicate, courts operate in this context within constraints that limit the redress they can offer.

Many of these eleven cases also offer stark reminders about how slow, costly, fragmented, and inaccessible legal proceedings can be for many individuals and communities. The time and cost challenges are evident in the progress of many of the cases in this volume. For example, *Smith v Fonterra* was filed in 2019, and after various appeals, the hearing on the merits is scheduled for fifteen weeks commencing in April 2027. *Mathur v Ontario* began in 2019, and after various appeals and finally a dismissal of Ontario's appeal to the Supreme Court of Canada in 2025, the case will be heard on the merits in 2026. These time and cost demands indicate the vital role of NGOs and pro bono legal assistance. Most of the cases in this volume, and many others, would not be possible but for the support and leadership of NGOs, legal clinics, and pro bono legal assistance.

Judicial interventions do not always support ambitious climate action. In a growing number of cases, litigation is being used to challenge or delay climate measures, such as emissions caps, renewable energy mandates, or

---

49 See for analysis of *Lliuya v RWE*, Maxim Bönnemann and Maria Antonia Tigre, 'What *Lliuya v. RWE* Means for Climate Change Loss and Damage Claims' (*Sabin Center*, 19 June 2025) <https://blogs.law.columbia.edu/climatechange/2025/06/19/what-lliuya-v-rwe-means-for-climate-change-loss-and-damage-claims/> accessed 29 August 2025; Jan-Erik Schirmer, 'Climate Liability after Lliuya v RWE: Causation and Duty Demystified' (2025) 7:3 J of European Tort L <https://ssrn.com/abstract=5153585> accessed 29 August 2025; Noah Walker-Crawford, Joy Reyes and Nicholas Petkov, 'Luciano Lliuya v. RWE: A major step forward for climate justice' (*Grantham Research Institute LSE*, 9 July 2025) <www.lse.ac.uk/granthaminstitute/news/luciano-lliuya-v-rwe-a-major-step-forward-for-climate-justice/> accessed 29 August 2025.

corporate disclosure requirements.[50] A further complexity lies in the tensions between climate mitigation or adaptation and other public interest objectives. *MK Ranjitsinh v Union of India* offers an example, with the Supreme Court weighing the rights of an endangered species against India's renewable energy expansion. Similar tensions arise around Indigenous rights, energy poverty, land use, and ecological protection. These emerging 'just transition' cases reflect a shift: as the need for climate action becomes politically unavoidable, litigation increasingly turns on *how* to act, not just whether to act. Courts are being asked to adjudicate competing claims of justice and to navigate ethical, distributive, and intergenerational dilemmas that defy easy resolution.[51]

Beyond their formal powers of adjudication and remedy, courts play a broader role. In an era marked by regulatory rollback, political inertia, and contested climate narratives, judicial venues can serve as platforms for public visibility by keeping the climate crisis on the agenda. By providing a structured space where evidence is tested and harms are discussed, courts help to anchor public discourse in science, rights, and intergenerational equity.

Climate litigation can also give voice to those sidelined in legislative and executive arenas, including youth, Indigenous communities, and other climate-vulnerable groups. Through their processes and their judgments, courts can amplify perspectives that are often marginalized in political and public debate. For many litigants, winning the case or securing a concrete remedy is only one objective; the legal process itself is often a strategy to generate public awareness, reshape narratives, and assert political and moral claims to justice. The origins and official mission statement of the NGOs involved in many of these cases indicate a goal of pursuing climate litigation for strategic purposes that transcend a particular case. Even where remedies are limited, judicial engagement can contribute to legitimizing demands for climate action and reinforcing democratic norms of accountability and transparency.

Across jurisdictions, we are witnessing efforts to dilute climate laws, stall transitions to renewable energy, or withdraw from international commitments. In this context, courts contribute to reinforcing constitutional and statutory commitments, supporting the legitimacy of civil society demands, and preserving continuity in climate governance amid shifting political tides. The cases in this volume offer evidence of those contributions.

Finally, measures of success in this context must extend beyond the traditional win/loss dichotomy. A case can fail (in the traditional sense) but still have significant impact. A loss can be overturned on appeal, or a loss can stand, but a case might have an educative function, catalyze other actors, or

---

50 Setzer and Higham (n 1) 42–46.

51 ibid 45.

make information available that had not previously been in the public domain. Statements made by a court on aspects of the case might clarify an issue, thus paving the way for further litigation with a greater chance of success. Dilling and Schaller capture this well in their concluding remarks about *Lliuya v. RWE*, a case that is a 'failure' but also a remarkable success for its innovative use of German property law and the message it enabled the Court to send to corporate actors.

> Beyond its legal implications, *Lliuya v RWE* exposed corporations to new financial and reputational risks, signaling to investors and insurers that climate litigation could impact profitability. The case has already inspired similar lawsuits in other jurisdictions, demonstrating how 'bottom-up' litigation can pressure companies to reduce emissions even where government regulations lag. Though dismissed, the ruling set a crucial precedent, paving the way for future claims that may succeed with stronger evidence of imminent harm. Ultimately, the case underscores the potential of strategic litigation to drive corporate accountability and shape the evolving landscape of climate justice.[52]

ACKNOWLEDGMENTS

The authors would like to acknowledge the financial assistance of the Schulich School of Law, Dalhousie University, and the Canadian Foundation for Legal Research.

52 Dilling and Schaller, ch 3.

APPENDIX

In Chapter 1, we noted that recent advisory opinions from international adjudicative bodies, most particularly the International Court of Justice advisory opinion on climate change (ICJ Opinion), have reinforced the importance of domestic courts in climate litigation. Since this chapter was finalized, we have already seen Canadian courts begin to interpret the ICJ Opinion and discuss its application in domestic cases. In *Lho'imggin v Canada*, the most recent motion to strike decision in the Misdzi Yikh matter discussed by Ralston in Chapter 9, the Federal Court of Canada recognized that 'while the ICJ Opinion is not binding in Canadian courts, [its] pronouncements may have significant legal implications in the Canadian context. The principles outlined in the Opinion can influence how courts interpret domestic laws, particularly in relation to constitutional rights and international obligations'.[53] The court then explained how a Canadian court could apply the ICJ Opinion to a section 7 Charter challenge: '[i]n a section 7 Charter challenge where it is argued that environmental harms arising from Canada's actions and omissions amount to an infringement of the right to life, liberty, and security of the person, the Court may use the ICJ's reasoning to inform its interpretation of the extent of climate protection guaranteed by the Charter'.[54] Given the ICJ Opinion's findings on the expansive measures states are required to take in the face of climate change, we expect claimants will argue that the ICJ Opinion supports the Charter guaranteeing broad climate protections.[55]

---

53 *Lho'imggin v Canada* 2025 FC 1586 [44].

54 ibid [64].

55 ibid [43].

2. *Mathur v Ontario*: youth-led constitutional challenge to government climate policy lights the path to future recognition of climate rights in Canada

Erin Dobbelsteyn and Nathalie Chalifour

1. INTRODUCTION

There are a growing number of rights-based climate lawsuits around the world asserting that governments have a legal duty to protect individuals and communities from the impacts of climate change, and Canada is no exception. *Mathur v Ontario*,[1] a ground-breaking case which continues to advance and offer hope for Canadians concerned about the climate crisis, was launched in 2019 by seven young climate leaders. The *Mathur* youth litigants argue that the greenhouse gas (GHG) emissions reduction target of the Ontario provincial government violates their rights to life, security of the person, and equality under the *Canadian Charter of Rights and Freedoms* (*Charter*).[2] *Mathur* is one of several rights-based climate cases grounded in *Charter* claims brought before Canadian courts since the late 2010s and the first to clear procedural hurdles

[1] *Mathur v His Majesty the King in Right of Ontario*, 2023 ONSC 2316 (*Mathur* ONSC), 2024 ONCA 762 (*Mathur* ONCA). Erin Dobbelsteyn and Nathalie Chalifour were pro bono counsel to Friends of the Earth Canada who intervened in the *Mathur* case before the Ontario Court of Appeal and the Ontario Superior Court of Justice. The chapter draws upon research that was supported by the Social Sciences and Humanities Research Council of Canada. The authors are grateful for the helpful feedback and support of the book's editorial team and Benjamin Ralston, the author of Chapter 9.

[2] Part I of the *Constitution Act, 1982*, being Schedule B to the *Canada Act 1982* (UK), 1982, c 11, ss 7, 15.

and proceed to a full hearing on the merits.[3] Although the case was unsuccessful after the initial hearing, the Ontario Court of Appeal overturned the lower court's decision in October 2024 and ordered that the matter be sent back for a new trial with guidance on the appropriate analytical framework to be applied to the claims. This revival of the case was hailed as a victory in part for its landmark confirmation that government climate policy is subject to review under the *Charter*. Ontario sought leave to appeal this decision to the Supreme Court of Canada, but their request was dismissed in May 2025. A new hearing before the Ontario Superior Court of Justice was scheduled for December 2025, but was put on hold after Ontario repealed the legislation on which the case turns just days before the hearing. In response, the youth litigants brought a motion to re-open the appeal before the Ontario Court of Appeal and, at the time of writing, were in the process of filing written materials.

This chapter begins with an overview of the case, identifying its impetus, the parties involved, the claims and the processes used to advance them, and the key findings in the judgments issued to date. The next two sections examine elements of the case that mark progress in Canadian human rights-based climate litigation. They also highlight issues that, if ignored, could leave Canada lagging behind other jurisdictions, especially regarding equality rights for youth disproportionately affected by climate change. We conclude by celebrating *Mathur*'s precedent-setting achievements which bring Canada closer to recognizing that governments have a duty to take scientifically sound action to reduce GHG emissions and protect the *Charter* rights of current and future generations.

---

3 The other climate cases grounded in *Charter* claims that have been advanced in Canada to date include: *Environnement Jeunesse v Attorney General of Canada*, 2021 QCCA 1871, leave to appeal to SCC ref'd (*ENJEU*); *La Rose v Canada*, 2020 FC 1008, 2023 FCA 241; *Misdzi Yikh v Canada*, 2020 FC 1059, 2023 FCA 241 (see Chapter 9 for a summary of the *Misdzi Yikh* case); *Dykstra et al v Saskatchewan Power Corporation et al* (31 May 2024), SKKB Court File No KB No 848 of 2023 (Amended Originating Application) <www.sustainablesask.ca/uploads/1/2/3/8/123825013/final_amended_application_may_31_2024.pdf> accessed 21 January 2025. Throughout this chapter, we use the terms 'rights-based climate litigation' and 'climate *Charter* litigation' to refer to these types of cases; they are a subset of the cases that fall under the broader heading of 'climate litigation', which we define as including any cases that have the issue of climate change at their core. We note that other scholars have defined Canadian rights-based climate litigation more expansively than we have to include procedural environmental human rights cases, among others, see Lisa Benjamin and Sara L Seck, 'Mapping human rights-based climate litigation in Canada' [2022] 13 Journal of Human Rights and the Environment 178.

2. OVERVIEW OF THE CASE

Canada is a federation – a union of ten provinces and three territories – with both a federal Parliament and a legislature in each province and territory. The Canadian legal system is predominantly based in the common law tradition, with a civil law system applying to private law matters in the province of Quebec.[4] As a settler-colonial nation, Aboriginal law and Indigenous law play an important role in the Canadian legal framework.[5] The *Charter*, which is part of Canada's Constitution, guarantees fundamental human rights and freedoms and must be taken into account by all Canadian governments when developing and applying law and policy.

While the *Charter* does not expressly contain a right to a healthy environment, this has not prevented litigants from arguing that existing *Charter* rights provide protection against the harms of environmental degradation and climate change. With the Ontario Court of Appeal's confirmation in 2024 that government climate policy is subject to scrutiny for compliance with the *Charter*, *Mathur* is leading the way on the issue of climate change and human rights in Canada. In this section, we explain the events that prompted the case ('why'), identify the applicants ('who'), set out the details of the claim ('what'), and summarize the judicial decisions taken in the case thus far ('where').

2.1 Why: A Newly Elected Government Weakens Ontario's Climate Target

From the time that Canada signed the United Nations Framework Convention on Climate Change, the politics around climate change policy in this country have been polarized. The divide is fuelled by diverging views about the fate of the fossil fuel industry upon which the Canadian economy has historically been highly dependent and which is amply subsidized by the Canadian government.[6] Canada has the third largest oil reserve in the world, most of which

---

4 Minister of Justice and Attorney General of Canada, *Canada's System of Justice* (Department of Justice Canada 2015) 4–5 <www.justice.gc.ca/eng/csj-sjc/just/img/courten.pdf> accessed 21 January 2025.

5 Indigenous law refers to the specific legal orders and traditions of Indigenous Nations. Aboriginal law refers to laws created by the Canadian government to apply to Indigenous Peoples and their relationship with the Canadian state, which encompasses the Aboriginal and treaty rights of Indigenous Peoples protected under section 35 of the Canadian Constitution, see *Constitution Act* (n 2).

6 In 2023, the Government of Canada provided at least $18.5 billion in financial support to the fossil fuel industry, see Julia Levin, 'Media Backgrounder: Canada's Fossil Fuel Funding in 2023' (*Environmental Defence Canada*, March

is located in the province of Alberta's oil sands.[7] Responsibility for climate policy is fragmented among federal, provincial, and territorial governments,[8] resulting in tensions between governments led by leaders with vastly different views about whether and how to address climate change.[9] This has certainly contributed to Canada's poor track record on climate change: Canada has failed to meet every climate target it has set to date and emissions are significantly higher than they were in 1990.[10] Canada is the 11th top GHG emitter in the world in absolute numbers (representing 1.5% of emissions), 2nd per capita

---

2024) <https://environmentaldefence.ca/wp-content/uploads/2024/03/Canadas-Fossil-Fuel-Subsidies.pdf> accessed 21 January 2025.

[7] Natural Resources Canada, 'Oil Resources' (*Government of Canada*, 16 December 2019) <https://natural-resources.canada.ca/our-natural-resources/energy-sources-distribution/fossil-fuels/crude-oil/oil-resources/18085> accessed 21 January 2025. Analysts have noted the mounting future liabilities of Canadian oil and gas companies to decommission oil and gas well, pipelines, and facilities; see Omar Mawji, 'Canada's Oil and Gas Decommissioning Liability Problem' (*IEEFA*, May 2022) <https://ieefa.org/sites/default/files/2022-05/Canadas%20Oil%20and%20Gas%20Decommissioning%20Liability%20Problem_May%202022_0.pdf> accessed 21 January 2025.

[8] See e.g. Nathalie Chalifour, 'Making Federalism Work for Climate Change: Canada's Division of Powers over Carbon Taxes' [2008] 22 NJCL 119 (discussing authority to implement carbon pricing at the federal and provincial level in Canada); Dayna Nadine Scott, 'The Environment, Federalism, and the Charter' in Peter Oliver, Patrick Macklem and Nathalie Des Rosiers (eds), *The Oxford Handbook of the Canadian Constitution* (OUP 2017) 493.

[9] For example, the province of Alberta's United Conservative Party voted in 2024 to abandon provincial GHG emissions reduction targets and classify carbon dioxide as 'a foundational nutrient for all life on earth'; see Jeremy Appel, 'Alberta's ruling party voted to dump remissions reduction plans and embrace carbon dioxide' (*Canada's National Observer*, 2 November 2024) <www.nationalobserver.com/2024/11/02/news/albertas-ruling-party-votes-emissions-reduction-carbon-dioxide> accessed 4 April 2025. Alongside the rise of populism, the conservative parties in Canada have swung increasingly to the right side of the political spectrum, opening up a divide about the response to climate change and increasing climate denialism, see Hannah Sladics, 'Climate Denialism in Mainstream Canadian Politics' (*Impact Ethics*, 21 November 2024) <https://impactethics.ca/2024/11/21/climate-denialism-in-mainstream-canadian-politics/> accessed 4 April 2025.

[10] Commissioner of the Environment and Sustainable Development, 'Report 6—Canadian Net-Zero Emissions Accountability Act—2030 Emissions Reduction Plan' (*Office of the Auditor General of Canada*, 2023), <www.oag-bvg.gc.ca/internet/English/parl_cesd_202311_06_e_44369.html> accessed 21 January 2025.

among these top emitting countries,[11] and the only G7 country whose emissions are still increasing.[12]

Much of the political wrangling over the last decade has centred on the use of carbon pricing, with the enactment by the federal government in 2018 of a national carbon pricing system becoming a lightning rod for political posturing and judicial challenges.[13] Ontario is Canada's most populated province and its emissions make it the second top provincial emitter; however, per capita Ontario's emissions are the third lowest in Canada and its overall emissions were 12% lower than 1990 levels in 2022 due largely to the progressive closure of coal-fired electricity plants between 2003 and 2014.[14] When Ontario Premier Doug Ford (leader of the Progressive Conservative Party of Ontario) was elected in 2018, his first act in office was to enact the *Cap and Trade Cancellation Act* (*CTCA*), a provincial law that dismantled the cap and trade program that had been implemented by the previous government.[15] The *CTCA* also rolled back Ontario's climate target for 2030 and eliminated the provincial targets for 2020 and 2050.[16] It is these actions that are challenged in

---

11 Environment and Climate Change Canada, 'Global greenhouse gas emissions: Canadian Environmental Sustainability Indicators' (*Government of Canada*, August 2024) <www.canada.ca/en/environment-climate-change/services/environmental-indicators/global-greenhouse-gas-emissions.html> accessed 21 January 2025.

12 Commissioner of the Environment and Sustainable Development (n 10).

13 See e.g. Paul Wells, 'A carbon tax? Just try them' (*MacLeans*, 7 November 2018) <https://macleans.ca/news/canada/a-carbon-tax-just-try-them/> accessed 21 January 2025 (the cover of the magazine shows five conservative leaders united in their opposition against the liberal federal Trudeau government implementing a carbon price).

14 Environment and Climate Change Canada, 'Greenhouse gas emissions' (*Government of Canada*, 19 December 2024) <www.canada.ca/en/environment-climate-change/services/environmental-indicators/greenhouse-gas-emissions.html> accessed 21 January 2025.

15 *Cap and Trade Cancellation Act, 2018*, SO 2018, c 13 (s 16 repeals the *Climate Change Mitigation and Low-carbon Economy Act, 2016*, SO 2016, c 7, which was the law that created the cap and trade program).

16 The prior *Climate Change Mitigation and Low-carbon Economy Act, 2016* (n 15) called for three GHG emissions reductions targets from a 1990 baseline: a 15% reduction by 2020; a 37% reduction by 2030; and an 80% reduction by 2050 [6(1)]. The new *Cap and Trade Cancellation Act, 2018* (n 15), sets a significantly weaker reduction target of 30% below 2005 by 2030, and omits any GHG reduction target for 2050. The revised 2030 target represents a 15% decrease from the earlier target, allowing for 30 megatonnes more in annual GHG emissions compared to the previous law. See Ontario Ministry of the Environment, Conservation

Mathur, though as elaborated upon below, the way in which the pleadings have been framed (e.g. around government action versus inaction) has been a central theme in judicial decisions to date.

2.2 Who: The Applicants and Interveners

The *Mathur* case is led by seven youth from Ontario (the Applicants): Sophia Mathur, Zoe Keary-Matzner, Shaelyn Wabegijig, Shelby Gagnon, Alexandra Neufeldt, Madison Dyck, and Beze Gray.[17] When the claim was filed in November 2019, the youth ranged in age from 12 to 24 years old.[18] The Applicants are represented on a pro bono basis by Ecojustice, the largest environmental law non-profit organization in Canada, with the support of a legal team from the Ontario civil litigation firm, Stockwoods LLP. The *Mathur* Applicants brought the claim as public interest litigants 'in solidarity with millions of youth in Ontario and around the world who are aware of the short period left to fight for their futures'.[19] Litigation costs in Canada are generally awarded to the successful party. Existing costs rules and high adverse costs awards have been identified as a barrier to accessing justice in public interest environmental litigation,[20] hence the importance of rights-based climate cases like *Mathur* being advanced on a pro bono basis.

All of the Applicants are actively engaged in a variety of political and community activities aimed at advancing climate justice and Indigenous rights and pushing governments to take stronger climate action. For example, the lead applicant, Sophia Mathur, has been a leader in the Fridays for Future movement in Ontario and was the first youth outside of Europe to strike from school in solidarity with Greta Thunberg. Three of the Applicants in *Mathur*

---

and Parks, 'Preserving and Protecting our Environment for Future Generations: A Made-in-Ontario Environment Plan' (*Queen's Printer for Ontario*, 2018) <https://prod-environmental-registry.s3.amazonaws.com/2018-11/EnvironmentPlan.pdf>. See also *Mathur v Her Majesty the Queen in Right of Ontario* (25 November 2019), ONSC Court File No CV-19-00631627-0000 (Notice of Application) [14]–[16].

17 Notice of Application (n 16) [2].

18 ibid [9].

19 ibid [6].

20 Canadian Environmental Law Association, Ecojustice and Environmental Law Centre, 'Costs and Access to Justice in Public Interest Environmental Litigation: Submissions to the Federal Court of Appeal and Federal Court Rules Committee' (23 November 2015) <https://cela.ca/wp-content/uploads/2019/07/Costs-Access-to-Justice-Public-Interest-Envl-Litigation.pdf> accessed 21 January 2025.

are Indigenous and grew up in different First Nations across the province.[21] Beze Gray, for example, is from the Aamjiwnaang First Nation, which has been subject to many environmental injustices over decades.[22]

The *Mathur* case has attracted a great deal of attention from environmental organizations, the media, and the Canadian public.[23] It has also been the subject of a significant amount of academic scholarship and commentary.[24]

---

21 Shaelyn Wabegijig grew up in Rama First Nation and is Caribou Clan. Shelby Gagnon is Anishinaabe and spent her early childhood in Aroland First Nation. Beze Gray is Anishinaabe and from Aamjiwnaang First Nation. See Ecojustice, 'Meet the young climate activists behind *Mathur et. al. v. His Majesty in Right of Ontario*' <https://ecojustice.ca/genclimateaction/> accessed 4 April 2025.

22 Notice of Application (n 16) [17]. Aamjiwnaang First Nation is located in Sarnia, Ontario, one of the most polluted places in Canada, often referred to as 'chemical valley'. There are over 60 large petrochemical, polymer, and other industrial facilities located within 25 kilometers of Aamjiwnaang First Nation that release millions of kilograms of toxic air pollution each year. Poor air quality and exposure to toxic substances have significantly harmed the physical and psychological health of the people living in this community, and governments have consistently failed to enforce adequate protection, see e.g. Environmental Commissioner of Ontario, 'Good Choices, Bad Choices: Environmental Rights and Environmental Protection in Ontario' (October 2017) 121–141 <www.auditor.on.ca/en/content/reporttopics/envreports/env17/Good-Choices-Bad-Choices.pdf>.

23 See e.g. 'Youth climate activists appeal Ontario court decision on greenhouse gas targets' (*CBC News*, 24 May 2023) <www.cbc.ca/news/canada/sudbury/climate-action-lawsuit-appeal-1.6853263> accessed 4 April 2025; Kate Helmore, 'From Montana to Ontario, youth take to the courts as a last resort to combat climate change' (*The Globe and Mail*, 4 September 2023) <www.theglobeandmail.com/business/article-from-montana-to-ontario-youth-take-to-the-courts-as-a-last-resort-to/> accessed 4 April 2025. Many members of the public attended the hearings before the Ontario Superior Court of Justice and the Ontario Court of Appeal.

24 See e.g. Camille Cameron, Riley Weyman and Claire Nicholson, 'Legal Hurdles and Pathways: The Evolution (Progress?) of Climate Change Adjudication in Canada' [2024] 47 Dal LJ 1; Stepan Wood, '*Mathur v Ontario*: Grounds for Optimism about the Recognition of a Constitutional Right to a Stable Climate System in Canada?' [2024] 69 McGill LJ 3; Benjamin and Seck (n 3); Sebastien Jodoin and Jean-Philippe Lemay, 'L'arbre vivant dans le conteste de la crise climatique: une analyse des litiges climatiques constitutionnels au Canada' (2022) 56 RJT ns 289; Camille Cameron and Riley Weyman, 'Recent Youth-Led and Rights-Based Climate Change Litigation in Canada: Reconciling Justiciability, Charter Claims and Procedural Choices' [2022] 34 J Env L 195; Lynda M Collins, 'Constitutional Eco-Literacy in Canada: Environmental Rights and Obligations

At the initial hearing before the Ontario Superior Court of Justice (Superior Court), there were six requests for intervention by a variety of organizations, which were all granted. Before the Ontario Court of Appeal, that number grew to twelve requests for intervention, on behalf of fifteen organizations (five of which were the same as at the Superior Court).[25] Once again, the requests for intervention were all granted. Many of these groups have been active in other climate litigation cases, including a case challenging the constitutionality of the Government of Canada's national carbon price (which was heard by the Supreme Court of Canada in 2020).[26]

2.3 What: The Application

The *Mathur* case was filed as an application instead of an action.[27] The main difference between an application and an action is that the former involves a hearing (rather than a trial) that generally proceeds based on a written record and with a judge sitting alone. Witnesses do not testify before the judge; instead, the parties provide their evidence by way of affidavits with limited use of oral testimony.[28] It provides a speedier, simpler (and ideally less expensive) procedure that avoids delays that could stem from lengthy discovery processes. In *Mathur*, the Applicants filed over a dozen affidavits (including those of the youth) and close to twenty reports providing expert evidence on the existing and projected impacts of climate change in Canada and Ontario. With the application process, the time from filing the Notice of Application to the hearing on the merits in *Mathur* was less than three years. This, in a Canadian context, is speedy. Not all jurisdictions in Canada have the option of advancing a claim in this way.[29]

---

in the Canadian Constitution' [2022] 26 Rev Const Stud 227; Dayna Scott and Garance Malivel, 'Intergenerational Environmental Justice and the Climate Crisis: Thinking with and Beyond the Charter' [2021] 17 J L & Eq 165.

25 See section 2.4 for further information on the interventions.

26 The Assembly of First Nations, Canadian Association of Physicians for the Environment, and Friends of the Earth Canada were among the 25 interveners before the Supreme Court of Canada in *References re Greenhouse Gas Pollution Pricing Act*, 2021 SCC 11.

27 See *Courts of Justice Act*, RSO 1990, c C.43; *Rules of Civil Procedure*, RRO 1990, Reg 194, Rule 14.05(3)(g.1).

28 The parties do have the opportunity to cross-examine the deponents of any affidavit served in the application.

29 Although speed is not the only relevant factor in comparing actions to applications, by way of contrast, the *La Rose* (n 3) case (a youth-led climate *Charter* case challenging the federal government's climate (in)action) is proceeding by action, which involves a trial with live witnesses. The *La Rose* case was also commenced

One of the downsides to proceeding as an application, however, is that the *Mathur* Applicants did not have the opportunity to share their stories about the way they've been impacted by climate change in front of a judge. Nor did the judge have the opportunity to hear directly from the Applicants or the experts regarding the climate science and other evidence relevant to the claims. While it is difficult to speculate on any potential impact the lack of *viva voce* evidence have on the final outcome of the hearing, it is important to note that oral testimony from those most directly impacted by climate harms has been a strategic priority for other rights-based climate cases around the world.[30]

For reasons that are explored further in section 3.1, the Applicants in *Mathur* were careful to frame their claim as targeting specific legislation. Specifically, the Applicants challenge the constitutionality of the 2030 GHG emissions reduction target set by Ontario under section 3(1) of the *Cap and Trade Cancellation Act* and the related *Made-in-Ontario Environmental Plan*, which is to reduce GHG emissions by only 30% below 2005 levels by 2030 (the Target). This distinguished *Mathur* from two other Canadian climate *Charter* cases, *La Rose* and *Misdzi Yikh*, which were launched around the same time and included broader allegations about the harms ensuing from a network of government laws, policies, and decisions.[31] While *La Rose* and *Misdzi Yikh* initially had difficulty clearing justiciability thresholds with the motion judges expressing concern about the broad and diffuse nature of the claims, the Federal Court of Appeal granted leave to amend the section 7 claims in these cases and proceed to trial.[32]

The *Mathur* Applicants also specifically chose not to root their challenge in the Ontario government's policy regression, but rather in the inadequacy of the Target relative to what is required from a scientific and legal standpoint. The Applicants allege that the current Target 'will lead to climate catastrophe' because it is incompatible with Ontario's share of the minimum level of GHG reductions necessary to limit global warming to the Paris Agreement standards and, as a result, violates Ontario's youth and future generations' rights to life, liberty, and security of the person under section 7 of the *Charter* and right to be free from discrimination under section 15 of the *Charter*. In order to establish a violation of section 7 of the *Charter*, a claimant has to show

---

in 2019, and took four years to clear pre-trial procedural hurdles and another year before the Federal Court of Canada scheduled the 8-week trial to begin in October 2026.

30 See e.g. Chapter 8 discussing the *Held v Montana* case.

31 See *La Rose* (n 3) and *Misdzi Yikh* (n 3). This aspect of *Mathur* distinguished it from earlier climate *Charter* claims, which each targeted a broader range of government action and inaction.

32 *La Rose v Canada*, 2023 FCA 241.

that the impugned government action interferes with or deprives them of their right to life, liberty, or security of the person and that this interference or deprivation is not in accordance with the principles of fundamental justice. For a section 15 violation, a claimant must show that the impugned government conduct directly or indirectly creates a distinction based on a protected ground and that this distinction is discriminatory.

The Applicants are seeking relief on behalf of their generation and future generations of Ontarians in the form of a series of declarations, namely that:

(a) the Target, and legislation that allows for the imposition of targets that are not set with regard to the Paris Agreement temperature standard or any kind of science-based process, violates the section 7 and section 15 *Charter* rights of Ontario youth and future generations;
(b) section 7 of the *Charter* includes the right to a stable climate system, capable of providing youth and future generations with a sustainable future; and
(c) the Target violates an unwritten constitutional principle that governments are prohibited from engaging in conduct that will, or reasonably could, result in the future harm, suffering, or death of a significant number of its citizens.

The Applicants also seek an order requiring the government to set a science-based GHG emissions reduction target and to implement a plan to meet that target.[33]

2.4 Where: Decisions in the *Mathur* Case to Date

Because the claim was brought against the provincial government of Ontario, the case was filed in the Ontario Superior Court of Justice (Superior Court), a court of general jurisdiction which hears matters ranging from civil to criminal proceedings. Appeals are generally made from the Superior Court to the Ontario Court of Appeal and then leave to appeal may be sought from the Court of Appeal to the Supreme Court of Canada. A summary of each stage of the proceeding that had occurred at the time of writing is set out in the following sections.

2.4.1 Motion to strike (denied)

As soon as the youth filed their Application, the government of Ontario filed a motion to strike the claim. These types of motions allow a judge to order the

33 Notice of Application (n 16) [8].

entirety or parts of the Application to be struck.[34] In order to do so, the motion judge must conclude that, assuming the facts as plead are true, it is plain and obvious that the pleading discloses no reasonable cause of action.[35]

In what can be considered the first human-rights based climate litigation victory in Canada, Justice Carole Brown of the Superior Court dismissed Ontario's motion to strike in November 2020, paving the way for the case to proceed to a full hearing.[36] Justice Brown's 55-page, 269-paragraph decision clarified that the government's preparation of a climate target and plan was reviewable by the courts for compliance with the *Charter* and that *Mathur* is a case 'of public interest' that 'transcends the interest of all Ontario residents, not just the Applicants' generation or the ones that follow'.[37] In other words, the matter was held to be justiciable. She also found that both the sections 7 and 15 claims had a reasonable prospect of success and that a case for positive obligations on the government could be made out in the context of climate change.[38] The characterization of a claim as 'positive' (i.e. seeking to impose a duty on the government to act) versus 'negative' (i.e. preventing the government from violating rights through its laws and conduct) is an important distinction in Canadian law that is explained further in section 3.3. Ontario tried to have this decision overturned, but their appeal request was dismissed in March 2021.[39]

2.4.2 Hearing before the Ontario Superior Court of Justice (application dismissed)

Canada's first full hearing of a human rights-based climate case was held over three days in September 2022 before a single judge (Justice Marie-Andrée Vermette). As noted earlier, the Superior Court allowed six organizations to intervene to provide the court with distinct and important perspectives, all of which generally supported the Applicants' positions on the issues of the case. These organizations were: Indigenous Climate Action, the Assembly of First Nations, the David Asper Centre for Constitutional Rights, the Canadian

---

34 *Rules of Civil Procedure* (n 27) Rule 21.01(1)(b).

35 *Mathur v Ontario*, 2020 ONSC 6918 [36]–[37] (*Mathur* Motion to Strike).

36 ibid.

37 ibid [61].

38 ibid [132], [139]–[140] (justiciability), [171], [189] (prospects of success), [228], [233], [234] (positive obligation).

39 *Mathur v Ontario*, 2021 ONSC 1624 (denying Ontario's request for leave to appeal the decision on motion to strike). See also Ecojustice, 'Youth climate case forges ahead after court affirms historic decision' (26 March 2021) <https://ecojustice.ca/news/youth-climate-case-forges-ahead-after-court-affirms-historic-decision/>.

Association of Physicians for the Environment, For Our Kids, and Friends of the Earth Canada.[40]

Although the application judge ultimately dismissed the case in April 2023, finding that there were no violations of *Charter* rights, she made several positive findings that were significant for this and future environmental rights cases in Canada. Notably, she held that the *Charter* claims raised by the Applicants are justiciable.[41] Justice Vermette also held that Ontario's Target falls severely short of the scientific consensus about what is required to limit global warming to 1.5 degrees Celsius and that this deficiency contributes to the increased risk of death and harm to security of the person faced by the Applicants and all Ontarians.[42] She rejected the government's defence that its emissions were insignificant in light of the global nature of the problem, joining courts around the world in holding that every emission contributes to the problem and that jurisdictions are accountable for what they emit.[43] Justice Vermette also acknowledged that climate change disproportionately impacts young people and Indigenous Peoples.[44] Despite the significance of these and other aspects of the case, which are discussed further in sections 3 and 4 below, Justice Vermette did not decide the case in favour of the Applicants.

2.4.3 Appeal to the Ontario Court of Appeal (appeal granted)

Undeterred by this setback, the Applicants appealed the Superior Court decision to the Ontario Court of Appeal. There were twelve interveners before the Court of Appeal, representing fifteen groups. The first five of the following organizations intervened at the initial hearing, while the rest were new: Assembly of First Nations, Canadian Association of Physicians for the Environment, David Asper Centre for Constitutional Rights, For our Kids, Friends of the Earth Canada, The Grand Council of Treaty #3, Greenpeace Canada and Stichting Urgenda, Canadian Lawyers for International Human Rights and Center for International Environmental Law, British Columbia Civil Liberties Association, the Canadian Civil Liberties Association, Citizens for Public Justice, Environmental Defence Canada Inc., and West Coast Environmental Law. The Ontario Court of Appeal heard arguments in January 2024 and released its decision in October 2024.[45]

Three judges of the Ontario Court of Appeal (Justices Lois Roberts, Steve Coroza, and Sally Gomery) unanimously concluded that the application judge

---

40 *Mathur* ONSC (n 1) [94].

41 ibid [106].

42 ibid [147].

43 ibid [149].

44 ibid [25], [178].

45 *Mathur* ONCA (n 1).

had failed to properly consider whether Ontario's Target was compliant with the *Charter* and sent the case back to the lower court for a new hearing to determine the section 7 and 15 *Charter* issues. In allowing the appeal and setting aside Justice Vermette's decision, the Court of Appeal declined to decide the *Charter* issues or to grant the relief sought by the appellants in their notice of appeal, which included directing Ontario to set a science-based GHG emissions reduction target and revise its climate change plan to implement this target.[46] The decision to remit the application for a new hearing was based largely on the Court of Appeal's view that the application judge had erred in her analytical approach by framing the Application as seeking to impose a freestanding positive obligation on Ontario to combat climate change and that important issues had been raised on appeal that were not considered by the application judge.[47] The Court of Appeal concluded that these serious matters needed to be considered afresh through the correct analytical lens and possibly with the benefit of further evidence.[48] In sections 3 and 4, we analyze key parts of this decision.

2.4.4 Appeal to the Supreme Court of Canada (dismissed)

In December 2024, the Ontario government sought leave to appeal the Ontario Court of Appeal decision to the Supreme Court of Canada. They submitted that the appeal raised three questions of national public importance.[49] The first was whether the *Charter* requires state actors in Canada to combat climate change (i.e. imposes a positive obligation on governments to do so). The second and third related to the situation in which a government voluntarily establishes a GHG emissions reduction target and asked whether the *Charter* requires this target to establish a minimum level of reduced GHG emissions and, if so, what level is required and what process must be followed to establish the target. Ontario's position was that there is uncertainty in Canadian jurisprudence regarding when it is appropriate for courts to adjudicate a claim in the climate change context, especially *Charter*-based claims like *Mathur*, and invited the Supreme Court to provide guidance about the 'parameters of justiciability in the climate change context and in the *Charter* context more generally'.[50] Ontario also took issue with the Court of Appeal's finding that the *Mathur* Applicants were not raising a claim for positive rights. The government argued that the Court of Appeal's decision creates an exception to

---

46 *Mathur v Ontario*, 2024 ONCA 762 (Notice of Appeal).

47 *Mathur* ONCA (n 1) [5], [6].

48 ibid [8].

49 *Mathur v Ontario*, SCC Court File No. 41596 (Application for Leave to Appeal) [25].

50 ibid [39].

the long-standing principle that section 7 of the *Charter* does not impose a positive obligation on Canadian governments to ensure that everyone enjoys life, liberty, or security of person but only restricts the state's ability to deprive people of these rights.

On May 1, 2025, the Supreme Court dismissed Ontario's leave to appeal, without reasons (as per standard practice for leave dismissals). So, what does this mean for *Mathur*? If the Supreme Court had heard the appeal, *Mathur* would have been the first climate *Charter* case to be reviewed by Canada's apex court. While this could have provided the country with much-needed clarity on the *Charter*'s role in confronting the climate crisis, the Supreme Court's decision has an upside worth celebrating. It preserves the Court of Appeal's decision, including findings that will be helpful for the Applicants as the case proceeds: that Canadian governments' GHG targets are subject to review for compliance with the *Charter*; that *Mathur* is not a positive rights case; and that Ontario's inadequate and unscientific target is exacerbating the risk to the lives and health of all people in Ontario. Days prior to the new hearing before the Superior Court, Ontario repealed the relevant sections of the Cap and Trade Canacellation Act upon which the case turns. Given the potential for the case to become moot, the Applicants requested that the Court of Appeal re-open the appeal in order to address unresolved issues and provide meaningful remedies for the Ontario government's violation of *Charter* rights.

3. GROUND-BREAKING ELEMENTS OF *MATHUR*

Many aspects of the *Mathur* case are novel, and the Superior Court and Court of Appeal decisions are notable successes for human rights-based climate litigation (and Constitutional litigation more broadly) in Canada. The Superior Court's adoption of a scientific standard against which to measure the adequacy of Ontario's Target, for instance, was a groundbreaking first. Additionally, the Superior Court's categorical rejection of the *de minimis* or 'drop in the bucket' defence by the Ontario government was a noteworthy success for the causation element of *Charter* climate litigation.[51] Clearing the justiciability threshold and the Court of Appeal's confirmation that a government's climate change laws must comply with the *Charter* is historic, as this had been an especially intractable roadblock in Canadian climate *Charter* litigation until very recently. We discuss these elements in more detail below, before turning to critiques of the judicial reasoning in the *Mathur* decision in section 4. It is

[51] Note that the Supreme Court of Canada also rejected this argument in the context of the division of powers analysis in the carbon pricing reference. See *References re Greenhouse Gas Pollution Pricing Act* (n 26) [188].

also worth noting that the *Mathur* case is the most successful negative rights climate case so far, which has the potential to be impactful in jurisdictions outside Canada where positive rights are not judicially recognized.

3.1 Clearing the Justiciability Threshold

One of the landmark aspects of *Mathur* is that it is the first climate *Charter* case in which it was acknowledged on a preliminary motion to strike the pleadings that Canadian courts have institutional capacity and authority to adjudicate this kind of lawsuit.[52] The justiciability of the Application, including the section 7 and 15 claims, was confirmed by the application judge after a full hearing on the merits.[53] Justice Vermette concluded that courts are required to review legislation and government action on climate change for *Charter* compliance even though the issues may be complex, contentious, and laden with social values, and this finding was upheld by the Ontario Court of Appeal.[54]

The question of how to frame climate *Charter* claims to avoid dismissal for non-justiciability has been at the centre of climate *Charter* litigation since its inception and continues to be plagued with difficulty and contradictions. Although *Mathur* is similar in several ways to the other human rights-based climate cases in Canada, and is ultimately alleging that Ontario's Target is inadequate, it successfully cleared the justiciability threshold largely because it was framed around specific government measures and actions: legislation (subsection 3(1) of the *CTCA*) and a related policy (the Target). In this way, Justice Vermette distinguished *Mathur* from other unsuccessful climate *Charter* case precedents at the time which had been framed more broadly as challenging a wide range of government laws, policies, and decisions related to climate change.[55] Justice Vermette explicitly acknowledged this when she

---

52 *Mathur* (n 1). Note that Justice Morrison found the claim by Environnement Jeunesse to be justiciable at first instance in *ENJEU* (though this finding was overturned by the Quebec Court of Appeal) and Justice Rennie for the Federal Court of Appeal recently overturned the decision to strike the claims in *La Rose* and *Misdzi Yikh*, finding the section 7 elements of those cases to be justiciable and allowing the applicants to amend their pleadings and proceed to a hearing on this issue. See *Environnement Jeunesse c Procureur général du Canada*, 2019 QCCS 2885; *La Rose* (n 3) and *Misdzi Yikh* (n 3) (heard together).

53 The question of determining Ontario's fair share of the global carbon budget, however, was held not to be justiciable, *Mathur* ONSC (n 1) [109].

54 See *Mathur* ONCA (n 1) [36]. The Court of Appeal took issue with other aspects of the application judge's decision, including her view that this was a positive rights case, which will be discussed later in the section.

55 See *ENJEU* (n 3); *La Rose* (n 3); *Misdzi Yikh* (n 3). See also (n 30).

stated that the Applicants challenged specific legislation 'very likely in order to avoid the dismissal of their Application as non-justiciable'.[56] This narrow framing, however, limited the scope of government action the Applicants could rely on to establish that the Target was causally connected to the alleged violation of section 7 of the *Charter*.[57] Justice Vermette disagreed with the Applicants that the Target authorizes, incentivizes, facilitates, and creates the level of dangerous GHG that imposes an increased risk of death and harm for all Ontarians in violation of section 7. In her view, the Target did no such thing and this was an attempt by the Applicants to shift the analysis away from the Target and the *CTCA* to Ontario's broad range of actions, programs, and policies that contribute to the overall level of GHG emissions in the province.[58] On this basis, she framed the Application as advancing positive rights because, in her view, what the Applicants were actually seeking was to restore an earlier statutory platform: the previous government's stronger GHG target.[59] The Court of Appeal concluded that this analytical approach, which was ultimately fatal to the Applicants' claims, was flawed.[60]

3.2 Adoption of Climate Science in Defining a Standard Against Which to Measure the Adequacy of Ontario's Target

The Superior Court judge in *Mathur* accepted the climate science as reported by the Intergovernmental Panel on Climate Change (IPCC) as relevant to climate litigation. Justice Vermette characterized the IPCC's reports as a 'reliable, comprehensive and authoritative synthesis of existing scientific knowledge about climate change and its impacts'.[61] Ontario tried to paint a different picture, using evidence from experts with connections to the fossil fuel industry and climate denialism to attack certain aspects of the Applicants' record, but the Superior Court rejected the evidence of Ontario's experts, noting that their credentials did not measure up to those of the IPCC.[62] In contrast, Justice Vermette accepted the expert evidence presented by the Applicants on the impacts of climate change, most of which was uncontested by Ontario.[63] This evidence documented a range of climate harms, from extreme heat events, the spread of infectious diseases, and mental health impacts to the increase

---

56 *Mathur* ONSC (n 1) [120].
57 ibid [120], [135].
58 ibid [135].
59 ibid [136].
60 *Mathur* ONCA (n 1) [4]–[5].
61 *Mathur* ONSC (n 1) [18].
62 ibid [19].
63 ibid [22].

in frequency and severity of wildfires, flooding, harmful algal blooms, and mercury toxicity in aquatic systems. The Superior Court also accepted evidence about climate tipping points and the disproportionate impacts of climate change on young people and Indigenous Peoples.[64]

With respect to global GHG emissions reduction goals, the Superior Court accepted the IPCC's finding (subsequently endorsed by countries, including Canada, in the *Glasgow Climate Pact*) that to limit warming to 1.5 degrees Celsius, global net anthropogenic $C0_2$ emissions must be reduced by approximately 45% below 2010 levels by 2030 and reach net zero by 2050 (the Scientific Target).[65] These scientific findings coupled with the evidence before the Superior Court allowed it to calculate that to reduce emissions in accordance with the Scientific Target, Ontario would need to reduce its emissions by 52% below 2005 levels by 2030.[66] Compared to this benchmark, the Superior Court concluded that Ontario's Target (30% below 2005 levels) was severely deficient.

The Superior Court appraised the adequacy of Ontario's Target by reference to the scientifically accepted international standard of global GHG emissions reductions. Justice Vermette, however, was not willing to adjudicate the issue of the proper approach to determining Canada's and Ontario's 'fair share' of the remaining global carbon budget, finding that the Court did not have the institutional capacity and legitimacy to make such a determination when it could not be based on a scientific formula.[67] She held that the factors identified by the Applicants' expert as relevant to the determination of 'fair share' – including population; historical responsibility; wealth; access to inexpensive, low carbon energy sources; equity; and common but differentiated responsibilities – lack a sufficient legal component to be justiciable.[68] On appeal, Ontario argued that the Applicants' request for a remedy that requires Ontario to set a Target that conforms to scientific standards, was vague, imprecise, and impossible for the courts to order. The Court of Appeal disagreed, noting that '[t]he unchallenged international standards and scientific consensus about

64 ibid [24–25].

65 ibid [20]. See United Nations Framework Convention on Climate Change, 'Glasgow Climate Pact' (13 November 2021) CP.26, art 22.

66 ibid [21].

67 ibid [109].

68 ibid. The Applicants maintained that the Superior Court could and should consider issues of equity and justice in determining Ontario's share of the global carbon budget. For more on the issue of 'fair share', see Wood (n 24) 24–27. See also Patricia Galvão Ferreira, '"Common but Differentiated Responsibilities" in the National Courts: Lessons from *Urgenda v The Netherlands*' [2016] 5 Transnational Environmental Law 329.

global warming and climate change and the remaining carbon budget in the evidence on this application is not imprecise' and if *Charter* rights violations are ultimately found, these standards can inform what a constitutionally compliant Target should look like.[69] The Court of Appeal did not directly address the issue of 'fair share', so it will be interesting to see whether the issue is revisited at the new hearing.

3.3 Climate Law and Policy is Subject to Review for Charter Compliance

At all stages of the proceedings, the Applicants carefully framed their case as one challenging actions taken by the Ontario government that lead to harmful levels of GHG emissions that exacerbate the climate crisis (a negative rights framing), as opposed to a failure of Ontario to take any action or to do enough to address the threat of climate change (a positive rights framing). While the distinction between positive and negative rights is often illusory, and is not so problematic in other jurisdictions, Canadian courts have long resisted interpreting section 7 of the *Charter* as conferring positive obligations on governments.[70] The Supreme Court of Canada recognized over 20 years ago that a positive obligation could be recognized under section 7 in 'special circumstances', and subsequent jurisprudence has identified climate change as a likely candidate.[71] Although Justice Vermette came very close in *Mathur*, no court has yet awarded a remedy on the basis of such a positive obligation. The 'positive versus negative rights' distinction has remained the subject of a longstanding, heated debate within the Canadian legal community.[72]

---

69 *Mathur* ONCA (n 1) [70].

70 See e.g. *La Rose* (n 3) [92]–[104] (surveying Canadian jurisprudence and scholarship on the positive/negative rights divide); Martha Jackman and Bruce Porter, 'Social and Economic Rights' in Peter Oliver, Patrick Macklem and Nathalie Des Rosiers (eds), *The Oxford Handbook of the Canadian Constitution* (OUP 2017) 84.

71 See *Kreishan v Canada (Citizenship and Immigration)*, 2019 FCA 223 [139] (Federal Court of Appeal Justices Webb, Rennie and Laskin offered climate change in *obiter dicta* as an example of the kind of circumstances that might qualify as special in the context of section 7 and positive obligations).

72 This is a simplified description of a longstanding debate within the Canadian judiciary and legal community. For more on positive rights, see e.g. Martha Jackman, '*Charter* Remedies for Socio-Economic Rights Violations: Sleeping under a Box?' in Justice Robert J Sharpe and Kent Roach (eds), *Taking Remedies Seriously* (Les Editions Yvon Blais 2009); Vanessa MacDonnell, 'The Protective Function and Section 7 of the Canadian Charter of Rights and Freedoms' [2012] 17 Rev Const Stud 53.

To avoid having to clear the hurdle of establishing 'special circumstances', the Applicants strategically framed their case as a negative rights claim, pointing to the *CTCA*, the Target, and Ontario's climate plan as government actions that have the effect of authorizing, incentivizing, facilitating, creating, and committing to dangerous levels of GHG emissions in Ontario. In other words, the Applicants argued that where the government acts, it must do so in a way that does not interfere with life, liberty, or security of the person. In the absence of a particular state action, Canadians cannot demand the government get involved to secure these same *Charter* interests.

Interestingly, Justice Vermette rejected the Applicants' characterization of the claim, finding both the section 7 and 15 claims assert a positive right since, in her view, the heart of the claim was that the Target was not 'sufficiently high'.[73] She went on to hold that the existential threat posed by climate change presents special circumstances that could justify the imposition of positive obligations under section 7.[74] She then held that the current section 7 test is not well-adapted to an analysis of positive obligations, and did not attempt to modify the test. Instead, applying the existing analysis, she found the Applicants had not shown that any deprivations of their rights were contrary to the principles of fundamental justice.[75] Justice Vermette similarly dismissed the section 15 claim, which she also characterized as a positive rights claim, even though the application was not presented as such.

The Ontario Court of Appeal overturned the Superior Court's characterization of the section 7 claim as one seeking positive rights. The Court of Appeal held that by enacting the *CTCA*, the Ontario government 'had voluntarily assumed a positive statutory obligation to combat climate change and to produce the Plan and the Target for that purpose' and, therefore, the ensuing Plan and Target must comply with the *Charter*.[76] Judicial recognition that government legislation aimed at reducing GHG emissions is subject to scrutiny for *Charter* compliance is an outcome of the Court of Appeal's decision to celebrate, however, it has potential negative implications that deserve mention. By 'effectively side-stepping'[77] the issue of whether the *Charter* imposes a positive obligation on governments to address climate change, the *Mathur* decision gave rise to the risk that governments might simply repeal

73 *Mathur* ONSC (n 1) [122].

74 ibid [138].

75 ibid [171].

76 ibid [5].

77 Martin Olzynski et al, 'A Landmark Decision in Canadian *Charter*-based Climate Litigation: *Mathur v Ontario*, 2024 ONCA 762' (*ABLawg*, 31 October 2024) <https://ablawg.ca/2024/10/31/a-landmark-decision-in-canadian-charter-based-climate-litigation-mathur-v-ontario-2024-onca-762/>.

existing climate laws and policies to avoid having to address allegations of *Charter* rights violations; a risk which did in fact materialize when Ontario took these steps just over a year after the Court of Appeal's decision.[78] While the Supreme Court's dismissal of Ontario's request for appeal confirmed the Court of Appeal's conclusion that *Mathur* is not a positive rights claim, it precluded the opportunity for timely, meaningful guidance from Canada's highest court on the positive/negative rights distinction for ongoing and future climate litigation in this country.[79] This guidance could have been particularly helpful to the Applicants as they respond to Ontario's attempt to evade accountability and seek to have the Court of Appeal re-open the appeal to clarify whether governments have any positive climate obligations under the *Charter*.

4. CRITIQUES OF *MATHUR*

In this final section, we discuss some of the areas that remain problematic in Canadian *Charter*-based climate litigation, as illustrated in the *Mathur* case to date. First, we discuss the way in which the Superior Court interpreted the principles of fundamental justice under section 7 of the *Charter*, critiquing the Superior Court's reliance on principles ill-suited to the climate context. Second, we address the Superior Court's restrictive analysis of section 15 equality rights, noting some internal logical inconsistencies in the decision with respect to positive rights and causation, and critiquing the Court's approach to age-related discrimination in the context of climate change. We also comment on the guidance provided by the Court of Appeal for the re-hearing of these issues by a lower court.

4.1 Principles of Fundamental Justice

The dismissal of the section 7 claims by the Superior Court turned largely on Justice Vermette's analysis of the principles of fundamental justice. Canadian courts have defined the principles of fundamental justice as 'a legal principle about which there is significant societal consensus that it is fundamental to the way in which the legal system ought fairly to operate, and it must be identified

[78] See e.g. Nathalie J Chalifour, Jessica Earle and Laura Macintyre, 'Coming of Age in a Warming World: The Charter's Section 15(1) Equality Guarantee and Youth-Led Climate Litigation' [2021] 17 JL & Equality 1, 52–54 (discussing the potential chilling effect on regulation if there is no positive obligation on governments to reduce GHG emissions).

[79] In *La Rose* (n 3), for instance, the plaintiffs use both a positive and negative rights framing. See *La Rose*, Court File No. T-1750-19, Amended Statement of Claim to the Defendants (31 May 2024).

with sufficient precision to yield a manageable standard against which to measure deprivations of life, liberty or security of the person'.[80] There are a number of established fundamental justice principles and, among them, three that are most commonly applied in section 7 claims: arbitrariness, overbreadth, and gross disproportionality.[81] New principles of fundamental justice may be recognized, but Canadian courts have rejected nearly every novel principle that has been proposed over the last few decades.[82]

In *Mathur*, the Applicants argue that the section 7 deprivations are contrary to the principles of arbitrariness and gross disproportionality. Additionally, they argue that the deprivations are contrary to a novel principle of 'societal preservation', which they defined by stating that 'a government cannot engage in conduct that will, or could reasonably be expected to, result in the future harm, suffering, or death of a significant number of its own citizens'.[83] Justice Vermette rejected both arguments, concluding that there was no violation of section 7 given that the Applicants had not demonstrated that any deprivation of their rights was contrary to the principles of fundamental justice. She was unable to find that the Target was arbitrary,[84] she held that gross disproportionality[85] did not apply in this case, and she refused to recognize the principle of societal preservation.[86]

The Court of Appeal concluded that Justice Vermette's mischaracterization of the *Mathur* Application as a positive rights case caused her to err

---

80 *Mathur* ONSC (n 1) [164].

81 Arbitrariness is the absence of any connection between the objective of an impugned law and its negative impact on section 7 rights. Overbreadth addresses the situation where the impugned law imposes limits on section 7 rights that go beyond what is required to achieve the objective. Gross disproportionality is where the effects of the impugned law are so extreme that they cannot be justified by its object. See *Canada (Attorney General) v Bedford*, 2013 SCC 72 [35], [93]–[123].

82 See Martha Jackman, 'Wizened Stump or Living Tree? Section 7 Principles of Fundamental Justice' in Howard Kislowicz, Richard J Moon and Kerri A Froc (eds), *Canada's Surprising Constitution: Unexpected Interpretations of the Constitution Act, 1982* (UBC Press 2024) 260.

83 *Mathur* ONSC (n 1) [163]. The Applicants also argued that societal preservation was an unwritten constitutional principle, which is discussed in section 4.2.

84 A law will be considered arbitrary where there is no connection between the effect of a provision of the law and its purpose or where there is no connection between the object of the law and the limit that it imposes on section 7 rights, see *Mathur* ONCA (n 1) [50].

85 A law will be considered grossly disproportionate when the seriousness of the section 7 deprivation is completely out of sync with the objective of the law, see *Mathur* ONCA (n 1) [52].

86 *Mathur* ONSC (n 1) [170].

in her analysis of the principles of fundamental justice.[87] This is among the issues that the Court of Appeal remitted for a new hearing to be considered through the correct analytical lens. Notably absent from the Court of Appeal's decision is a consideration of Justice Vermette's analysis with respect to the Applicants' proposed novel principle of societal preservation despite the fact that it contained a number of flaws challenged by the Applicants on appeal. When the case is re-heard by the Superior Court, the focus should go beyond the established principles of arbitrariness and gross disproportionality to consider the application of novel principles of fundamental justice such as societal preservation.[88]

4.2 Unwritten Constitutional Principles

Unwritten constitutional principles are the principles that are foundational to the creation and operation of the Canadian Constitution, yet not explicitly set out in its text. The established unwritten constitutional principles include democracy, the rule of law, federalism, parliamentary sovereignty, constitutionalism, the separation of powers, judicial independence, and the honour of the Crown.[89] As explained by the Superior Court in *Mathur*, unwritten principles serve a gap-filling purpose in constitutional analysis and may be used by Canadian courts to assist with interpretation of the Constitution or to develop new doctrines that help to support the structure of the Constitution.[90]

The Applicants in *Mathur* argue that societal preservation, in addition to being a principle of fundamental justice, is an unwritten constitutional principle and that this should be used to interpret their *Charter* claims.[91] Friends of the Earth, an intervener, argued for recognition of another novel unwritten constitutional principle: ecological sustainability.[92] Neither of these arguments were successful in the Superior Court. Justice Vermette concluded that it was

---

87 *Mathur* ONCA (n 1) ibid [50].

88 For an in-depth analysis of the challenges and opportunities for the principles of fundamental justice analysis in climate *Charter* litigation, see Erin Dobbelsteyn, 'Fundamental (In)justice and the Existential Threat of Climate Change' [2025] Can J Women & Law (submitted).

89 See e.g. Vanessa A MacDonnell, 'Rethinking the Invisible Constitution: How Unwritten Constitutional Principles Shape Political Decision-Making' [2019] 65 McGill LJ 175, 178–79; Mari Galloway, 'The Unwritten Constitutional Principles and Environmental Justice: A New Way Forward?' 52 Ottawa L Rev 199, 205–14.

90 *Mathur* ONSC (n 1) [186].

91 ibid [185].

92 *Mathur* ONSC (n 1) (Factum of the Intervener Friends of the Earth Canada); *Mathur* ONCA (n 1) (Factum of the Intervener Friends of the Earth Canada).

unnecessary to consider whether either was in fact an unwritten constitutional principle, as their only role would be to assist in interpreting sections 7 and 15 and no such assistance was warranted in this case.[93] She did not provide clarification as to why no such assistance was necessary, even though the Applicants and Friends of the Earth specifically argued they should be recognized for this very purpose.

Friends of the Earth argued that the Constitution includes a binding obligation to sustain the ecological foundations of the Canadian state, including a stable climate, upon which all of the rights, interests, and powers of the Constitution depend.[94] With respect to the Applicants' *Charter* claims, Friends of the Earth stated that this unwritten principle of ecological sustainability supports an interpretation of sections 7 and 15 that aligns with the Applicants' arguments. For example, ecological sustainability would support an interpretation of section 7 that recognizes the positive ecological dimensions of the principles of fundamental justice, which require governments to set GHG emissions reduction targets that are at a minimum aligned with the international scientific consensus.

The Court of Appeal found it unnecessary to fully review or consider the issue of whether societal preservation and/or ecological sustainability are unwritten constitutional principles. In allowing the appeal and remitting the Application for a new hearing, the Court of Appeal stated, however, that 'how this issue may inform the question of whether the Target and Plan are *Charter* compliant … is another question that may require reconsideration at the new hearing'.[95] The Court of Appeal also noted that the interveners had raised 'relevant, important issues' that were not determined by Justice Vermette, including the recognition and impact of the unwritten constitutional principles of societal preservation and ecological sustainability.[96] We are hopeful the Superior Court will engage with this issue at the new hearing, given that the recognition of a novel unwritten constitutional principle – particularly ecological sustainability – has the potential to fundamentally alter the Canadian constitutional landscape by acknowledging that protection of the long-term

---

93 *Mathur* ONSC (n 1) [187].

94 *Mathur* ONSC (n 1) (Factum of the Intervener, Friends of the Earth Canada) [34]–[49].

95 *Mathur* ONCA (n 1) [77].

96 ibid [6], [77]. Note that the Court of Appeal cautioned that the Applicants may have to obtain leave to amend their pleadings and amplify the evidentiary record in order to have the additional issues raised by the interveners adjudicated, see ibid [7].

viability and well-being of ecological systems is essential to upholding *Charter* rights.[97]

4.3 A Restrictive, Formalistic Reading of Equality Rights in the Climate Context

Canada's constitutional guarantee of equality is found in section 15(1) of the *Charter*. While courts have paid lip service to the importance of substantive equality, equality claims have struggled since the *Charter*'s inception, especially in cases where the discriminatory treatment is indirect (adverse impacts cases).[98] In claims that allege both section 15(1) and section 7 violations, courts have often given the equality claim comparatively less judicial attention, sometimes disposing of the case under only section 7. The Supreme Court of Canada recently cautioned courts against this possibility, holding that section 15 claims deserve to be properly analyzed.[99]

The Superior Court's decision in *Mathur* was no exception to this general trend. Despite recognizing that climate change has disproportionate impacts on young people and Indigenous Peoples, the *Mathur* decision did not advance the state of jurisprudence on equality-based climate claims in Canada.[100] To establish a section 15(1) violation in an adverse impacts case, the Applicants would need to show that the government's conduct creates or contributes to a disproportionate impact on the claimant group based on a protected ground (which includes age and race), as compared to other groups. The government conduct does not need to be the only or even the dominant cause of the disproportionate impact.[101]

The Application judge dismissed the section 15(1) claim on the basis that it was a positive rights claim (and thus not captured by section 15) and also that it failed on causation. The Superior Court curiously applied a stricter causation

[97] See Lynda Collins and Lorne Sossin, 'In Search of an Ecological Approach to Constitutional Principles and Environmental Discretion in Canada' [2019] 52 UBC L Rev 293; Lynda Collins, *The Ecological Constitution: Reframing Environmental Law* (Routledge 2021), 18.

[98] See Jennifer Koshan and Jonnette Watson Hamilton, '"Clarifications" or "Wholesale Revisions"? The Last Five Years of Equality Jurisprudence at the Supreme Court of Canada' [2023] Supreme Court Law Review (forthcoming).

[99] See *Canadian Council for Refugees v Canada*, 2023 SCC 17 [180].

[100] For a broader discussion of climate equality claims under the *Charter*, see Nathalie J Chalifour, 'Equality Rights in Canadian Climate Litigation – Still a Struggle even with the World on Fire' [2025] Can J Women & Law (submitted).

[101] *Mathur* ONSC (n 1) [174] citing *Re Sharma*, 2022 SCC 39 [42], [45].

standard to section 15 than section 7, producing a judgment that is internally inconsistent on causation.

4.3.1 Misconstruing positive obligations in the section 15 context

Justice Vermette agreed with the Applicants that young people are especially susceptible to the physical health impacts of climate change and bear a disproportionate impact of its mental health impacts.[102] She also agreed that young people and future generations will bear the brunt of the catastrophic impacts of climate change as they worsen over time.[103] However, while she showed willingness to interpret section 7 as conferring positive obligations on the Applicants in light of the special circumstances presented by climate change, she was categorical in stating that section 15(1) does not impose a 'general, positive obligation on the state to remedy social inequalities or enact remedial legislation'.[104] Justice Vermette held that courts are under no positive obligation to remedy inequalities and that leaving gaps between protected groups and others does not infringe section 15. Justice Vermette relied on a recent decision of the Supreme Court of Canada (*R v Sharma*), which narrowed the approach to section 15(1) by stating that the section does not impose a 'general positive obligation'.[105] The Applicants were not seeking recognition of such an obligation, but rather a finding that the government's Target would lead to disproportionate impact on the Applicants. They distinguished *Sharma*, noting that it left intact previous Supreme Court of Canada jurisprudence recognizing positive obligations in section 15 claims in certain situations.[106] Justice Vermette did not address this argument nor equality case law beyond *Sharma*.

The Ontario Court of Appeal overturned the Superior Court, holding that Justice Vermette had once again, like in the section 7 analysis, mischaracterized the case as one seeking positive rights. Instead, the Court should have focused on whether Ontario's voluntarily assumed statutory obligation to address climate change under the *CTCA* was compliant with section 15.[107]

4.3.2 Internally inconsistent approach to causation

One of the more curious elements of the Superior Court's decision is that it applied a more stringent causation test for the section 15 claim as compared to section 7.

---

102 ibid [177].
103 ibid [177].
104 ibid [178].
105 *Sharma* (n 101).
106 *Mathur* ONSC (n 1) [87].
107 *Mathur* ONCA (n 1) [57].

Justice Vermette's causation analysis for section 7 was robust and aligned with the jurisprudence in other jurisdictions accepting the global, collective action nature of climate change in assessing the link between government action and the harms from climate change. Justice Vermette explicitly stated that 'the "collective action problem" has to be taken into consideration when assessing causation' and concluded that Ontario's severely deficient Target contributes to an increased risk of death and risks to security of the person faced by the Appellants and others.[108]

In her section 15 analysis, Justice Vermette ignored her own directive that the global, collective nature of climate change should be taken into consideration when assessing causation. Instead, she concluded that there is no causal link between this disproportionate impact of climate change on young people and Ontario's deficient Target.[109] Not only did this produce an internal inconsistency, but it also imposes a higher standard than the section 15(1) jurisprudence requires. The Supreme Court of Canada recently confirmed that section 15(1) requires claimants to show that the impugned state conduct 'creates or contributes to' the disproportionate impact but need not be the 'only or the dominant cause of the disproportionate impact'.[110] The causal connection may be established by a 'reasonable inference' and may not even require evidence when causation is obvious. The Superior Court did not offer an explanation to justify applying this stricter analysis for the equality claim. Not only did doing so create an internally inconsistent decision, but it once again imposed a higher bar for equality claimants, contrary to the Supreme Court of Canada's recent direction that the '*Charter* should not be treated as if it establishes a hierarchy of rights in which s. 15 occupies a lower tier' than section 7 claims.[111]

The Ontario Court of Appeal overturned the Superior Court's decision on section 15, holding that the lower court mischaracterized the equality claim as a positive rights case. According to the Court of Appeal, the lower court should have assessed whether 'Ontario committed itself to a level of greenhouse gas emissions that will create or contribute to a disproportionate impact on the basis of an enumerated or analogous ground'.[112] The Ontario Court of Appeal also called out the lower court for its inconsistent causation analysis, noting that '(t)he application judge's conclusion about causation under s. 15 … is difficult to reconcile with her conclusion about causation under s. 7 …' and calling out the Court for not explaining this apparent inconsistency.[113] Rather

108 *Mathur* ONSC (n 1) [149], [150].
109 ibid [178].
110 *Sharma* (n 101) [42], [45], [49].
111 *Canadian Council for Refugees* (n 99) [180].
112 *Mathur* ONCA (n 1) [58].
113 ibid [65].

than decide the issues itself, the Court of Appeal remanded the matter for rehearing by the Superior Court.

4.3.3 Temporal versus age-related distinctions

Another issue which seems to have troubled the Superior Court in *Mathur* is the temporal aspect of climate change. Temporal distinctions are not generally actionable under equality rights in Canada. This includes distinctions related to the date an event happened, such as the date a relationship ends due to a partner's death,[114] the date of remarriage,[115] or the date someone acquires a disease.[116] In her section 15 analysis, Justice Vermette characterized the distinction between young and older people as temporal.

The Superior Court seems to be conflating date-related temporal distinctions with age-related distinctions. While age-related distinctions may have a temporal aspect, in that they implicate the passage of time, this is quite different from distinctions based on age.[117] Young people will suffer disproportionately from climate change due to their young age in at least three ways. First, young people experience distinct physical and mental health harms from climate change because of their physiology and stage of life that make them uniquely vulnerable to impacts. On the motion to strike, Justice Brown correctly held that young people will suffer the most of all generations and that climate change impacts will 'exacerbate their pre-existing vulnerability and disability'.[118] Second, young people under the age of 18 are unable to vote in Canada, meaning that they cannot participate in the democratic process of selecting leaders that will take serious action on climate change. The Superior Court characterized this claim as an attack on the age limit to vote and held that the evidentiary record was insufficient to proceed on this basis.[119]

The third way in which young people suffer disproportionately from climate change is due to the *duration* of the impacts over their lifetimes, the *compounding magnitude* of those impacts, and the *increasing uncertainty and destabilization* associated with those impacts over time. While this set of interrelated impacts implicates the passage of time, the claim about disproportionate harm is not a temporal distinction. It is the fact of young age that causes the disproportionately longer duration of exposure to the irreversible, urgent,

---

114 *Canada (Attorney General) v Hislop*, 2007 SCC 10 [37].

115 *Bauman v Nova Scotia (Attorney General)*, 2001 NSCA 51 [65].

116 *Guild v Canada (Attorney General)*, 2006 FC 1529 [13], aff'd 2007 FCA 311.

117 See Nathalie J Chalifour, Jessica Earle and Laura Macintyre (n 78) 55–61 (discussing enumerated ground of age in the climate context).

118 *Held v Montana*, No CDV-2020-307, 2023 WL 5229257 (Mont 1st Dist Ct Aug 14, 2023) [8], [104]–[107], [111], [119]–[124], [127], [133]–[136].

119 *Mathur* ONSC (n 1) [181].

and existential features of the climate crisis.[120] As an American court recently found in *Held v Montana*, children born in 2020 will experience a two- to seven-fold increase in extreme events compared to people born in 1960.[121]

Justice Vermette's mischaracterization of the distinction as temporal may have stemmed from her decontextualized comparison of young and older persons in a given year. She chose 2050 as a snapshot year and held that everyone will be impacted by climate change at that time. This obscured the real question of whether young people are disproportionately impacted. Using a given 'snapshot year' disregarded the essential existential and irreversible features of climate change. It also obscured the fact that young age is the key factor that determines young people's physiological and psychological vulnerabilities as well as the magnitude and duration of their exposure to harms.

The Court's reliance on a 'snapshot year' also failed to account for the urgency in reducing GHG emissions given the decades-long delays between emissions and climate impacts. These time-lags have largely been ignored in Canadian climate jurisprudence to date.

Justice Vermette also did not factor in the burden created for young people and future generations who will need to undertake more radical future GHG reductions, something the German Constitutional Court held justifies imposing stricter reductions now.[122]

Interestingly, the Court of Appeal did not comment on the Superior Court's characterization of the distinction as temporal. When Justice Vermette rehears the case in December 2025, she will have the opportunity to address this shortcoming.

5. DISCUSSION AND CONCLUSION

This chapter provided a summary of *Mathur* and an overview of why it is a landmark case in the recent wave of Canadian rights-based climate litigation. Although it has yet to result in a finding of human rights violations, *Mathur* is leading the way towards holding Canadian governments accountable for the human rights implications of their inadequate responses to the climate

120 *Mathur* Motion to Strike (n 35) [187] (Justice Brown held that '[the] adverse effects of climate change on younger generations – who presumably would have more years to live than current generations – may be considered self-evident, especially if the Applicants are able to present evidence of historical or sociological disadvantage the Applicants have experienced as a result of their age').

121 *Held* (n 118) [136]. The Montana Supreme Court recently upheld the decision, see Chapter 8.

122 *Neubauer et al v Germany*, 1 BvR 2656/18, Order of 24 Mar 2021 (Federal Constitutional Ct) [192].

crisis. Before concluding, we want to comment on a few additional aspects of the case, including its reliance on and role in transnational climate change jurisprudence.

Given Canada's unique constitutional framework and the legal complexities of the hurdles faced and overcome so far, *Mathur* may be limited in its translation to foreign jurisdictions. However, the confirmation from the Ontario Superior Court of Justice and Court of Appeal that Canadian courts have the institutional capacity and competence to judicially review government climate laws and policy brings Canadian courts in alignment with what many other courts around the world have already recognized and adds to the influential ammunition for the jurisdictions where judicial reticence to adjudicating climate change remains. Additionally, *Mathur* is another example of courts taking the collective nature of the global warming problem seriously in rejecting national and subnational governments' arguments that their contribution to climate change is so minimal that it is meaningless for courts to require them to take any action at all.

In terms of *Mathur*'s reliance on foreign jurisprudence, it is worth noting that public interest organizations intervening before both the Superior Court and Court of Appeal focused their submissions on the successful judicial outcomes of rights-based climate litigation in other jurisdictions and the applicable lessons for Canada.[123] The application judge referred to the Netherlands Supreme Court decision in *Urgenda* as providing persuasive support of her conclusion that she could use the international scientific consensus on global GHG emissions reductions to conclude that Ontario's Target falls short of what is required.[124] The Ontario Court of Appeal noted the relevance and importance of issues raised by interveners, but not determined by the application judge, including the application of international law and international environmental law in the interpretation of *Charter* rights, and indicated they may need to be addressed at a fresh hearing.[125] This demonstrates the importance of public engagement with rights-based climate litigation, but it remains to be determined what role, if any, foreign and international cases will have in a potential successful outcome in *Mathur*.

---

[123] See e.g. *Mathur* ONSC (n 1) (Factum of the Intervener Friends of the Earth Canada); *Mathur* ONCA (n 1) (Factum of the Interveners Greenpeace Canada and Stichting Urgenda).

[124] *Mathur* ONSC (n 1) [146] ('While this Court is obviously not bound by decisions of foreign courts made in the context of different legal systems, I note that the Supreme Court of the Netherlands adopted a similar reasoning in a similar case, which I find persuasive').

[125] *Mathur* ONCA (n 1) [6].

More than six years have passed since the courageous young leaders in *Mathur* filed their legal challenge against Ontario's weak climate target. With a new hearing before the Supreior Court currently on hold pending the outcome of a motion before the Court of Appeal to re-open the case to consider unresolved issues, it will be at least a few more years before the case is finally decided, at which time many of the youth Applicants will no longer be youth. The delays inherent in pursuing *Charter* litigation and the judicial conservatism regarding climate change in Canada are in conflict with the urgent need for ambitious government action to mitigate GHG emissions. Every day counts as the climate clock ticks towards irreparable harm, imperiling the future for all young people and future generations. Despite these challenges, *Mathur* continues to advance the needle closer to a successful rights-based climate case in Canada, giving hope for all Canadians concerned about the climate crisis.

3. *Lliuya v RWE*: introducing polluter pays via transnational climate litigation?

Olaf Dilling and Romina Luz Schaller

1. INTRODUCTION

The increasing concentration of greenhouse gases (GHG) in the atmosphere is widely regarded as the principal driver of the rising global temperature, which in turn is contributing to the acceleration of climate change.[1] The impacts of climate change are profound, affecting both natural ecosystems and human systems in ways that threaten fundamental human rights.[2] Vulnerable populations, particularly in the Global South, bear the brunt of these effects, suffering disproportionate loss and damage from climate-related events, such as extreme weather events, rising sea levels, and disrupted agricultural systems.[3]

The largest historic contributors to GHG emissions are the countries with the most developed economies,[4] and it is largely due to the inaction or insufficient

[1] Nicholas Stern, *The Economics of Climate Change: The Stern Review* (CUP 2007); Ove Hoegh-Guldberg and others, '2018: Impacts of 1.5°C Global Warming on Natural and Human Systems' in Change Intergovernmental Panel on Climate (eds), *Global Warming of 1.5°C: IPCC Special Report on Impacts of Global Warming of 1.5°C Above Pre-industrial Levels in Context of Strengthening Response to Climate Change, Sustainable Development, and Efforts to Eradicate Poverty* (CUP 2022).

[2] Hoegh-Guldberg and others, '2018: Impacts of 1.5°C Global Warming on Natural and Human Systems' (n 2).

[3] Marina Romanello and others, 'The 2023 Report of the Lancet Countdown on Health and Climate Change: The Imperative for a Health-Centred Response in a World Facing Irreversible Harms' (2023) 402 The Lancet 2346. See also Sarah Schöngart and others, 'High-Income Groups Disproportionately Contribute to Climate Extremes Worldwide' (2025) 15 Nature Climate Change 627.

[4] H Damon Matthews and others, 'National Contributions to Observed Global Warming' (2014) 9 Environmental Research Letters; Matthew W Jones and

measures taken by these countries and corporations that the crisis continues to deepen.[5] Despite widespread scientific consensus on the need for drastic reductions in emissions, many developed economies and corporations have demonstrated a reluctance to assume responsibility for their role in climate change.[6] This lack of accountability has spurred growing demands for reparations to address both past and ongoing harm. While most climate-related legal actions have traditionally targeted states, an increasing number of civil claims are now being brought against corporations, holding them directly accountable for their emissions and the resulting harm.[7]

---

others, 'National Contributions to Climate Change Due to Historical Emissions of Carbon Dioxide, Methane, and Nitrous Oxide since 1850' (2023) 10 Scientific Data 155; Lea Beusch and others, 'Responsibility of Major Emitters for Country-Level Warming and Extreme Hot Years' (2022) 3 Communications Earth & Environment 7; UNEP, *Emissions Gap Report 2024: No more hot air ... please! With a Massive Gap Between Rhetoric and Reality, Countries Draft New Climate Commitments* (2024).

5 While emerging economies, most notably China, are currently among the highest GHG emitters in absolute terms, their per capita emissions remain significantly lower than those of most developed countries. Moreover, their cumulative historical contributions to climate change are still substantially smaller. In line with the principle of *common but differentiated responsibilities and respective capabilities* (CBDR-RC), established under the UNFCCC (art 3(1) and reaffirmed in the Paris Agreement (art 2(2)), developed countries should continue to take the lead in combatting climate change. This includes deeper emissions reductions and enhanced support for developing countries through finance, capacity building, and technology transfer. See Matthews and others, 'National contributions to observed global warming' (n 5); Michael Grubb and others, 'Introduction and Framing' in IPCC (eds), *Climate Change 2022: Mitigation of Climate Change* (CUP 2022) 151, 158.

6 Romanello and others, 'The 2023 report of the Lancet Countdown on health and climate change: The imperative for a health-centred response in a world facing irreversible harms' (n 4) 2379; Simone Cenci and others, 'The alignment of companies' sustainability behavior and emissions with global climate targets' (2023) 14 Nature Communications 7831; Yann Robiou du Pont and Malte Meinshausen, 'Warming assessment of the bottom-up Paris Agreement emissions pledges' (2018) 9 Nature Communications 4810; UNGA 'Report of the Special Rapporteur on the promotion and protection of human rights in the context of climate change' UN Doc A/77/226 (2022); UNGA 'Report of the Special Rapporteur on the promotion and protection of human rights in the context of climate change, Elisa Morgera' UN Doc A/HRC/56/46 (2024).

7 UNGA 'Report of the Special Rapporteur on the promotion and protection of human rights in the context of climate change, Ian Fry' UN Doc A/78/255 (2023). Joana Setzer and Catherine Higham, 'Global Trends in Climate Change

The case of *Saúl Lliuya v RWE*[8] is a striking example of this new wave of climate litigation, highlighting efforts to hold corporations accountable for their role in climate change. Saúl Lliuya is a claimant from a climate-vulnerable region of the Peruvian Andes who sued RWE AG (RWE), a German energy company with global operations, for its contribution to GHG emissions and the associated climate risks that threatened his property. Specifically, Lliuya argued that climate change, exacerbated by emissions from companies like RWE, has accelerated the melting of glaciers above his hometown of Huaraz, significantly increasing the risks of a catastrophic glacial lake flood. Lliuya was seeking a contribution from RWE to the costs of safety measures to protect his property from the tangible risks associated with climate change, proportional to RWE's alleged 0.38% share of global emissions.

In May 2025, the Higher Regional Court of Hamm dismissed Lliuya's claim against RWE, ruling that the flood risk (estimated at 1% over 30 years) did not meet the threshold of imminent or concrete harm required under German private law (Section 1004(1) of the German Civil Code).[9] Despite this, the Court made two landmark findings: first, that corporations, like RWE, could in principle be held proportionately liable for climate-related damages abroad; and second, that German private law could apply extraterritorially in such cases. While the dismissal was a setback for Lliuya, the judgment affirmed the justiciability of private climate claims, setting a precedent for future lawsuits based on scientifically grounded attribution evidence.

Against this backdrop, this chapter examines the application of the 'polluter pays' principle through the lens of *Lliuya v RWE*, analyzing its significance within the broader context of transnational climate litigation. We explore the legal and normative implications of holding corporations accountable for their contributions to climate change and how such cases challenge traditional boundaries of corporate liability. In this context, litigation such as *Lliuya v RWE* can influence regulatory frameworks and international agreements by highlighting gaps in current regulations and driving demands for stronger

---

Litigation: 2025 Snapshot' (Grantham Research Institute LSE 2025) 18, place this type of litigation among what they call 'polluter pays' cases.

8 LG Essen, 2 O 285/15, NVwZ 2017, 734; OLG Hamm, judgement of 28 May 2025 – 5 U 15/17.

9 Civil Code BGB as promulgated on 2 January 2002 (Bundesblattgesetz (BGBl., Federal Law Gazette) I p. 42, 2909; 2003 I p. 738), last amended by Article 1 of the Act of 17 July 2025 (Federal Law Gazette 2025 I No. 163). An English translation provided by the Federal Ministry of Justice and Consumer Protection is available at <www.gesetze-im-internet.de/englisch_bgb/englisch_bgb.html #p4705> accessed 6 August 2025.

corporate liability. These developments contribute to the evolving landscape of global climate governance.

To this end, we first contextualize Germany's and RWE's contribution to climate change with a short overview of GHG emissions in Germany (section 2). We then provide a summary of the case and the court proceedings (section 3), followed by an analysis of the claim within the framework of German legal doctrine (section 4). Particular attention is given to how the Lliuya plaintiffs strategically shaped the case to promote the bottom-up integration of the polluter pays principle into climate governance (section 5). The subsequent sections examine the financial (section 6) and transnational legal (section 7) implications of the case and provide concluding remarks on *Lliuya v RWE*'s contribution to global climate change litigation (section 8).

2. GERMANY'S AND RWE'S ROLE IN CLIMATE CHANGE

To understand the relevance of the case, it is important to situate RWE's role within the broader context of Germany's carbon footprint: Germany, where RWE is headquartered, ranks among the top ten global emitters of CO_2.[10] RWE itself has referred to the company as 'Europe's largest single emitter of CO_2'.[11] During the court proceedings, its cumulative contribution to global historic industrial CO_2 emissions was estimated as 0.38% and just under 0.24% of all CO_2 emissions.[12] According to RWE's own business reports, the company was emitting approximately 150 million tons of CO_2 annually at the time the action was filed in 2014. As of 2023, RWE was still ranked as the 109th largest emitter of CO_2 globally, accounting for approximately 0.15% of total annual global emissions.[13]

Also as of 2023, Germany's net GHG emissions per capita stood at 8.0 tons of CO_2 equivalent (tCO_2-eq), exceeding the EU average of 6.9 tCO_2-eq. The country accounts for approximately 1.3% of global emissions, positioning it as

[10] Emissions Database for Global Atmospheric Research (EDGAR), 'Greenhouse Gases Emissions and Climate' <https://edgar.jrc.ec.europa.eu/climate_change> accessed 6 August 2025.

[11] RWE AG, 'Our Responsibility. Report 2009' (2010) 31 <https://t1p.de/57i6k> accessed 6 August 2025.

[12] See the Higher Regional Court judgment in the *Lliuya v RWE* case, OLG Hamm, judgment of 28 May 2025, 5 U 15/17 <https://openjur.de/u/2521471.html> accessed 6 August 2025.

[13] Carbon Majors Database, <https://influencemap.org/briefing/The-Carbon-Majors-Database-2023-Update-31397> accessed 6 August 2025.

the 13th largest emitter worldwide.[14] The energy sector remains the primary source of emissions (29%), followed by industry (22%) and transport (21%).[15] Notably, between 2005 and 2023, Germany reduced its net GHG emissions by 30%, largely due to reforms in the energy and building sectors. Compared to 1990 levels, net emissions, including those from the Land Use, Land-Use Change and Forestry (LULUCF) sector, declined by 47% by 2023.[16] As a party to the Paris Agreement, Germany has committed to achieving climate neutrality by 2050.[17] Furthermore, as an EU member state, it is also bound by supranational climate governance frameworks such as the European Green Deal and European Climate Law,[18] which mandates a legally binding target of net-zero GHG emissions by 2050[19] and an interim net GHG reduction of at least 55% by 2030.[20]

At the national level, Germany has implemented a series of policy and regulatory measures to meet these obligations. In 2016, the federal government introduced the Climate Action Plan,[21] later supplemented by the Climate Action Programme 2030,[22] to operationalize the long-term 2050 objectives. Legislatively, Germany has enacted the Federal Climate Action Law (*Bundes-*

---

14 Emissions Database for Global Atmospheric Research (EDGAR), 'GHG emissions of all world countries 2024 report' <https://edgar.jrc.ec.europa.eu/report_2024?vis=ghgtot#emissions_table> accessed 4 August 2025.

15 European Union, 'Climate Action Progress Report 2024 – country profile Germany' <https://t1p.de/v4qp4> accessed 6 August 2025.

16 ibid 1. For more details on LULUCF (Land Use, Land-Use Change, and Forestry), see the IPCC, 2022: Annex I: Glossary, 1806. For specific data on LULUCF emissions in Europe, refer to European Environment Agency, available at <https://www.eea.europa.eu/en/analysis/indicators/greenhouse-gas-emissions-from-land> accessed 4 August 2025.

17 Paris Agreement, adopted 12 December 2015, entered into force 4 November 2016 55 ILM 740, art 4(1).

18 Council Regulation 2021/1119 of 30 June 2021 establishing the framework for achieving climate neutrality and amending Regulations (EC) No 401/2009 and (EU) 2018/1999 (European Climate Law) (2021) OJ L243/1.

19 Council Regulation 2021/1119 of 30 June 2021 establishing the framework for achieving climate neutrality and amending Regulations (EC) No 401/2009 and (EU) 2018/1999 (European Climate Law) (2021) OJ L243/1.

20 European Climate Law, art 1, art 4.1.

21 Nature Conservation Federal Ministry for the Environment, Building and Nuclear Safety (BMUB), 'Climate Action Plan 2050 – Principles and goals of the German government's climate policy' <https://t1p.de/r3aed> accessed 6 August 2025.

22 Federal Ministry of Agriculture, Food and Regional Identity (BMLEH), 'Klimaschutzprogramm 2030' <https://t1p.de/yg2eb> accessed 6 August 2025.

Klimaschutzgesetz),[23] which sets a more stringent timeline than the EU framework, by mandating net GHG neutrality by 2045 (five years earlier than the EU target) and net negative emissions by 2050. Recognizing that meeting these targets requires not only substantial emission reductions but also active carbon dioxide removal options (CDR) strategies,[24] the law incorporates specific provisions for annual emissions accounting in the land use sector[25] and acknowledges the role of technological solutions.[26]

German courts, particularly the constitutional and administrative spheres, have played a key role in clarifying and reinforcing the country's climate obligations under existing legal frameworks. A landmark example is the 2021 *Neubauer* decision, in which the Federal Constitutional Court ruled that intergenerational equity is a constitutional imperative. The Court held that the state must distribute emissions reductions fairly over time to avoid shifting disproportionate burdens on future generations and limiting their opportunities.[27] Similarly, administrative courts have issued rulings mandating that the federal

---

23 Federal Climate Action Act of 12 December 2019 (Federal Law Gazette I, p. 2513), as last amended by Article 1 of the Act of 15 July 2024 (Federal Law Gazette I No. 235) s 3.

24 Priyadarshi Shukla and others, 'IPCC 2022: Summary for Policymakers' in IPCC (ed), *Climate Change 2022: Mitigation of Climate Change Contribution of Working Group III to the Sixth Assessment Report of the Intergovernmental Panel on Climate Change*, C.11.1; Federal Ministry for Economic Affairs and Energy, 'Langfriststrategie Negativemissionen zum Umgang mit unvermeidbaren Restemissionen (LNe) – Eckpunktepapier' <https://t1p.de/b9bgi> accessed 4 August 2025; Johannes Förster and others, 'Framework for Assessing the Feasibility of Carbon Dioxide Removal Options Within the National Context of Germany' (2022) 4 Front Clim 758628; Malgorzata Borchers and others, 'A Comprehensive Assessment of Carbon Dioxide Removal Options for Germany' (2024) 12 Earth's Future e2023EF003986.

25 Federal Climate Action Law, s 3a. By 2030, emissions should reach a minimum of -25 million tCO2-eq; by 2040, emissions should achieve at least -35 million tCO2-eq; and by 2045, emissions should be reduced to no less than -40 million tCO2-eq.

26 Federal Climate Action Law, s 3b.

27 1 BvR 2656/18 NJW 2021, 1723. The English version is available at <https://t1p.de/62yc> accessed 6 August 2025. See Mathias Hong, 'Intertemporal Freedom in the Historic Climate Protection Ruling of the German Federal Constitutional Court: An Inspiration for Strasbourg's European Court of Human Rights?' (*VerfBlog*, 8 July 2023) <https://verfassungsblog.de/intertemporal-freedom-in-the-historic-climate-protection-ruling-of-the-german-federal-constitutional-court/> accessed 6 August 2025.

government adopt more stringent actions to meet national climate targets.[28] Together, these judicial interventions reflect an emerging constitutional and administrative jurisprudence that reinforces the binding nature of Germany's climate commitments.

Despite this comprehensive legal and policy architecture, and significant judicial rulings on state obligations, major corporate emitters like RWE have continued to report high annual emissions. This persistent trend calls into question the effectiveness of existing regulatory mechanisms and suggests that holding (major) emitters legally accountable through private litigation may be necessary. Against this backdrop, the *Lliuya v RWE* case emerges as a critical test: whether large-scale polluters can be held liable under private law in civil law jurisdictions for their proportional contribution to climate-related harms.

3. CASE SUMMARY

The plaintiff is the co-owner of a residential building in the town of Huaraz, Peru, which he purchased from his parents in the spring of 2014 through a notarized agreement. Huaraz, the capital of the Ancash Region, is situated in the Callejón de Huaylas valley at an altitude of 3,050 m, on the right bank of the Santa River, at the base of the Cordillera Blanca mountain range, whose peaks rise above 6,000 m. Above the town, at an elevation of 4,562 m, lies Lake Palcacocha, a glacial lake dammed by a natural moraine. The lake is fed by meltwater from surrounding glaciers but has only limited drainage capacity due to the region's geological characteristics. In the late 1930s, the lake held between 10 to 12 million cubic meters of water.

In the past, glacial lake outburst floods triggered by earthquakes or landslides have posed significant risks to the area. In 1941, an earthquake caused an ice avalanche to plunge into the lake. The resulting flood devastated large portions of Huaraz, claiming approximately 1,800 lives and causing widespread destruction.[29] Since then, various protective measures, including

[28] OVG Berlin-Brandenburg, judgment of 16 May 2024, Az. OVG 11 A 22/21, OVG 11 A 31/22.

[29] Various publications estimate that between 3,000 and 5,000 people lost their lives in this earthquake. However, contemporaneous records from the Peruvian Red Cross report a significantly lower death toll of approximately 1,800. For more details see: Carlos Batalla Sotelo, *Una idea, una acción: 125 años de la Cruz Roja Peruana* (Tarea Asociación Gráfica Educativa 2004) 213; Steven Wegner, *Lo que el agua se llevó – Consecuencias y lecciones del aluvión de Huaraz de 1941. Nota técnica n° 7.* (Ministerio del Ambiente. Viceministerio de Desarrollo de Recursos Naturales, Dirección General de Cambio Climático y Desertificación 2014) 3, 40;

artificial drainage systems and dams, have been implemented to prevent the risk of future catastrophes by controlling the lake's water volume.

Despite these efforts, the water volume of Lake Palcacocha increased to 17.3 million cubic meters by 2009. Additional measures failed to significantly reduce the water level, and by February 2016, the volume had continued to rise. Scientific studies have indicated that, in the event of another glacier lake outburst flood, the area where the plaintiff's property is located would be at high risk of inundation.[30] However, the extent and imminence of this risk became the most contested issue in *Lliuya v RWE*.[31]

The defendant, RWE, is the parent company of various subsidiaries engaged in energy generation and supply. The plaintiff alleges that RWE's activities have contributed to global warming, which in turn has accelerated the melting of the Cordillera Blanca glaciers and increased the risk of flooding from Lake Palcacocha. Based on scientific estimates, the plaintiff attributes 0.47% of global warming to RWE, corresponding to its share of global GHG emissions since the beginning of industrialization (revised down to 0.38% at the concluding stages of the case, based on the Carbon Majors Study reported by Richard Heede).[32] The plaintiff argues that scientific models allow for the quantification of the defendant's specific contribution to climate change and to the associated flood risk.

Building on this attribution analysis, the plaintiff bases his legal claim on Section 1004(1) of the German Civil Code (*Bürgerliches Gesetzbuch* – BGB), which governs protection against interference with property. He asserts that RWE qualifies as a *Störer* (disturber)[33] because its emissions have materially contributed to a risk that threatens his legally protected property interests. In

Mark Carey, *Glaciares, cambio climático y desastres naturales: Ciencia y sociedad en el Perú* (Institut français d'études andines 2021) 48–80.

30 See Christian Huggel and others, 'Anthropogenic climate change and glacier lake outburst flood risk: Local and global drivers and responsibilities for the case of Lake Palcacocha, Peru' (2020) 20 Nat Hazards Earth Syst Sci 2175; Alina Motschmann and others, 'Losses and damages connected to glacier retreat in the Cordillera Blanca, Peru' (2020) 162 Climatic Change 837, 845; Jan Klimes and Adam Emmer, 'Comments on Landslide Occurrence under Changing Climate in Earthquake-Prone Cordillera Blanca, Peru' (2024) 9 Revista Científica de Glaciares y Ecosistemas de Montaña 97, 103.

31 See OLG Hamm, judgement of 28 May 2025 – 5 U 15/17.

32 Richard Heede, 'Carbon Majors: Accounting for carbon and methane emissions 1854–2010' (*Climate Mitigation Services* 2014).

33 According to an official translation provided by the Federal Ministry of Justice and Consumer Protection, the term used for *'Störer'* is *'disturber'*. The English translation is available at <https://www.gesetze-im-internet.de/englisch_bgb/englisch_bgb.html#p4705> accessed 6 August 2025.

line with established German nuisance doctrine, he seeks a declaratory judgment that RWE is proportionally liable (0.47%, later 0.38%) for the risk of harm to Lliuya's property; and proportional compensation for preventive measures (e.g., flood protection infrastructure) to avoid future interference with his property rights.[34] Because of the complex scientific and probabilistic nature of climate attribution, the plaintiff invokes Section 287 of the German Code of Civil Procedure (*Zivilprozessordnung* – ZPO),[35] which allows the court to estimate the extent of harm and apportion liability where precise quantification is not possible. The plaintiff added three auxiliary remedies to the action, requesting the Court to:

- order the defendant to take appropriate measures to ensure that the water volume in Lake Palcacocha is reduced in accordance with the defendant's contribution to the cause, with the exact amount to be determined by the court in accordance with German civil procedure rules.[36]
- order the defendant to pay their share of EUR 17,000 to the municipal association Waraq for appropriate protective measures suitable to safeguard the plaintiff.
- order the defendant to pay the plaintiff EUR 6,384 to compensate for measures he had already taken to protect his house. This remedy is based on Section 1004(1) in combination with Section 683, 684 BGB, which govern the reimbursement of expenses in cases of *negotiorum gestio* (performance on another's behalf without prior consent). Under this framework, a person who incurs expenses while acting on their own account, such as in the exercise of a right of removal, may only be entitled to reimbursement, rather than further compensation.[37]

RWE requested that the Court dismiss the action, contesting both the admissibility and the merits of the claims. The company argued that there is no causal link between its emissions and the flood risk alleged by the plaintiff. RWE

---

34 An English translation of plaintiff's claim is available at <https://climatecasechart.com/non-us-case/lliuya-v-rwe-ag/> accessed 6 August 2025.

35 Code of Civil Procedure as promulgated on 5 December 2005 (Federal Law Gazette) I p. 3202; 2006 I p. 431; 2007 I p. 1781), last amended by Article 1 of the Act of 24 October 2024 (Federal Law Gazette 2024 I No. 328). An English translation provided by the Federal Ministry of Justice and Consumer Protection is available at <https://www.gesetze-im-internet.de/englisch_bgb/englisch_bgb.html#p4705> accessed 6 August 2025.

36 Section 287 ZPO.

37 Mai-Lan Tran, *Grenzüberschreitende Klimaklagen* (Mohr Siebeck 2024) 90.

further denied that it qualifies as a *Störer* (disturber) under German property law and maintained that there is no legal basis for holding a private actor liable for harms resulting from ubiquitous environmental pollution.[38]

The Essen Regional Court dismissed the action at first instance, finding it to be partly inadmissible and partly unfounded,[39] on the following grounds:

- *Inadmissibility of the main claim and first two auxiliary remedies*: The Court deemed the main claim and the first two auxiliary remedies inadmissible due to a lack of specificity under Section 253(2) No. 2 ZPO. These remedies were deemed unenforceable because they did not articulate a clearly defined and sufficiently precise request. Additionally, the Court noted that the municipal association Waraq, mentioned in the remedy, was not a legally recognized entity, rendering enforcement of the proposed protective measure impossible.
- *Unfounded third auxiliary remedy*: While the third auxiliary remedy, which covered a specified sum of money, was considered admissible, the Court found it unfounded. It ruled that the plaintiff had not established a sufficient causal link between the defendant's emissions and the alleged flood risk. Given the vast number of global emitters, the Court concluded that the defendant's emissions were not significant enough to be considered a material cause of the flooding risk.

After the dismissal at first instance, the case was appealed to the Higher Regional Court of Hamm. Although the appeal was filed in 2017, procedural delays, including a motion for judicial recusal, meant that a final decision was only issued in May 2025.[40] The Court ultimately dismissed the appeal. The key issue in the dismissal revolved around the evidence demonstrating the probability of the plaintiff's property being damaged by a glacial lake outburst flood or mudslide. Based on court-commissioned expert evidence, the risk of such an outburst was found to be minimal: less than 1% within the next 30 years. Even if an outburst were to occur, the Court concluded it would be unlikely to directly destroy the plaintiff's home.[41] Despite the unsuccessful

---

[38] Additionally, the defendant raises a statute of limitations defense, which was dismissed by the Higher Regional Court in Hamm. According to the plaintiff, the significant facts that led to his claim were relatively new such as a warning by public authorities and the significant rise of the lake's level. See OLG Hamm.

[39] LG Essen, 2 O 285/15 NVwZ 2017, 734.

[40] OLG Hamm, I-5 U 15/17, ZUR 2018, 118.

[41] Paul Grimm and Jan-Erik Schirmer, 'Success Without Victory: Lliuya v. RWE Decided' (*VerfBlog*, 1 Juny 2025) <https://verfassungsblog.de/lliuya-rwe-climate-judgment/> accessed 6 August 2025.

appeal, many legal and climate experts viewed the ruling as a success because for the first time, a German court acknowledged that GHG emitters could, in principle, be held liable for climate change-related damages.[42] This sets a significant precedent, paving the way for future climate litigation.

4. LEGAL ANALYSIS

This section explains why *Lliuya v RWE* was a strategically chosen and path-breaking case. Every aspect of its design, from legal framing and selection of parties to the choice of forum, was deliberately structured to test the boundaries of German civil law. In general, the possibilities for environmental litigation under German private law are quite limited. However, this case innovatively repurposed a property law provision, traditionally applied for disputes between neighboring landowners, to address global climate-related harm. What made this approach remarkable was its expansion of a legal instrument originally confined to local conflicts, applying it instead to transboundary environmental damage.[43]

The case has high precedential value, as the Court's legal reasoning was thoroughly substantiated. This is evident in its detailed analysis of the claim.[44] Although the case was dismissed on factual grounds, as the Court concluded that the plaintiff's property did not face a sufficiently imminent hazard, this limitation was tied to the specific circumstances of Lliuya's claim.[45] Essentially, the Court's reasoning leaves the door open for similar legal arguments to succeed in future cases where the factual basis is stronger.

In *Lliuya v RWE*, the plaintiff strategically focused on threatened harm rather than damage that had already occurred. This approach is important from the perspective of German legal doctrine. The claim was not grounded in tort law but instead relied on property law, specifically the 'neighbor law'

---

[42] Jan-Erik Schirmer, 'Erfolgreich gescheitert: Warum Großemittenten grundsätzlich für Klimaschäden haften können' (*VerfBlog*, 30 May 2025) <https://verfassungsblog.de/lliuya-rwe-klimaklage-olg-hamm/> accessed 6 August 2025.

[43] Jannika Jahn, 'In der Klimakrise sind alle Nachbarn' (30 May 2025) <https://www.mpg.de/24809085/rwe-klimaklage> accessed 6 August 2025; see also the acknowledgement as 'innovative' by Maria Antonia Tigre and Margaretha Wewerinke-Singh 'Beyond the North-South Divide: Litigation's Role in Resolving Climate Change Loss and Damage Claims', 32 Review of European, Comparative & International Environmental Law 439, 452.

[44] A thorough analysis of the legal soundness of the claimant's argumentation is given in the reasons of the judgement, OLG Hamm, judgement of 28 May 2025 – 5 U 15/17, pp. 39–91.

[45] OLG Hamm, judgement of 28 May 2025 – 5 U 15/17, pp. 91–134.

(*Nachbarschaftsrecht*) under Section 1004(1) BGB. This legal pathway allows for injunctive relief to prevent or remove interferences with property rights, even in the absence of actual damage.[46]

In German private law, neighbor law, including Section 1004(1) BGB, is considered a subset of property law, distinct from tort law or contract law. This area typically includes issues involving private nuisances, boundary fencing, encroaching structures, or overhanging branches and tree roots.[47] While tort law (governed by Section 823 BGB and following) typically addresses compensation for damages caused by intentional or negligent acts, neighbor law deals with ongoing or imminent infringements on property rights. Section 1004(1) BGB states:

> (1) If the ownership is interfered with by means other than removal or retention of possession, the owner may demand that the disturber remove the interference. If there is concern that further interferences will ensue, the owner may seek injunctive relief.
> (2) The claim is excluded if the owner is obliged to tolerate the interference.[48]

This legal provision grants property owners the right to seek injunctive relief against disturbances (*Störungen*) that do not involve the loss of possession. It applies when an interference hinders the lawful use of their property (e.g., pollution, noise, smoke, water runoff, or in this case, increased flood risk). In other words, the property owner may demand that the 'disturber' (*Störer*) cease the interference or remove its effects. If future interference is anticipated, the owner may also file a pre-emptive lawsuit for an injunction.

However, this right does not apply if the owner is legally obliged to tolerate the interference (e.g., under statutory toleration duties or public interest exceptions). These obligations are further specified in Section 906 BGB, which excludes minor impairments (*unwesentliche Beeinträchtigung*) or those considered customary for the area (*ortsübliche Benutzung*).

---

46 For an anthropological perspective on the 'neighborly relations' of *Lliuya v RWE*, see Noah Walker-Crawford, 'Climate Change in the Courtroom: An Anthropology of Neighborly Relations' (2022) 23 Anthropological Theory 76.

47 See Gerhard Wagner and Arvid Arntz, 'Liability for Climate Damages under the German Law of Torts' in Wolfgang Kahl and Marc-Philippe Weller (eds), *Climate Change Litigation – A Handbook* (C.H. Beck 2021) 423.

48 An English translation provided by the Federal Ministry of Justice and Consumer Protection is available at <www.gesetze-im-internet.de/englisch_bgb/englisch_bgb.html#p4705> accessed 6 August 2025. Although the official translation uses the term '*prohibitory injunction*', the expression '*injunctive relief*' is more commonly employed.

From a legal doctrinal standpoint, the plaintiff in *Lliuya* is not seeking compensation for past damages but rather the removal or prevention of a material risk to his property. Importantly, his claim does not demand that RWE stop its emissions directly. Instead, he argues that the defendant should contribute proportionally to the costs of preventive measures necessary to avoid a glacial lake outburst and the resulting damages to his property. Because claims under Section 1004(1) BGB are forward-looking and aimed at stopping or mitigating ongoing or expected harm, they are not subject to the same stringent requirements as tort claims under Section 823 BGB. Notably, the plaintiff need not to prove fault (intent or negligence) or demonstrate specific past damages.

Typically, Section 1004(1) BGB has been applied to disputes between neighboring landowners, where spatial proximity leads to frequent interactions and direct physical interferences with property rights. However, in *Lliuya v RWE*, this legal instrument is being elevated to the scale of the 'global village': the defendant is not a physical neighbor but a major corporation whose emissions are alleged to contribute to a distant, yet foreseeable, climate-related harm. This constitutes a legally innovative application of German property law to address transboundary climate harm.

This approach offers strategic advantages. First, as previously noted, the evidentiary burden is lower: while causation must still be demonstrated, there is no need to prove fault.[49] This is a significantly easier threshold to meet. Second, under German tort law, defendants often invoke the 'permit defense', arguing that emissions compliant with regulatory standards exempt them from civil liability. However, this defense does not apply in property-based claims under Section 1004(1) BGB. By framing the case as a property dispute, the plaintiff's legal strategy avoided some of the key hurdles inherent in a conventional tort claim.

Another aspect of the *Lliuya v RWE* case is the role of the parties involved. Under German civil procedure law, the plaintiff must be a natural person asserting a legally protected right; in this case, a property owner facing climate-related risks such as flooding or mudslides. However, the lawsuit was not pursued solely as an individual claim. The plaintiff received substantial support from Germanwatch, an independent environmental and human rights NGO advocating for sustainable global development. Germanwatch covered all legal and procedural costs, signaling that the case extends beyond the plaintiff's personal interests.

---

49 Ingo Wolfgang Sarlet and Gabriel Wedy, 'Notas sobre os assim chamados "litígios climáticos" na Alemanha – O Caso Lliuya Vs. Rwe' (2020) 21 Espaço Jurídico Journal of Law [EJJL] 277.

While the plaintiff's legal standing is based on the threat to his private property, the broader objective is to safeguard a public good: the protection of Huaraz from climate-induced disasters. Reflecting this, the plaintiff has agreed contractually to transfer any awarded compensation to the provincial government of Ancash if the case succeeds.[50] From a legal perspective, grounding the claim in individualized property rights provides a clearer path to standing and justiciability. Yet in substance, the case targets a public harm, demonstrating how strategic litigation can leverage private law mechanisms to advance broader accountability.

The selection of RWE as the defendant was strategic. While RWE is a single corporate entity, and from a purely legal perspective, any other GHG emitter could have been chosen, it serves as a representative case for similar polluters (*pars pro toto*). Under German legal doctrine, any significant emitter could theoretically be held liable under the principle of equivalence in causation (*Äquivalenztheorie*). This doctrine, applied in both criminal and civil law, treats any contributing factor as legally relevant if the harm would not have occurred without it. This is also known as the *conditio sine qua non* principle. In this case, the plaintiff argued that RWE's historical GHG emissions, though accounting for only 0.38% of global GHG emissions, were a necessary contributing factor to increased flood and landslide risks in Huaraz. Under causation doctrine, even a minor contribution can constitute a legally relevant cause.

This aspect of the case highlights broader challenges in establishing causation under German law, particularly in environmental litigation involving diffuse harms like climate change.[51] The Essen District Court's reasoning reflects lingering doctrinal uncertainty in such complex cases. A relevant precedent is the Federal Court of Justice (*Bundesgerichtshof – BGH*) 1980s ruling on *Waldsterben*, the widespread decline of German forests due to acid rain caused by coal power plants and vehicle emissions of sulphur dioxide (SO_2) and nitrogen oxides (NO_x). In that case, forest owners sued the German Federal government for failing to protect the environment (state liability),[52] but the BGH dismissed the claims, ruling that the sources of forest damage were too diffuse and the causal link too indirect to establish legal responsibility.

---

50 Sebastian Balzter, 'Warum ein peruanischer Kleinbauer RWE verklagt' *Frankfurter Allgemeine Zeitung* (Frankfurt am Main, 2 November 2021) <www.faz.net/aktuell/wirtschaft/klima-nachhaltigkeit/rwe-wird-von-kleinbauer-aus-peru-auf-schadensersatz-verklagt-17609050.html> accessed 6 August 2025.

51 Nicolas de Sadeleer, *Environmental Principles – From Political Slogans to Legal Rules* (OUP 2020) 66.

52 So called 'forest damages judgements' (*Waldschadensurteile*), BGH, Judgement of 10 December 1987, Az III ZR 220/86, see also BVerfG NJW 1998, 3264.

In *Lliuya v RWE*, both the Essen District Court and the Higher Regional Court of Hamm addressed the issue of 'diffuse and distant causation' in markedly different ways. The District Court relied heavily on *Waldsterben* precedent, where courts rejected liability due to the difficulty of linking specific emissions to localized forest damage. However, that analogy overlooks critical scientific distinctions. Acid rain produces spatially variable effects shaped by local ecological and meteorological factors, which complicate attribution to individual emitters. Climate change, by contrast, stems from the cumulative impact of well-mixed GHGs, especially CO_2, which contribute uniformly to global warming regardless of their source. The plaintiff invoked attribution science, including the *Carbon Majors Report*,[53] to demonstrate RWE's proportional contribution to global emissions and the resulting climate risks. The Higher Regional Court of Hamm recognized that such proportional causation could, in principle, support liability under Section 1004(1) BGB and allowed the case to proceed.

The District Court also downplayed the evidentiary strength of climate attribution science and questioned the causal link between global emissions and local flood risks. Notably, it did not cite any climate science literature or scientific expert studies to support its skepticism, diverging from the approach taken by most courts worldwide. Instead, the Court relied almost exclusively on a single legal academic article, selectively quoting and paraphrasing passages without engaging with the broader scientific consensus.[54] The reasoning of the Essen District Court was very basic: it cited only legal literature (without quoting correctly) while asserting, without basis, that climate change remains scientifically contested. By contrast, the Higher Regional Court of Hamm addressed the causation issue with greater doctrinal rigor and scientific precision.

That said, this development does not constitute a legal innovation in the doctrinal sense. Rather, it reflects an evolution in how German courts assess causation in the light of advances in climate science. The Higher Regional Court acknowledged the well-established anthropogenic causes of global warming, framing the central question not as *whether* human emissions cause climate warming, but as *how* this warming contributes to glacial melt and the specific flood risks threatening the plaintiff's property. A substantial part of the Court's decision focused on this scientific question, engaging directly with expert reports and probabilistic risk assessments.[55]

---

53 Heede, 'Carbon Majors: Accounting for carbon and methane emissions 1854–2010' (n 33).

54 District Court of Essen, see above (n39), Chatzinerantzis, NJOZ 2010, 594.

55 For a comprehensive account of the attribution science and methodologies involved, see the work of Rupert F Stuart-Smith and others, 'Liability for

Despite its more rigorous engagement with climate science, the Higher Regional Court's approach was not exempt from criticism. The case was ultimately dismissed based on court-appointed experts' conclusions that the probability of a glacial lake outburst flood affecting the plaintiff's property within 30 years was below 1%, with likely minimal damage. Critics argue that this assessment was overly restrictive. As Grimm and Schirmer point out, the experts relied exclusively on historical local data, failing to incorporate forward-looking climate risks, contrary to current scientific standards, and excluded the impacts of rockfalls, which IPCC reports indicate will increase in frequency and intensity due to climate-induced permafrost melt.[56]

A private expert for the plaintiff proposed applying a general 'climate factor' to account for the increased likelihood of trigger events, but this suggestion was dismissed due to a lack of localized permafrost data. The Court accepted the position of its appointed experts without further inquiry, despite procedural obligations to engage seriously with private expert evidence that is scientifically substantiated. As Grimm and Schirmer note, this raises questions about whether the legal threshold for imminence was applied too narrowly considering evolving climate risks.

5. BOTTOM-UP CLIMATE JUSTICE: A STRATEGY TO INTEGRATE POLLUTER PAYS IN CLIMATE GOVERNANCE

Most strategic climate litigation to date has aimed to push states toward more ambitious, science-based emission reduction targets, often invoking the 1.5°C goal set by the Paris Agreement. These 'top-down' lawsuits often challenge weak regulations or lack of climate ambition but overlook emissions that fall below the threshold set by these targets. In contrast, lawsuits like *Lliuya v RWE* follow a 'bottom-up' approach. Instead of targeting government policy, they directly apply the polluter pays principle to hold large corporate emitters accountable for their proportional share of climate-related harm.

Climate Change Impacts: The Role of Climate Attribution Science' in Elbert R De Jong et al (eds), *Corporate Responsibility and Liability in Relation to Climate Change* (Intersentia 2022) 206–235; Elisabeth A Lloyd and Theodore G Shepherd, 'Climate Change Attribution and Legal Contexts: Evidence and the Role of Storylines' (2021) 167 Climatic Change 28; Tobias Pfrommer and others, 'Establishing Causation in Climate Litigation: Admissibility and Reliability' (2019) 152 Climatic Change 67.

[56] Paul Grimm and Jan-Erik Schirmer, 'Success Without Victory: Lliuya v. RWE Decided' (*VerfBlog*, 1 June 2025) <https://verfassungsblog.de/lliuya-rwe-climate-judgment/> accessed 6 August 2025.

These two strands of climate litigation are often treated as distinct. One aims to pressure states to change laws and policies, while the other seeks to enforce direct accountability for emissions. In practice, they are deeply interlinked. Polluter pays cases like *Lliuya v RWE* do more than pursue redress for individual harm. By attributing a share of climate-related damage to specific emitters, these cases expose corporations to financial and reputational risks (see section 6). Such risks, in turn, can influence corporate behavior and investment strategies.

We argue that *Lliuya v RWE* is driving a conceptual shift toward private climate litigation against corporations. To clarify this shift, it is helpful to distinguish between two approaches to the polluter pays principle in climate governance: the *target-consistent approach* and the *social cost of carbon approach*.[57]

At its core, the 'polluter pays' principle asserts that those responsible for environmental harm should bear the costs of preventing, mitigating, or remedying it. In climate governance, this principle can be applied in two main ways:

- *The target-consistent approach* holds polluters accountable only to the extent required to meet collective emissions goals, such as the 1.5°C target. This logic underpins most emissions trading systems and climate mitigation regulatory frameworks, where emissions reduction benchmarks are shaped by scientific evidence but also subject to political negotiation. Under this approach, many emissions remain unregulated or legally permissible, as the focus is on achieving predefined targets rather than accounting for all harm caused.
- *The social cost of carbon approach*, in contrast, seeks to make polluters pay for the full marginal damage of their emissions, regardless of whether those emissions occur above or below a given target. Each ton of CO_2 is assigned a monetary value reflecting its estimated societal harm (e.g., crop losses, sea-level rise, health impacts). This approach prioritizes actual and expected damage over policy targets, supporting not only carbon pricing mechanisms but also liability claims for attributable climate harm. Polluters are expected either to compensate for these costs directly (e.g., through taxes or liability claims) or to internalize them into their decision-making.

57 Hans Christian Bugge, 'The polluter pays principle: Dilemmas of justice in national and international contexts' in Jonas Ebbesson and Phoebe Okowa (eds), *Environmental Law and Justice in Context* (CUP 2009) 411, 416.

The social cost of carbon approach reframes CO_2 and other GHG emissions not as unavoidable by-products of industrial activity, but as environmental pollution-harmful waste or externalities that impose costs on society. Under this framework, such emissions must be (a) eliminated at the source, by reducing or avoiding emissions, or (b) safely contained or removed by, for example, carbon capture and storage or other carbon dioxide removal technologies, or (c) financially compensated for, given the damage or risk of harm they cause.[58]

In other words, when emissions are treated as a form of harmful waste or as externalities, they may be considered a legally and economically relevant interference when they harm the property rights of others.

While widely used in economic modelling,[59] the social cost of carbon approach has rarely been explicitly invoked in climate litigation. Yet, this framing aligns with strict liability concepts, because even if emissions are legally permitted or compliant with regulatory regimes, emitters may still bear civil responsibility for the harm caused. This mirrors the logic applied to toxic waste or hazardous substances, where polluters are held accountable for cleanup or compensation, regardless of fault or regulatory compliance. For example, the EU's Environmental Liability Directive (ELD) imposes liability for environmental damage even when activities are permit-compliant.[60]

To date, there is broad scientific and economic consensus that the costs of climate loss and damage and climate adaptation far exceed those of timely mitigation. Given this, integrating the full social cost of carbon into climate governance is not just reasonable but imperative. This does not require a literal 'zero-tolerance' policy for emissions, but rather a systematic effort to internalize the external costs of emitting activities. In this framing, GHG emissions, like other hazardous industrial byproducts, should be treated as cost-generating pollutants, for which emitters bear responsibility even where emissions are legally permissible.

This shift in perspective is exemplified by *Lliuya v RWE*, where plaintiffs used climate litigation to promote a social cost of carbon approach, moving beyond mere target-setting. Even if CO_2 emissions comply with national or EU targets, they can still be framed as harmful activity. Courts may then assess

---

[58] Markus Petersen, 'CO_2 ist Müll' (*Klimareporter*, 25 October 2022) <www.klimareporter.de/gesellschaft/co2-ist-muell> accessed 6 August 2025; Brian F Snyder, 'Beyond the Social Cost of Carbon: Negative emission technologies as a means for biophysically setting the price of carbon' (2020) 49:9 Ambio 1567; William D Nordhaus, 'Revisiting the Social Cost of Carbon' (2017) 114 Proceedings of the National Academy of Sciences 1518.

[59] See Stern (n 2).

[60] See Sadeleer (n 52) 69.

liability not based on illegality but on foreseeable damage to property and the emitter's proportional contribution to it.

6. FINANCIAL IMPLICATIONS OF *LLIUYA V RWE*

Lliuya v RWE challenged traditional property law by attempting to hold corporations accountable for climate-related damages, illustrating the growing pressure to internalize the social cost of carbon into corporate decision-making. Though dismissed on its specific facts, the ruling's broader implications, particularly its acknowledgment that major GHG emitters could face proportional liability for climate adaptation costs, are significant. *Lliuya v RWE* was seen as a 'narrow escape' for the defendant, highlighting the reputational, operational, and financial risks that major emitters and their associated stakeholders face as climate litigation proliferates.[61]

A successful claim in a similar case could expose corporations to substantial outlays, including adaptation measures, compensation orders, and related costs.[62] Such liability may compromise a company's financial stability, depress share prices, and increase the risk of stranded assets, meaning those rendered obsolete or unprofitable by market and regulatory transitions to decarbonization.[63] Reputational risks further compound these exposures, as stakeholders grow more attuned to climate accountability and stranded assets, potentially leading to a decline in the company's share prices.[64] Negative publicity surrounding climate litigation can tarnish the defendant's brand, erode consumer trust, and deter potential investors.[65]

The ripple effects are not confined to defendants. Financial institutions, such as banks and insurers, face indirect exposure through their relationships

---

[61] Network for Greening the Financial System, 'Climate-Related Litigation: Raising Awareness About a Growing Source of Risk' (2021); UNEP FI, 'Liability Risk and Adaptation Finance' (2021); Javier Solana, 'Climate Change Litigation as Financial Risk' (2020) 2 Green Finance 344.

[62] UNEP FI (n 62) 32; Ben Caldecott and others, 'Stranded Assets: Environmental Drivers, Societal Challenges, and Supervisory Responses' (2021) 46 Annual Review of Environment and Resources 417, 423.

[63] Javier Solana, 'Climate Litigation in Financial Markets: A Typology' (2020) 9 Transnational Environmental Law 103, 131; UNEP FI (n 62) 20; Caldecott and others (n 63) 425, 431.

[64] Solana, 'Climate Change Litigation as Financial Risk' (n 62) 355.

[65] Jonathan M Karpoff, 'Does Reputation Work to Discipline Corporate Misconduct?' in Timothy G Pollock and Michael L Barnett (eds), *The Oxford Handbook of Corporate Reputation* (OUP 2012) 361; Solana, 'Climate Litigation in Financial Markets: A Typology' (n 64) 28.

with high-emission clients. For insurers, a successful claim could trigger large payouts under liability policies, potentially affecting reserves and premiums system-wide.[66] For lenders, the risk of default increases if the borrower's balance sheet is strained by environmental liability.[67] This interlinkage underscores the systemic risks posed by environmental liability, as the financial repercussions ripple through the broader economy.

The financial and reputational challenges associated with environmental liability can serve as a deterrent, incentivizing corporations to adopt more sustainable practices.[68] Faced with the prospect of costly litigation, major emitters are compelled to reduce GHG emissions and align their operations with the precautionary principle. This shift not only mitigates future litigation risks but also contributes to broader climate goals, reinforcing the polluter pays principle.[69] In this context, climate litigation can act as a catalyst for corporate behavioral change.[70] By internalizing the financial risks of climate-related damage, corporations are driven to prioritize emissions reduction and invest in sustainable technologies, thereby driving positive environmental outcomes. Lawsuits such as *Lliuya v RWE* may thus function not only as tools for redress but also as mechanisms of deterrence, making the financial risks of climate-related harm more visible and immediate. They give practical effect to the polluter pays principle within a transnational, bottom-up framework.

This evolving litigation landscape coincides with a broader regulatory shift. The European Union has recently adopted binding instruments such as the Corporate Sustainability Reporting Directive (CSRD)[71] and the Corporate Sustainability Due Diligence Directive (CSDDD),[72] which mark a transition

---

[66] UNEP FI (n 62) 17; Task Force on Climate-related Financial Disclosures, 'Recommendations of the Task Force on Climate-related Financial Disclosures' (2017) 10.

[67] Network for Greening the Financial System (n 62) 7; UNEP FI (n 62) 32, 53.

[68] Peter Gailhofer, 'Functions and Objectives of Corporate Liability for Transboundary Environmental Harm' in Peter Gailhofer and others (eds), *Corporate Liability for Transboundary Environmental Harm: An International and Transnational Perspective* (Springer International Publishing 2023) 12.

[69] ibid 12.

[70] N K Dubash and others, 'National and Sub-National Policies and Institutions' in IPCC (eds), *Climate Change 2022: Mitigation of Climate Change* (n 6) 1375.

[71] Directive (EU) 2022/2464 of the European Parliament and of the Council of 14 December 2022 amending Regulation (EU) No 537/2014, Directive 2004/109/EC, Directive 2006/43/EC and Directive 2013/34/EU, as regards corporate sustainability reporting (Text with EEA relevance) [2022] OJ L322/15.

[72] 2024/1760 of the European Parliament and of the Council of 13 June 2024 on corporate sustainability due diligence and amending Directive (EU) 2019/1937 and Regulation (EU) 2023/2859 (Text with EEA relevance) [2024] OJ L.

from voluntary ESG[73] standards to mandatory environmental and human rights compliance. These frameworks require companies to assess, disclose, and address environmental risks, potentially easing evidentiary burdens in future litigation and reinforcing expectations of climate accountability.

As these financial and regulatory dynamics evolve, the influence of *Lliuya v RWE* is already extending beyond Germany, inspiring similar claims in other jurisdictions and contributing to a growing body of transnational climate litigation, which we explore in the next section.

7. TRANSNATIONAL IMPLICATIONS AND THE SPREAD OF CLIMATE LIABILITY CLAIMS

Lliuya v RWE represents a pivotal development in efforts to establish corporate liability for climate change, significantly contributing to the development of transnational climate litigation jurisprudence. As one of the first cases to seek compensation from a major emitter for climate-related damages across borders – linking emissions from a German energy major to risks faced in the Peruvian Andes – it challenges legal boundaries, such as corporate responsibility, extraterritorial jurisdiction, and the polluter pays principle.

The significance of this case extends beyond its legal implications. It represents a milestone in access to climate justice. By demonstrating that civil law systems may adjudicate transnational climate claims even in the absence of specific liability statutes,[74] *Lliuya v RWE* broadens the scope of liability under private law, especially through property-based claims. This illustrates how bottom-up litigation can serve as a tool for enforcing climate accountability where regulatory frameworks remain insufficient. While challenges like proving causation remain, the case has the potential to mitigate future climate-related harms and spur stricter regulations. Its influence is already evident in cases like *Asmania et al v Holcim*, currently pending before the Swiss civil courts.[75]

---

[73] Marc Helmold and others, *ESG, CSR und SDG als langfristiger Wettbewerbsvorteil* (Springer Gabler Wiesbaden 2024) 1–9.

[74] Malte-Christian Gruber, 'Escaping the Epistemic Trap' in Klaus Mathis and Bruce Huber (eds), *Environmental Law and Economics* (Springer International Publishing 2017) 107, 114.

[75] In the case of *Asmania et al v Holcim*, four inhabitants of the Indonesian island of Pari have sued the Swiss building materials company Holcim. The plaintiffs request both CO_2 reduction and proportional compensation for climate change-related damages in Pari, as well as a financial contribution to adaptation measures in Pari. See Sabine Center for Climate Change, 'Asmania et al. vs. Holcim' (*Global Climate Change Litigation Database*) <https://climatecasechart

Lliuya v RWE sets a potential precedent for holding major polluters legally accountable for global climate damage, paving the way for transnational 'polluter pays' litigation. A successful ruling could significantly impact private emitters like RWE, increasing their legal and financial risks. This could trigger a rise in similar claims globally, with broad economic and political consequences and also place pressure on judicial systems. The prospect of procedural overload in courts may have transformative effects, pressuring legislators to strengthen corporate accountability regulations or incentivizing companies to adopt stricter self-regulation. Thus, litigation like *Lliuya v RWE* can indirectly shape climate governance by driving legal, political, and market-based responses to systemic risks.

8. CONCLUSION

The case of *Saúl Lliuya v RWE* marks a significant step in climate litigation, testing whether corporations can be held liable for their contributions to climate change under private law. Although the German court dismissed the claim, finding the 1% risk of a glacial flood damaging Lliuya's property over 30 years too speculative, it made two landmark rulings: first, that companies like RWE could, in principle, be held proportionally responsible for climate-related damages abroad; and second, that domestic private law could apply extraterritorially in such cases.

The lawsuit innovatively used German property law (rather than tort law) to argue that RWE's emissions interfered with Lliuya's property rights, sidestepping the need to prove negligence or noncompliance with legal standards. While unsuccessful, the case advanced the 'polluter pays' principle by treating emissions as harmful waste requiring financial accountability, even if legally permitted. It also highlighted the growing role of climate attribution science in establishing corporate liability.

Beyond its legal implications, *Lliuya v RWE* exposed corporations to new financial and reputational risks, signaling to investors and insurers that climate litigation could impact profitability. The case has already inspired similar lawsuits in other jurisdictions, demonstrating how 'bottom-up' litigation can

.com/non-us-case/four-islanders-of-pari-v-holcim> accessed 6 August 2025; Laura A Duarte Reyes and Nina Burri, 'Transnational corporate liability in the era of loss and damages: The case of Asmania et al. v Holcim' in Stefano Zirulia, Lidia Sandrini and Cesare Pitea (eds), *What future for environmental and climate litigation? Exploring the added value of a multidisciplinary approach from international, private and criminal law perspectives* (Milano University Press 2024) 97, 107.

pressure companies to reduce emissions even where government regulations lag. Though dismissed, the ruling set a crucial precedent, paving the way for future claims that may succeed with stronger evidence of imminent harm. Ultimately, the case underscores the potential of strategic litigation to drive corporate accountability and shape the evolving landscape of climate justice.

4. Common law at the coal face: *Smith v Fonterra and others*

Vernon Rive

1. INTRODUCTION

The February 2024 New Zealand Supreme Court judgment in *Smith v Fonterra Co-operative Group Ltd* (*Smith v Fonterra*)[1] joins a sizable and growing pool of judicial decisions grappling with the capacity of tort law in common law jurisdictions to respond to threats posed by rising greenhouse gas (GHG) emissions. *Smith v Fonterra* has received widespread attention and coverage globally: in the media,[2] civil society community[3] and legal academy.[4] Opinions about the case are sharply divided—while Greenpeace Aotearoa Executive Director Russell Norman welcomed the Supreme Court's decision as a 'hugely

1 *Smith v Fonterra Co-Operative Group Ltd* [2024] NZSC 5, [2024] 1 NZLR 134 (*Smith v Fonterra (Supreme Court)*).

2 See e.g. Eva Corlett, 'The Māori Climate Activist Breaking Legal Barriers to Bring Corporate Giants to Court' *The Guardian* (6 March 2024) <www.theguardian.com/world/2024/mar/06/mike-smith-maori-climate-activist-right-to-sue-companies> accessed 4 January 2025; Sharon Bretkelly, 'Extraordinary Climate Claim Has the World Watching' *RNZ* (21 February 2024) <www.rnz.co.nz/programmes/the-detail/story/2018926685/world-first-climate-action-in-nz-s-top-court> accessed 4 January 2025.

3 See e.g. Paula Reisdorf, 'Māori Activist Wins Right to Sue Major New Zealand Polluters' (*CorpWatch*, 16 February 2024) <www.corpwatch.org/article/maori-activist-wins-right-sue-major-new-zealand-polluters> accessed 4 January 2025.

4 See e.g. Sam Bookman, '*Smith v Fonterra* and the Climatisation of Tort Law' [2024] *Modern L Rev* 1468; Douglas A Kysar, 'The Duty of Climate Care' (2024) 73 *DePaul L Rev* 487; Jacob Lerner, 'Some Objections to *Smith v Fonterra*' [2024] *NZ Law J* 185–189; Caroline E Foster, 'Tort Law and New Zealand's Corporate Greenhouse Gas Emissions' (*Verfassungsblog,* 17 March 2024) <http://verfassungsblog.de/tort-law-and-new-zealands-corporate-greenhouse-gas-emissions/> accessed 4 January 2025.

significant win' which has 'confirmed polluters like Fonterra could be held legally responsible for the climate devastation they've caused',[5] others were less enthusiastic. A conservative New Zealand think tank commentator described the decision as 'absurd', suggesting that the New Zealand emissions trading scheme (ETS) 'makes any common law tort toothless and superfluous'.[6]

The true effect of the judgment is more restrained than either of these early reactions. The Supreme Court did not 'confirm' legal responsibility for any climate polluter, nor did it make any conclusive assessment on the roles of regulatory tools such as ETSs and tort mechanisms to control GHG emissions. Instead, the Court reversed lower court decisions on applications to strike the plaintiff's claims, ruling that a lawsuit against seven large New Zealand companies brought by Michael Smith, a Māori elder and climate change spokesperson, seeking to establish tort liability for their contributions to global GHG emissions, could proceed to trial.

The outcome of the case—yet to commence its substantive stage—will likely not be known for many years. Its ultimate import will depend on how the evidence and legal arguments of the claimants and defendants fare before trial and (almost inevitably) appellate judges. However, the Supreme Court's decision heralds a new phase of accountability for corporate entities involved in emissions intensive activities in New Zealand. Whatever the ultimate result for Smith, the ripples of this case will be felt widely in climate litigation and tort law contexts in New Zealand, and, potentially, worldwide.

1.1 New Zealand Climate Litigation Landscape

The Supreme Court delivered its decision in *Smith v Fonterra* just under 33 years after New Zealand's first climate change case, a 1991 judicial review of a planning decision by a local authority approving a subdivision and residential development on low-lying coastal land without considering the implications of climate change-related sea level rise.[7] In that case, which predated the adoption of the 1992 United Nations Framework Convention on Climate

---

5 Russel Norman, 'Huge Win for the Climate in *Smith v Fonterra* Supreme Court Ruling' (*Greenpeace Aotearoa*, 7 February 2024) <www.greenpeace.org/aotearoa/press-release/smith-vs-fonterra-supreme-court-ruling-win/> accessed 4 January 2025.

6 Oliver Hartwich, 'Absurd: NZ Courts Can Now Decide on Climate Change', *The Australian* (15 February 2024) <www.theaustralian.com.au/commentary/absurd-new-zealand-courts-can-now-decide-on-climate-change/news-story/b336698c31b25b69eaf174fbbe61efdb> accessed 4 January 2025.

7 *Maruia Society Inc v Whakatane District Council* (1991) 15 NZTPA 65 (*Maruia*).

Change[8] (UNFCCC) by one year, the High Court judge regarded the existence of 'climatic change' (or at least its anthropogenic character) as an open question—one that he would not consider.[9] From the mid-1990s, litigants brought a steady stream of cases centred on climate change issues before New Zealand national and regional authorities,[10] the specialist Environment Court,[11] the High Court[12] and appellate courts.[13] Early judicial hesitation towards climate change science gave way, generally, to acceptance of the content in Intergovernmental Panel on Climate Change (IPCC) reports.[14] In 2010, during a planning appeal under the *Resource Management Act 1991* (RMA), the High Court signaled its unwillingness to 'enter into a discussion of [climate change's] causes, directions and magnitude', on the basis that the 2004 inclusion of the term 'climate change' in the RMA[15] 'reflect[s] the statutory assumption that climate change is occurring'.[16]

---

8 United Nations Framework Convention on Climate Change 1771 UNTS 107 (opened for signature 4 June 1992, entered into force 21 March 1994) (UNFCCC).

9 *Maruia* (n 7) [21] (Doogue J): 'Whether the evidence at present available in respect of matters such as the "greenhouse" effect is anything more than conjectural I do not know'.

10 For an example of a national authority see Hon Simon Upton, 'Air Discharge Permit Taranaki Combine Cycle Power Station: Decision of Hon Simon Upton Minister for the Environment' (Ministry for the Environment, 23 March 1995). For an example of an appeal from a decision of a regional authority see e.g. *Environmental Defence Society v Taranaki Regional Council* [2002] NZEnvC 441.

11 See e.g. *Environmental Defence Society v Auckland Regional Council* [2002] NZRMA 492; [2002] NZEnvC 315; (2003) 9 ELRNZ 1.

12 See e.g. *Greenpeace New Zealand Inc v Northland Regional Council* (2006) 12 ELRNZ 377, [2007] NZRMA 87.

13 See e.g. *Greenpeace New Zealand Inc v Genesis Power Ltd* [2008] NZSC 112. For an overview of climate change litigation in New Zealand to 2020, see eg Davey Salmon, 'Thoughts on Climate Litigation in New Zealand' in Simon Mount and Max Harris (eds), *The Promise of Law: Essays Marking the Retirement of Dame Sian Elias as Chief Justice of New Zealand* (LexisNexis NZ Limited 2020) 152–157.

14 See e.g. *Fore World Developments Ltd v Napier City Council* Environment Court 13 April 2006 ENC Wellington W29/06, [2006] ELHNZ 143 [73]–[74] (IPCC global estimates for sea-level rise accepted by the Environment Court in proceedings concerning zoning of land in coastal hazards area).

15 Section 4 of the *Resource Management (Energy and Climate Change) Amendment Act 2004* inserted a statutory definition of 'climate change' into s 2 of the *Resource Management Act*.

16 *Meridian Energy Ltd v Central Otago District Council* (2011) 1 NZLR 482 [157] (Chisholm and Fogarty JJ).

The pattern of New Zealand climate litigation in the two decades between the mid-1990s and mid-2010s roughly followed the model of what Peel and others describe as 'first generation' cases.[17] These cases generally involved planning appeals or administrative law challenges 'seeking to incorporate climate change within the scope of decision-making on a project'.[18] As Peel and others discuss in an Australian context, 'next-generation' climate lawsuits represent 'a shift away from this project-level focus and environmental statutory basis' towards cases 'founded on an accountability model, whereby legal interventions are designed to hold governments and corporations *directly* to account for their contributions to climate change'.[19] 'Next-generation' lawsuits became more frequent in New Zealand after the mid-2010s.

One of New Zealand's most prominent next-generation cases was a judicial review brought in 2015 by law student Sarah Thomson seeking declarations and associated orders concerning government decisions about domestic and international emissions reduction targets.[20] With pro bono assistance from trial and appellate litigator Davey Salmon (appointed as Queen's Counsel in 2021, lead counsel for the plaintiff in *Smith v Fonterra*) and others, Thomson succeeded in identifying legal shortcomings in the central government decision-making process. The High Court declined to order relief as, after a change in government since the claim had been filed, a new Minister had confirmed the government's intention to set new targets.[21] Despite this, the case established a model for legal proceedings challenging government entities about aspects of climate change policy.

Various next-generation proceedings have been brought by indigenous Māori. In the 2016 action *Mataatua District Māori Council v New Zealand*, representatives of a statutory Māori organisation lodged a claim in the Waitangi Tribunal (a quasi-judicial commission of inquiry established to hear and make recommendations about claimed breaches of the Treaty of Waitangi).[22] The

---

17 Jacqueline Peel and others, '"Next Generation" Climate Change Litigation in Australia' in Jolene Lin and Douglas A Kysar (eds), *Climate Change Litigation in the Asia Pacific* (1st edn, Cambridge University Press 2020) 176.

18 ibid. A high-profile example is the appeal dismissed in a majority judgment by the Supreme Court of an unsuccessful application for a declaration of the relevance of indirect emissions from the overseas burning of coal mined in New Zealand: *West Coast ENT Inc v Buller Coal Ltd* (2014) 1 NZLR 32; [2013] NZSC 87.

19 Jacqueline Peel and others (n 17) 181.

20 *Thomson v Minister for Climate Change Issues* [2017] NZHC 733, [2018] 2 NZLR 160.

21 ibid [178].

22 *Mataatua District Māori Council v New Zealand* [2016] WAI 2607.

Mataatua claim that the New Zealand government was in breach of its Treaty obligations towards Māori as a result of its failure to implement adequate policies to address the threats posed by global climate change was the forerunner of subsequent Māori-led climate change-focused proceedings in the Waitangi Tribunal. Subsequent hearings commenced in November 2024.[23] Other notable cases in the next-generation category include: *All Aboard Aotearoa Inc v Auckland Transport* (2022);[24] *Students for Climate Solutions Incorporated v Minister of Energy and Resources* (2022, 2024);[25] *Lawyers for Climate Change Action NZ Incorporated v Minister of Climate Change* (2023);[26] *Smith v Attorney-General* (2022, 2024);[27] and, of course, the subject of this chapter, *Smith v Fonterra.*

1.2 New Zealand's Climate Change Profile

New Zealand has a climate change profile which reflects its position as an Annex I and II 'developed country' party under the UNFCCC in some respects, but is more characteristic of a small, developing state in other respects. As of 2023, per capita GHG emissions were 15.3 metric tons of carbon dioxide equivalent (MT CO_2e) in New Zealand, the fourth highest per capita emissions of Organization for Economic Cooperation and Development countries.[28] However, its territorial share of global emissions is small (a mere 0.16% in 2023),[29] reflecting both a low population base (5.3 million in 2023)[30] and

23 Waitangi Tribunal *Climate Change Priority Inquiry*, WAI 3325; see <www.waitangitribunal.govt.nz/en/inquiries/kaupapa-inquiries/climate-change-priority-inquiry>.

24 *All Aboard Aotearoa Inc v Auckland Transport* [2022] NZHC 1620.

25 *Students for Climate Solutions Incorporated v Minister of Energy and Resources* [2022] NZHC 2116; [2024] NZCA 152.

26 *Lawyers for Climate Change Action NZ Incorporated v Minister of Climate Change* [2023] NZHC 1835.

27 *Smith v Attorney-General* [2022] NZHC 1693 (*Smith v Fonterra (High Court)*); [2024] NZCA 692 (*Smith v Fonterra (Court of Appeal)*).

28 Statista, 'Per capita greenhouse gas emissions in OECD countries in 2023' <statista.com/statistics/478783/leading-countries-based-on-per-capita-greenhouse-gas-emissions/> accessed 2 January 2024.

29 Emissions Database for Global Atmospheric Research (EDGAR), 'GHG emissions of all world countries 2024 report' <edgar.jrc.ec.europa.eu/report_2024?vis=ghgtot#emissions_table> accessed 12 January 2025.

30 Stats NZ 'National population estimates: At 30 September 2023' <www.stats.govt.nz/information-releases/national-population-estimates-at-30-september-2023/> accessed 12 January 2025.

relatively high proportion of renewable electricity generation (88% in 2023).[31] A distinctive feature of New Zealand's emissions inventory more commonly seen in non-Annex I parties is the unusually high contributions of its agricultural sector. In 2022, agricultural emissions (including biogenic methane, largely from dairy cattle, and nitrous oxide from agricultural fertiliser) were by far the largest source of emissions by sector, contributing 53% of the country's total emissions.[32]

Successive New Zealand governments have announced, accepted or adopted various domestic and international emissions reductions targets since as early as 1990.[33] In 2021, the government updated its Nationally Determined Contribution (NDC), communicated under the Paris Agreement to include a 'headline' target of a 50% reduction in net emissions from the 2005 gross emissions baseline by 2030 (including contributions from land use, land use change and forestry).[34] Alongside international commitments, the *Climate Change Response (Zero Carbon) Amendment Act 2019* provides a statutory framework for the following domestic targets:

(i) net zero carbon emissions (excluding biogenic methane) by 2050;
(ii) a reduction in biogenic methane emissions by 24–47% from a 2017 baseline by 2050;[35] and
(iii) a five-yearly interim emissions budget and associated emissions reductions plan to be set by the government, taking into account advice from an independent Climate Change Commission.[36]

---

31 Ministry of Business, Innovation and Employment, *Energy in New Zealand 2024* (2024) 6.

32 New Zealand's Greenhouse Gas Inventory, *New Zealand's Greenhouse Gas Inventory 1990–2022* (Ministry for the Environment 2024) 36.

33 For summary of early targets adopted or accepted between 1990 and 2011, see Vernon Rive, 'New Zealand Climate Change Regulation' in Alastair Cameron (ed), *Climate Change Law and Policy in New Zealand* (LexisNexis NZ Ltd 2011) 178–179.

34 New Zealand Submission under the Paris Agreement. 'New Zealand's Nationally Determined Contribution' updated 4 November 2021 <https://unfccc.int/sites/default/files/NDC/2022-06/New%20Zealand%20NDC%20November%202021.pdf>.

35 *Climate Change Response Act 2004,* s 5Q.

36 *Climate Change Response Act 2004,* s 5X.

In 2008, the government introduced an ETS, making New Zealand the second nation to enact such a scheme after the European Union.[37] Originally designed as an 'all gases, all sectors' scheme, the New Zealand ETS has been the subject of intense political lobbying and criticism for delivering far less than its architects intended (or signaled).[38] An initial grace period for the agricultural sector stretched to a 16-year deferral before legislation was passed in November 2024 formally excluding agricultural emissions from the scheme.[39] It remains the case that no carbon price exists for agricultural biogenic methane or nitrous oxide.

In his amended statement of claim tabled with June 2022 submissions to the Supreme Court, Smith pleaded: 'despite enacting the *Climate Change Response Act*, since 2002 New Zealand's net and gross greenhouse gas emissions have increased and have not reduced'.[40]

New Zealand's Greenhouse Gas Inventory for the period from 1990–2020 (published shortly before Smith's written submissions were tabled) recorded that, as of 2020, the country's gross emissions had increased by 21% from 1990 levels, with net emissions increasing by 26% 'due to the underlying increase in gross emissions'.[41] The 2024 Greenhouse Gas Inventory (reporting on emissions for the period from 1990–2022) reported that (as of 2022) net emissions had increased by 33% (gross emissions by 14%) above 1990 levels. The report also observed that both gross and net emissions had peaked in 2006, and that although gross and net emissions had increased between 2016 and 2019, gross emissions had declined each year since 2019.[42]

---

37 Alastair Cameron and Vernon Rive, 'Emission Trading: Setting the Scene' in Alastair Cameron (ed), *Climate Change Law and Policy in New Zealand* (LexisNexis NZ 2011).

38 See e.g. Catherine Leining and others, 'The New Zealand Emissions Trading Scheme: Critical Review and Future Outlook for Three Design Innovations' (2020) 20 *Climate Policy* 246. For a recent blunt assessment of the ETS from the outgoing Chair of the New Zealand Climate Change Commission, see Shannon Williams, '"The Emissions Trading Scheme is Broken": Carr' *Carbon News* (2024).

39 *Climate Change Response (Emissions Trading Scheme Agricultural Obligations) Amendment Act 2024*.

40 Draft Amended Statement of Claim, 15 June 2022, attached to Davey Salmon KC and others, 'Appellant's Synopsis of Submissions on Appeal' (15 June 2022) [69].

41 New Zealand's Greenhouse Gas Inventory, *New Zealand's Greenhouse Gas Inventory 1990–2020* (Ministry for the Environment 2022) xxiv.

42 New Zealand's Greenhouse Gas Inventory (n 32).

1.3 New Zealand's Legal Framework and Culture

New Zealand's constitutional framework has been variously described as 'unwritten',[43] 'derived from eclectic sources',[44] 'inherited' from Britain,[45] and having 'relatively open-textured constitutional arrangements compared with many other countries'.[46] Following the signing of Te Tiriti o Waitangi/the Treaty of Waitangi by representatives of the British Crown and a selection of Māori chiefs on 6 February 1840, Britain assumed sovereignty over the South Pacific territory. Then, British representatives took further steps to formally establish New Zealand as a separate colony from (Australian) New South Wales and establish a system of representative government (initially including a bicameral Parliament).[47] Through a succession of 'English Laws' acts from 1854, the 'law of England' (as of 14 January 1840) was legislated to apply 'so far as applicable to the circumstances of the said Colony of New Zealand'.[48]

Since the abolition in 1950 of the Legislative Council (operating as an upper house), New Zealand has had a unicameral Parliament, operating under a 'mixed-member proportional' (MMP) system since 1996.[49] Under MMP, governments have typically consisted of multiple political parties, either in formal coalition or confidence and supply arrangements. The current (January 2025) coalition government comprises members from the centre-right National Party of New Zealand, right-wing libertarian ACT Party, and nationalist populist New Zealand First Party.[50] Other major political parties include the centre-left

43 Matthew Palmer, 'Constitution – What is a constitution?', Te Ara – the Encyclopedia of New Zealand' <www.TeAra.govt.nz/en/constitution/page-1> (accessed 8 January 2025). But see Janet McLean, 'The Unwritten Political Constitution and Its Enemies' (2016) 14 *Intl J Constitutional L* 119, 120.

44 Bruce Harris, *New Zealand Constitution: An Analysis in Terms of Principles* (Thomson Reuters New Zealand 2018) 18.

45 Jacinta Ruru and others, *The Aotearoa New Zealand Legal System: Structures and Processes* (7th LexisNexis NZ Limited 2022) 125.

46 Geoffrey Palmer and Matthew Palmer, *Bridled Power: New Zealand's Constitution and Government* (4th edn, Oxford University Press 2004) 5.

47 See Jacinta Ruru and others (n 45) 114–124.

48 *English Laws Act 1858*, s 1. The 'Law of England' includes, of course, *common* law. See Philip Joseph, *Joseph on Constitutional and Administrative Law* (5th edn, Thomson Reuters New Zealand 2021) 32–33. The current position on the application of the 'common law of England' in New Zealand is statutorily regulated in the *Imperial Laws Application Act 1988*, s 5.

49 See Matthew Palmer and Dean Knight, 'Parliament' in *Constitution of New Zealand: A Contextual Analysis* (Hart Publishing 2021) 107–136.

50 A useful overview of the process and outcomes of New Zealand's 2023 election is in Stephen Levine (ed), *Back on Track?: The New Zealand General Election of 2023* (Te Herenga Waka University Press 2024).

Labour Party, the left-wing environmental and social justice-focused Green Party and the Māori-led Te Pāti Māori.

In the absence of a codified, 'higher law' constitution (New Zealand is one of only three countries in the world with this state of constitutional affairs),[51] Parliament acts as the supreme lawmaking body.[52] As Palmer and Knight observe, acts of Parliament override any other law to the contrary. This includes all regulations, the common law, exercises of the royal prerogative or any earlier Act that is, in its terms, inconsistent with the later Act.[53]

In 1990, a legislative Bill of Rights—*the New Zealand Bill of Rights Act 1990 (NZBORA)*—was passed, designed (among other objectives)[54] to give effect in domestic law New Zealand's obligations under the International Covenant on Civil and Political Rights, which it had ratified in 1978. Consistent with the legislative supremacy noted prior, New Zealand courts have no powers under the *NZBORA* (or any other statutes or sources of law) to strike down primary legislation.[55] However, in 2018, the New Zealand Supreme Court recognised and implemented a 'declaration of inconsistency' as a formal judicial declaration that a statutory provision represents an unjustified breach of a right affirmed by *NZBORA*.[56] A structured process for bringing declarations of inconsistency to the attention of Parliament was incorporated in the *NZBORA* through a 2020 amendment passed with cross-party support.[57]

1.3.1 Legal pluralism and *tikanga* Māori

The assumption of British sovereignty over New Zealand in 1840 was followed by a rapid increase in European settlement, conflict over land and natural resources, and, by the early 1860s, outright war.[58] For a period of time after 1840, courts in the new colony recognised (and, on occasion, applied in

---

[51] The other two countries without a codified constitution are the United Kingdom and Israel.

[52] David McGee, *Parliamentary Practice in New Zealand* (4th edn, Oratia Media Ltd 2021) para 41.1.1.

[53] Matthew Palmer and Dean Knight, *Constitution of New Zealand: A Contextual Analysis* (Hart Publishing 2021) 126.

[54] See Andrew S Butler and Petra Butler, 'Background to and History of the New Zealand Bill of Rights Act 1990' in *The New Zealand Bill of Rights Act: A Commentary* (2nd edn, LexisNexis NZ Limited 2015) 15–47.

[55] *New Zealand Bill of Rights Act 1990*, s 4.

[56] *Attorney-General v Taylor* [2018] NZSC 104, (2019) 1 NZLR 213.

[57] *New Zealand Bill of Rights (Declarations of Inconsistency) Amendment Act 2022.*

[58] See Vincent O'Malley, *The New Zealand Wars | Ngā Pakanga o Aotearoa* (Bridget Williams Books 2019).

a limited way) what was described under British law as 'custom',[59] a subset of a 'set of [Māori] values, principles, understandings, practices, norms and mechanisms' known as *tikanga*.[60] A widely cited description of *tikanga* is offered by Professor Sir Hirini Moko Mead:[61]

> Tikanga embodies a set of beliefs and practices associated with procedures to be followed in conducting the affairs of a group or an individual…Tikanga are tools of thought and understanding. They are packages of ideas which help to organise behaviour and provide some predictability in how certain activities are carried out.

With few exceptions, by the 1880s, neither the terms of Te Tiriti, nor the standing of *tikanga* or Māori customary rights were recognised by the judicial or parliamentary branches of the emerging settler government.[62] Despite the common law's ostensible recognition and preservation of customary law so far as it was not inconsistent with statute, there had been minimal recognition of *tikanga* as a relevant source of law by any of the branches of government until the 2000s. This position began to change with the publication of an important Law Commission report in 2001.[63] By the 2010s, *tikanga* Māori was being recognised regularly in court decisions[64] and was increasingly referenced in legislation.[65] As will be discussed following, the role of *tikanga* in relation to Smith's claim, and to the evolution of tort law in New Zealand generally, emerged as a majorly contested issue in those proceedings.

---

59 An example of an early case cited by the Supreme Court is *James Reynolds v Simon Tuangau* West Coast Times (New Zealand, 8 August 1866): *Smith v Fonterra (Supreme Court)* [184]–[185]. See generally Te Aka Matua o te Ture | Law Commission, *He Poutama* (NZLC SP24 2023) [5.27]–[5.32].

60 ET Durie, 'Will the settlers settle? Cultural conciliation and law' (1996) 8 *Otago L Rev* 449 at 452.

61 Hirini Moko Mead, 'The Nature of Tikanga' (paper presented to Mai i te Ata Hāpara Conference, Te Wānanga o Raukawa, Otaki, 11–13 August 2000) at 3–4.

62 *Wi Parata v Bishop of Wellington* (1878) 3 NZ Jur (NS) SC 72 [74] (Prendergast CJ) holding that pre-European settlement Māori were 'without any kind of civil government, or any settled system of law'.

63 Law Commission Te Aka Matua o te Ture, *Māori Custom and Values in New Zealand Law* (NZLC SP9 2001).

64 Notable examples include: *Takamore v Clarke* [2012] NZSC 116, (2013) 2 NZLR 733 (*Takamore*); *Proprietors of Wakatū v Attorney-General* [2017] NZSC 17, (2017) 1 NZLR 423; *Ngāti Whātua Ōrākei Trust v Attorney-General* [2018] NZSC 84, (2019) 1 NZLR 116; *Trans-Tasman Resources Ltd v Taranaki-Whanganui Conservation Board* [2021] NZSC 127 (*Trans-Tasman*); (2021) 23 ELRNZ 47; *Peter Hugh McGregor Ellis v R* [2022] NZSC 114 (*Ellis*).

65 See 'Tikanga and Statute Law' in Law Commission Te Aka Matua o te Ture, *He Poutama* (NZLC SP24, NZLC SP24 2023) ch 6.

2. OVERVIEW AND SUMMARY OF THE CASE

2.1 Plaintiff

The plaintiff in this case was Michael Smith, a 67-year-old Māori *kaumatua* (elder) of two *iwi* (tribes) from the north of New Zealand, Ngāpuhi and Ngāti Kahu. In written submissions presented to the Supreme Court, Smith's counsel introduced him as follows:[66]

> Mr Smith (Ngāpuhi, Ngāti Kahu) is the climate change spokesperson for the *Iwi* Chairs forum. He has an interest, according to *tikanga* Māori, in the Mahinepua C block near Kaeo, in Northland. Mahinepua C is situated on the coast at Wainui Bay. On that land, and around it, are numerous sites of deep customary, cultural, historical, nutritional and spiritual significance to Mr Smith and his *whānau*. Mr Smith pleads that Mahinepua C, and these surrounding resources and sites of significance have been, and will be, harmed by climate change. Mr Smith sues to restrain the respondents' ongoing contributions to this harm.

Smith has a long history as a Māori and environmental activist.[67] In 1992, he represented Ngāpuhi at the Global Earth Summit in Rio de Janeiro, subsequently assuming a leadership role on climate change issues within a pan-Māori organisation (*Iwi* Chairs Forum) whose purpose includes advocating for 'the collective priorities of the [Forum] *iwi* in discussions with others, recognising the rangatiratanga/independence of *iwi*'.[68]

Alongside his claim in this case, Smith has also started proceedings against the Attorney-General of New Zealand, summarised in a December 2024 decision of the Court of Appeal as an action:[69]

> brought for an alleged breach of a common law duty to avoid dangerous interference with the climate system, breach of the right to life and the right to culture affirmed in the New Zealand Bill of Rights Act 1990 (NZBORA), and for breach of the Treaty of Waitangi | te Tiriti o Waitangi (the Treaty | te Tiriti) and of fiduciary duties owed to Māori.

---

66 Davey Salmon KC and others (n 40) [5].

67 For useful background on Mike Smith and the case, see Pete McKenzie, 'How an Old British Law May Hold New Zealand's Biggest Polluters to Account', *New Zealand Geographic* (2024) <www.nzgeo.com/stories/how-an-old-british-law-may-hold-new-zealands-biggest-polluters-to-account/> accessed 30 December 2024.

68 Iwi Chairs Forum Secretariat, 'Sharing The Vision of Kotahitanga' (*Iwi Chairs Forum*, no date) <https://iwichairs.maori.nz/tikanga/> accessed 12 January 2025.

69 *Smith v Attorney-General* [2024] NZCA 692 [2] (Mallon J) (*Smith v Attorney General (Court of Appeal)*).

As with his claim against Fonterra and others, Smith's claim against the Attorney-General was the subject of a strike-out application. In July 2022, the High Court (Grice J) struck the claim out in its entirety.[70] Smith's appeal of that decision was dismissed in a unanimous decision by the Court of Appeal in December 2024.[71] While upholding the High Court's overall position that the claim against the government was untenable, Mallon J (writing for Wylie J and Cooper P) acknowledged the potential for *NZBORA* and Treaty of Waitangi-based judicial review actions against the executive branch of government with respect to specific decisions or exercises of power.[72]

2.2 Defendants

Smith's claim is made against seven New Zealand-based corporate defendants, described in submissions to the Supreme Court as 'some of the biggest emitters and fossil fuel suppliers in New Zealand', said to be responsible in total for approximately one-third of the country's GHG emissions.[73] The defendants are:

(i) Fonterra Co-operative Group Ltd (Fonterra), whose assets and operations include eight New Zealand-based diary factories using coal for energy generation;
(ii) Genesis Energy Limited, operator of the country's largest thermal power station fueled by coal and natural gas;
(iii) Dairy Holdings Ltd, operator of 59 dairy farms and a significant source of biogenic methane and nitrogen dioxide released from nitrogen-based fertiliser used on the farms;
(iv) New Zealand Steel Ltd, operator of the country's largest steel mill primarily fueled by coal;
(v) Z-Energy (successor to some of the assets and fuel retail operations of Shell New Zealand and Chevron New Zealand);

[70] *Smith v Attorney-General* [2022] NZHC 1693.
[71] *Smith v Attorney-General (Court of Appeal)* (n 69).
[72] ibid [102], [132], [151].
[73] Davey Salmon KC and others (n 40) [7]. Commenting on the emitter status of the defendants, the Supreme Court noted 'it is here beyond question that the respondents are either very substantial emitters of GHGs or are (or have been) very substantial suppliers of fossil fuels that release GHGs when burned by others': *Smith v Fonterra Co-Operative Group Ltd (Supreme Court)* (n 1) [155].

(vi) Channel Infrastructure NZ Ltd, formerly the operator of a major oil refining facility (closed in March 2022), also the operator of the country's 'largest fuel import terminal, storing and distributing 40% of New Zealand's fuel, including 80% of New Zealand's jet fuel';[74] and

(vii) BT Mining Ltd, owner and operator of the largest opencast mine in New Zealand, producer of bituminous, coking and thermal coal (most of which is exported to overseas markets).

2.3 Overview of the Pleaded Claims

In a 2024 interview, Smith explained the tactical decision to initiate what represented a departure from typical climate litigation strategy (at least in New Zealand), a *private* law action against corporate emitters:[75]

> Our *Iwi* Chairs climate group has been petitioning the government over the years to take firmer action, but to no avail…So what that tells us is that, while we can expect and demand that the government do things, quite often they won't. So we have to take other avenues—like taking on the companies themselves, directly.
>
> You'd think responsible organisations would be taking urgent action to remedy their impacts and their culpability, but they're not. So they have to be compelled to do it.

Smith's claim relies on three distinct but related causes of action, two of which are established categories of tort law, and the third, largely novel. The first is a claim that the actions of the corporate defendants amount to a public nuisance. The second is a claim in negligence. The third seeks to fashion a novel form of tort liability, described in submissions as 'a duty to avoid contribution to dangerous anthropogenic interference in the climate system' and given a short form label by the Supreme Court of 'the proposed climate system damage tort'.[76] Each cause of action, and its treatment in the New Zealand courts so far, is briefly explained in the following.

---

74 Channel Infrastructure, 'We are Chanel Infrastructure' <https://channelnz.com/> accessed 12 January 2025.

75 Siena Yates, 'Michael versus the Goliaths' *E-Tangata* (18 February 2024) <https://e-tangata.co.nz/comment-and-analysis/michael-versus-the-goliaths/> accessed 30 December 2024.

76 *Smith v Fonterra (Supreme Court)* (n 1) [58].

2.3.1 Public nuisance claim

In his June 2022 amended statement of claim, Smith pleaded:[77]

> By releasing Greenhouse Gases into the atmosphere (or producing or exporting coal, in the case of BT Mining; and producing or supplying Fuel Products in the case of Channel NZ Refining and Z Energy), the defendants have interfered with, or contributed to interference with, and will in the future interfere with or contribute to interference with, the rights of the public.

Introducing the Court of Appeal's analysis of this part of Smith's claim, French J described 'the law of public nuisance as lacking some precision'.[78] Following a framework articulated by the House of Lords in *R v Rimmington*,[79] the Court of Appeal identified four key issues (or 'questions') for assessing whether, in circumstances such as this, a public nuisance claim could be established:[80]

(a) whether actionable public rights were pleaded;
(b) whether independent illegality was required;
(c) whether the special damage rule was met or required; and
(d) whether there was a 'sufficient connection' between the pleaded harm and the respondents' activities.

In the High Court, Wylie J focused his analysis on the second, third and fourth, questions outlined. On each of those matters, Wylie J ruled that Smith had not tenably established a basis for the cause of action. He struck out the public nuisance claim.[81] The Court of Appeal said that it did 'not agree with all of the reasons given by the [High Court] and/or advocated by the respondents for striking out Mr Smith's claim'.[82] However, it still reached the same conclusion: 'the claim in public nuisance is clearly untenable and should be struck out. To allow it to proceed would not extend the existing law but distort it'.[83]

The key difference decided between the High Court and Court of Appeal on the public nuisance claim concerned item (b); whether (as Wylie J had considered) 'it was a necessary ingredient of the tort that the interference with the public right must be independently unlawful, in addition to interfering with public rights'.[84] Having considered the (in parts, conflicting) caselaw and

---

77 Draft Amended Statement of Claim (n 40) [84].
78 *Smith v Fonterra (Court of Appeal)* (n 27) [58].
79 *R v Rimmington* [2005] UKHL 63, [2006] 1 AC 459.
80 *Smith v Fonterra (Court of Appeal)* (n 27) [60]–[93].
81 *Smith v Fonterra (High Court)* (n 27) [56]–[73].
82 *Smith v Fonterra (Court of Appeal)* (n 27) [59].
83 ibid [93].
84 ibid [69].

commentary, the Court of Appeal concluded that '[t]he weight of authority is . . . it is not necessary for the act or omission to be in itself a legal wrong separate from nuisance'. 'What matters', it held, 'is that the act or omission causes common injury'.[85]

The Court of Appeal did not regard issues (a) or (b) as insurmountable barriers to the public nuisance cause of action. However, in its view, (c) and (d) were fatal to the claim. On the matter of the 'special damage rule' (discussed further following), the Court conceded that the position was more nuanced than understood by the High Court. But even adopting 'the most liberal formulation' of the rule, 'Mr Smith did not satisfy it'.[86] Considering a submission from Smith that the time had come to abolish the rule, the Court of Appeal acknowledged that 'the special damage rule has been criticised by some commentators'[87] but stated that it was not necessary to decide whether it should be retained 'because we are satisfied that even were the [rule] abolished...a claim in public nuisance is still doomed to fail and should therefore be struck out'.[88]

The Court of Appeal's analysis of issue (d)—whether there was a 'sufficient connection' between the pleaded harm and the respondents' activities—included a 'causation analysis'.[89] Here, the Court acknowledged that in a number of cases where there were multiple potential tortfeasors, courts had departed from a conventional 'but for' causation test, preferring instead 'the proposition that a defendant will not be exempted from liability on the grounds that they were simply one of many causing a nuisance'.[90] Still, that was no help to Smith as 'all of the cases which have invoked this aggregation principle have involved a finite number of known contributors to the harm, all of whom were before the Court'.

In the Supreme Court decision, public nuisance was the only one of Smith's three pleaded causes of actions to receive a detailed analysis.[91] The Court accepted that each cause 'has its own doctrinal underpinning', but noted that the 'same facts' applied to each one and 'the deeper questions of necessary

[85] ibid [72].

[86] ibid [75], [82], [78].

[87] ibid [84].

[88] ibid [87].

[89] ibid [88].

[90] ibid [89]–[90].

[91] During the Supreme Court hearing, Kós J remarked that he regarded public nuisance as 'the real battleground in this case'. While the Judge took care to clarify that this was *his* view, the focus on public nuisance largely to the exclusion of an analysis of negligence and the proposed climate system tort suggests that the rest of the Court may have been inclined towards this view: *Smith v Fonterra (Supreme Court)* (n 1) transcript 9. See also transcripts 213, 240.

relationship, proximity, causation, disproportionality and indeterminacy raise[d] issues common to all'.[92] Having reached the position that a claim in public nuisance had been pleaded tenably (discussed following), and that '[a]ny added burden the respondents may be required to bear in confronting two additional causes of action will not be significant',[93] the Court concluded that 'it [was] neither necessary nor appropriate that this Court traverse the remaining claims struck out in the Courts below'.[94]

The Supreme Court's discussion on public nuisance started with an account of the tort's origins. These origins lie in early English common law criminalisation of 'socially objectionable actions and omissions' resulting in harm which, suffered by a community as a whole, could not be addressed through *private* nuisance actions.[95] The offence of public nuisance was 'traditionally used to deal with the obstruction of public highways and rivers, and activities that caused a loss of amenity in a neighbourhood (for example through noises or smells)'.[96] As a common law crime, public (or 'common') nuisance proceedings were brought in the name of the Attorney-General. However, in the 16th century, English courts began to recognise public nuisance as a civil suit, actionable by private persons, provided plaintiffs could show that they had 'suffered greater hurt or inconvenience than the generality have' (this became the so-called 'special damage rule').

Tracing the evolution of the tort of public nuisance in New Zealand, the Court noted that common law crimes were abolished in the colony in 1893 when criminal nuisance became an exclusively statutory offence. Civil actions for public nuisance continued, affirmed in cases such as the 'leading authority in New Zealand on public nuisance',[97] *Attorney-General v Abraham and Williams Ltd* (1949).[98]

Turning to Smith's claim of public nuisance, the Supreme Court agreed that the Court of Appeal had aptly identified the four key questions. It rejected submissions of the defendants challenging the Court of Appeal's findings on items (a) (actionable public rights tenably pleaded)[99] and (b) (independent illegality not required).[100] It then turned to the more contentious findings on the

---

92 *Smith v Fonterra (Supreme Court)* (n 1) [175].
93 ibid.
94 ibid [176].
95 ibid [103].
96 ibid [105].
97 ibid [108].
98 *Attorney-General v Abraham and Williams Ltd* [1949] NZLR 461 (CA).
99 *Smith v Fonterra (Supreme Court)* (n 1) [144].
100 ibid [147].

third and fourth items, on both of which the Court of Appeal had ruled against Smith.

The third 'question' concerned the 'special damage rule'. As previously noted, the Court of Appeal had side-stepped the issue of whether a standing requirement for public nuisance claims remained appropriate in the contemporary New Zealand context.[101] The Supreme Court had no qualms addressing the issue directly: it concurred with academic commentary that, in light of the departure from criminalisation of public nuisance and modern approaches to class actions and case management, 'the rationales for the standing rule are not now particularly convincing'.[102] The Court stated:[103]

> We consider the special damage rule requires reconsideration in a 21st century context, in which the implications of ubiquitous harms such as pollution (including from GHGs) are more evident and better understood, and in which class actions and active judicial case management have developed and are better able to meet fears of an oppressive multiplicity of actions.

The Court rejected the earlier courts' findings that Smith could not satisfy the special damage requirement, however formulated. It held:[104]

> Mr Smith has a tenable claim to meeting its present requirements because of his pleading of damage to coastal land at Mahinepua C in which he and others he represents claim both a legal interest and distinct tikanga interests. If the interests of many others, whether proprietary or tikanga, are likewise affected, that may say more about the gravity of the alleged tort than the propriety of entertaining it.

The fourth and final question in the public nuisance analysis of 'sufficient connection, or causation' may well be the most contentious element of the entire claim. This overarching issue concerned potential limitations of the common law to respond to the described 'existential crisis' of climate change. At a more granular level, the Court needed to address doctrinal matters such as the relevance of and alternatives to conventional 'but for' causation requirements (including addressing the implications of multiple contributors to harm), and the question of relief.

The Supreme Court's position on the capacity of tort law to address climate change is a major departure from the earlier court findings. Implicitly rejecting

---

101 *Smith v Fonterra (Court of Appeal)* (n 27) [87].

102 ibid [149] citing Carolyn Sappideen and Prue Vines (eds), *Fleming's The Law of Torts* (10th edn, Thomson Reuters 2011) 490.

103 ibid [151].

104 ibid [152].

the Court of Appeal's view that 'the issue of climate change cannot be effectively addressed through tort law',[105] the Supreme Court responded:[106]

> It may indeed be beyond the capacity of the common law to resolve climate change in fact, but we are not presently convinced, at this stage of the proceeding, addressing only strike out, that the common law is incapable of addressing tortious aspects of climate change.

The Court noted that the common law has previously faced and responded to other crises. The industrial revolution was of a 'lesser scale' to the climate crisis,[107] but catalysed common law responses to 'new, widespread risk and damage caused by air and water pollution and the escape of biohazards'.[108] Therfore, Smith should be permitted to proceed to trial where the Court can assess, with the benefit of evidence and full argument, whether the common law might contribute, as it has done in the past, to responses to 'newer technologies and newer harms'.[109]

On causation, substantiality, unreasonableness and remedy, the Court observed that 'logic and experience suggest' these issues represent the 'fundamental battleground between the parties'.[110] At the strike-out stage, it did not consider that these matters were necessarily fatal to the claim. On causation, it noted that it was at least arguable that the 'but for' test is not appropriate for a public nuisance claim.[111] On the requirement that tortfeasors' contributions to a public nuisance be 'substantial' and 'unreasonable' (a test not questioned in the judgment),[112] 'this remains a fundamental issue of fact for trial. We do not prejudge that issue here'.[113] On the question of remedy, the Court attached some weight to the fact in this case, Smith has not sought damages, but rather, *injunctive* relief ('an equitable remedy involving a substantial measure of discretion in the calibration of remedial impact').[114] Nor, at this stage, did it consider that the utility of declaratory relief should be dismissed.[115]

105 *Smith v Fonterra (Court of Appeal)* (n 27) [28].
106 *Smith v Fonterra (Supreme Court)* (n 1) [154].
107 *Smith v Fonterra (Supreme Court)* (n 1) [156].
108 ibid [157].
109 ibid [166].
110 ibid [170].
111 ibid [165].
112 See e.g. ibid [111], [147], [168].
113 ibid [169].
114 ibid [171].
115 ibid.

The Supreme Court's overall conclusion on the tenability of the public nuisance claim was that 'in sum', it did 'not consider the obstacles are so overwhelming as to meet the standard for strike out':

> The consequence, therefore, is that [the defendants] must now submit to argument, and evidence, at trial. In this area, the common law must develop, if at all, in the fertile fields of trial, not on the barren rocks of a strike out application.[116]

2.3.2 Negligence

Smith's second cause of action is outlined in the amended statement of claim in the following terms: 'the defendants owe the plaintiff (and persons like him) a duty to take reasonable care not to operate their business in a way that will cause the plaintiff loss by contributing to dangerous anthropogenic interference in the climate system'.[117]

As noted by Kysar, negligence claims against corporate emitters face a range of particular doctrinal challenges which, to date, have impeded success in common law courts.[118] The first instance decision of the Federal Court of Australia in *Sharma v Minister for the Environment* was a rare example of judicial acceptance of a broadly configured duty of care on the part of a ministerial decision-maker assessing an application to approve the extension of an opencast coal mine. Greenhouse gas emissions from the mine extension were said to contribute materially to adverse impacts on vulnerable young people in Australia.[119] However, the decision was overturned by the full Federal Court of Australia on appeal.[120]

At the High Court, Wylie J held that a negligence action would be 'clearly untenable'.[121] He ruled that the cause of action had no reasonable prospects of success on the grounds of lack of foreseeability, causation, proximity and having regard to policy considerations.[122] The Court of Appeal differed from the High Court on the question of foreseeability (it accepted that it could be

116 ibid [173].

117 Draft Amended Statement of Claim (n 40) [90].

118 See Douglas A Kysar, 'What Climate Change Can Do about Tort Law' [2011] *Environmental L* 1. For a more recent analysis including commentary on the *Smith v Fonterra* litigation, see Douglas A Kysar, 'The Duty of Climate Care' (2023) 73 *DePaul L Rev* 487.

119 See *Sharma v Minister for the Environment* [2021] FCA 560 [76] (Bromberg J), [513].

120 *Minister for the Environment v Sharma and others* (2023) 4 LRC 1.

121 *Smith v Fonterra (High Court)* (n 27) [100].

122 ibid [74]–[100].

an issue for trial),[123] but affirmed the first instance findings on proximity,[124] causation[125] and policy considerations.[126]

For the reasons explained previously, the Supreme Court decision said very little about the viability of negligence as a cause of action in Smith's case. A review of submissions for and against the strike-out (together with the record of oral argument before the Supreme Court)[127] shows that the central (although not only) point of contention concerned the element of proximity. Smith's primary argument about proximity was that it was a matter to be decided at trial. At this stage of the proceedings, it was sufficient to acknowledge that the plaintiff was 'part of an identifiable group (coastal Māori in Northland) subject to some "particular" or "distinctive" risk to that faced by the general public'.[128] Counsel for the first to fifth defendants submitted that the Court of Appeal had correctly concluded that 'the degree of proximity was insufficient to establish a duty of care: there was no physical or temporal proximity; no direct relationship; and no causal proximity'.[129] Developing that point in oral submissions, he argued that the 'very legal concept of proximity…in its express form in negligence, but also in its implicit form in public nuisance' required 'a real relationship between plaintiff and tortfeasor'; a relationship that did not exist between Smith and the defendants in the case.

The Supreme Court's decision to allow all three causes of action to proceed to trial means that a negligence claim remains 'live'. As previously noted,[130] oral commentary by at least one Supreme Court judge indicate a preliminary view that public nuisance would appear to better suit the claim. Legal, conceptual and evidentiary challenges concerning proximity (present most

123 *Smith v Fonterra (Court of Appeal)* (n 27) [100] (French J).
124 ibid [103].
125 ibid [113].
126 ibid [116].
127 Davey Salmon KC and others (n 40) [119]–[131], [160]; DR Kalderimis and others, *Smith v Fonterra (Supreme Court)*, 'Synopsis of submissions for first to fifth respondents' (20 July 2022) [118]–[120]; *Smith v Fonterra (Supreme Court)* (n 1) transcript 28, 45, 81, 89, 156, 170, 225, 231, 238–239, 246, 249, 253, 374.
128 Davey Salmon KC and others (n 40 [28]). Counsel for Smith also submitted at [123] that 'the pleaded knowledge of actual risk is a significant indicator of sufficient proximity', at [125] that 'vulnerability is a factor particularly relevant to the proximity inquiry', and at [160] that 'the *tikanga* approach to proximity between the perpetrator and victim of "*hara*" [an issue or cause] is more expansive than that traditionally adopted by common law torts'.
129 DR Kalderimis and others (n 127) [118].
130 *Smith v Fonterra (Court of Appeal)* (n 27).

obviously—but not necessarily exclusively—in a negligence claim) seem to support this view,[131] but its fate remains to be determined at trial.

2.3.3 Climate system damage tort

Smith's third cause of action is presented in his amended statement of claim as follows:[132]

> The defendants owe a duty, cognisable at law, to cease materially contributing to damage to the climate system, dangerous anthropogenic interference with the climate system, and the Adverse Effects of climate change through their emission of Greenhouse Gases into the atmosphere (or their production or exportation of coal in the case of BT Mining; and their production and supply of Fuel Products in the case of Channel NZ Refining and Z Energy).

Wylie J's first instance decision to allow this cause of action (resting on what he described as an 'inchoate duty') to proceed to trial reflected his 'reluctan[ce] to conclude that the recognition of a new tortious duty which makes corporates responsible to the public for their emissions, is untenable'.[133] The judge clearly had doubts about its prospects, but was unwilling to rule out the potential for the 'evolution of the law of tort',[134] including the possibility that 'the special damage rule in public nuisance could be modified' and 'that climate change science will lead to an increased ability to model the possible effects of emissions'.[135] The Court of Appeal shared neither Wylie J's reluctance to prevent the opportunity for tort law's engagement with climate change to evolve, nor his openness to doctrinal or evidential innovation. In its words, 'the mere fact of novelty cannot be enough. Otherwise any claimant would be able to proceed to trial simply by asserting a new tort'.[136]

---

131 This appears to be acknowledged in a 2022 academic publication by David Bullock, one of Smith's co-counsel in the *Fonterra* proceeding. Writing in the *Modern Law Review*, Bullock conceded that 'climate change poses a *plaintiff problem* and a *defendant problem* for the application of tort law', and posed the rhetorical question 'can tort law draw conceptual boundaries as to parties that enable it to be both workable and relational?' The article proceeds to outline how both plaintiff and defendant 'problems' may be overcome through *public nuisance* doctrine: David Bullock, 'Public Nuisance and Climate Change: The Common Law's Solutions to the Plaintiff, Defendant and Causation Problems' (2022) 85 *Modern L Rev* 1136, 1137.

132 Draft Amended Statement of Claim (n 40) [98].

133 *Smith v Fonterra (High Court)* (n 27) [103].

134 ibid [102], [103].

135 ibid [103].

136 *Smith v Fonterra (Court of Appeal)* (n 27) [124].

As with the negligence claim, the Supreme Court included little specific commentary on the proposed climate system damage tort, content to allow it to remain 'live' on the basis that to do so would not significantly add to the defendants' burden at trial.[137] However, some indications of how arguments and evidence might develop at trial can be gleaned from written submissions and the record of oral argument before the Supreme Court.

A feature of the proposed tort that arose in oral arguments concerned its potential to operate on the basis of an entirely different approach to conventional tort law proximity, instead reflecting an 'ecocentric as opposed to anthropocentric way of looking at harm in relation to the environment'.[138] Making submissions in support of the relevance of *tikanga* to all three causes of action, Smith's co-counsel Natalie Coates observed that '*tikanga* provides an alternative lens through which to look at wrongs and harms'. Acknowledging that *tikanga* is involved in responses to human-to-human harms, counsel continued:[139]

> But we say pollution and harm involving the environment and the natural world are a different category of [wrongs] to te ao Māori...pollution directed, or harms directed at that environment do not tend to focus or have this focus on individualized relational proximity between two people.

In an interchange between counsel for the first to fifth respondents and Supreme Court judges, counsel relayed his understanding that the previous noted submissions contemplated 'a sort of duty to the environment generally, not really a duty to Mr Smith as such', and that (in respect of the '*tikanga* element'), 'the appellant's case for the third cause of action was a discussion free of proximity'.[140] A novel climate change tort that dispenses of conventional conceptions of human-to-human relational proximity, refashioned (as counsel for BT Mining put it) 'a tort for injury to Gaia or the like' would have far-reaching implications, the consideration of which is beyond the scope of this chapter.[141] However, if the submissions before the Supreme Court are an indication of the future directions of the proceeding, the next phase promises to cover fascinating territory.

---

137 *Smith v Fonterra (Supreme Court)* (n 1) [175].

138 *Smith v Fonterra (Supreme Court)* (n 1) transcript 89 (Natalie Coates).

139 ibid [87]–[88].

140 *Smith v Fonterra (Supreme Court)* transcript 248–249 (Daniel Kalderimis).

141 *Smith v Fonterra (Supreme Court)* transcript 374 (RJ Gordon).

2.3.4 Relief

As explained in submissions,[142] Smith seeks a range of remedies. First, a declaration that the defendants' activities have 'unlawfully caused or contributed to' the relevant tort.[143] Second, an injunction requiring the defendants take specified emissions reductions steps including achieving a net zero position by 2050.[144] In the alternative, third is an injunction 'which may be suspended' requiring the defendants to 'immediately cease emitting net emissions'.[145] Fourth, a general request for 'other relief as the Court determines appropriate to enable the mitigation of or adaptation to damage to climate systems contributed to by the defendants'.[146]

Relief occupied some time and attention at the Supreme Court hearing. Consistent with its preference for judicial economy in dealing with strike-out applications, (recognising that 'the matter...may yet return to [it]')[147] the Court addressed the issue of relief only briefly in its decision. The one paragraph on relief (at the end of its analysis on public nuisance) hints at an inclination towards a flexible form of relief (that would appear consistent with a suspended injunction)[148] as well as the option of declaratory relief.[149]

2.3.5 *Tikanga*

The role of *tikanga* in the proceeding developed as the case progressed through the courts. In the High Court, *tikanga* received only confined attention. There was no explicit reference to *tikanga* in the original statement of claim,[150] but the matter was addressed in oral submissions as part of an explanation of Smith's connection with the land in Northland pleaded as adversely impacted by climate change (including as a result of emissions associated with

142 Davey Salmon KC and others (n 40) [164].

143 Draft Amended Statement of Claim 15 June 2022, (n 40) [89(a)], [97(a)], [99(a)].

144 ibid [89(b)], [97(b)], [99(b)].

145 ibid [89(c)], [97(c)], [99(c)].

146 ibid [89(d)], [97(d)], [99(d)].

147 *Smith v Fonterra (Supreme Court)* (n 1) [143].

148 ibid [171], referring to 'an equitable remedy involving a substantial measure of discretion in the calibration of remedial impact'.

149 ibid: 'the utility of the declaration of inconsistency jurisdiction in public law suggests the court should not dismiss the power of purely declaratory relief in private law'.

150 In submissions before the Supreme Court, counsel for Smith explained 'Mr Smith did not initially "plead" *tikanga* Māori on the understanding that, as it forms part of the laws of New Zealand, there was no need to do so' (Davey Salmon KC (n 40) [51]).

the defendants' activities).[151] *Tikanga* received two brief mentions in the High Court decision, simply recording that it formed part of the basis of Smith's association with the land affected.[152]

By the Court of Appeal hearing, *tikanga* had assumed a more prominent role. Having received an indication from the Court that '*tikanga* Māori needed to be pleaded…if he intended to argue that it could inform the principles of tort law',[153] Smith's counsel proposed a new clause in the Statement of Claim:[154]

> *Kaitiakitanga* as a principle of *tikanga* Māori incorporates concepts of guardianship, protection and stewardship of the natural environment including recognising that a right in a resource carries with it a reciprocal obligation to care for its physical and spiritual welfare as part of an ongoing relationship.

The Court of Appeal interpreted Smith's argument on *tikanga* as 'a principle and value that should infuse the court's consideration of the issues in relation to all three causes of action'.[155] Alongside that broader submission, Smith's counsel argued that *tikanga* was relevant when considering whether Smith had experienced damage above and beyond that experienced by the general public: the 'special damage rule'.[156]

The Court of Appeal was not moved by the amended *tikanga* pleading or argument. On the relevance of *tikanga* to the public nuisance special damage requirement, the Court rejected the proposition that Smith could claim harm above and beyond 'many other people in New Zealand (or elsewhere in the world) who suffer the same interference, including landowners, other *iwi* and *hapū*'.[157] The position was 'not something that can be overcome by re-pleading or invoking concepts of *tikanga*'.[158] Furthermore (and surprisingly, given that no evidence had been introduced on the issue), the Court provided its view on the implications of *kaitiakitanga* (a *tikanga* principle often translated as 'guardianship', but whose meaning is more complex than the English concept or term)[159] to the claim. It stated 'we consider that controlling climate change through regulatory means [as opposed to tort mechanisms] is consistent with *kaitiakitanga*'.[160]

---

151 *Smith v Fonterra (High Court)* (n 27) [5].
152 ibid [5], [60].
153 Davey Salmon KC and others (n 40) [51].
154 *Smith v Fonterra (Court of Appeal)* (n 27) [7].
155 ibid [8].
156 ibid [57].
157 ibid [82].
158 ibid.
159 See eg Law Commission Te Aka Matua o te Ture (n 65) [3.116—3.119].
160 *Smith v Fonterra (Court of Appeal)* (n 27) [34].

The role and relevance of *tikanga* was addressed at length in oral argument before the Supreme Court, and in its decision.[161] A significantly more detailed clause about *tikanga* was included in an updated proposed Statement of Claim, referencing additional *tikanga* principles beyond those included in the amended Claim before the Court of Appeal.[162] A detailed analysis is beyond the scope of this chapter, but two points are highlighted following.

First, departing from the approach taken in the lower courts, the Supreme Court had accepted, without hesitation, the relevance of *tikanga* to the matters raised in the pleading, and to the development of tort law in New Zealand generally (in the context of climate change). Justices Williams and Kós reached into New Zealand legal history, citing decisions from the 19th and early 20th centuries where New Zealand courts had acknowledged the relevance of *tikanga* in tort claims, and to the common law more generally.[163]

Unlike the Court of Appeal, the Supreme Court considered *tikanga* to be directly relevant to the question of standing and special damages in the context of the public nuisance claim. The Court accepted that 'the specific loss pleaded by Mr. Smith in this case is in part *tikanga*-based'. The Court held that 'the trial court will be required to engage with *tikanga*'.[164]

In broader terms, the Court also regarded *tikanga* as relevant to other causes of action, including the proposed climate system damage tort:[165]

> [T]he trial court will need to grapple with the fact that Mr Smith purports to bring proceedings not merely as an alleged proprietor who has suffered loss, but as a kaitiaki [guardian] acting on behalf of the whenua [land], wai [water] and moana [ocean]—distinct entities in their own right. And it must consider some tikanga conceptions of loss that are neither physical nor economic. In other words, addressing and assessing matters of Tikanga simply cannot be avoided.

---

[161] At the Supreme Court, leading Māori lawyer Natalie Coates was part of the legal team for Smith and addressed the court about the *tikanga* dimensions of the case. Te Hunga Rōia Māori Aotearoa | The Māori Law Society was granted leave to participate as an intervener, and senior Māori barrister Matanuku Mahuika delivered submissions focusing on Māori *tikanga*: see MK Mahuika and HK Irwin-Easthope, *Smith v Fonterra (Supreme Court)*, 'Submissions for Te Hunga Rōia Māori o Aotearoa (The Māori Law Society)' (22 June 2022). For the Court's analysis of *tikanga*, see especially: *Smith v Fonterra (Supreme Court)* (n 1) [59]–[61], [151], [169], [177]–[189].

[162] Draft Amended Statement of Claim 15 June 2022, above (n 40) [82].

[163] *Smith v Fonterra (Supreme Court)* (n 1) [177]–[189].

[164] ibid [182].

[165] ibid [188] (footnote omitted).

Having accepted that *tikanga* was, in principle, legally relevant, the Court observed that evidence would need to be heard on the topic, potentially utilising 'analytical methodolog[ies]' employed in previous Supreme Court hearings.[166] It concluded that in this context, 'Mr Smith's ultimate prospects at trial will depend, in part, on the quality of the evidence, including that in relation to *tikanga*'.[167]

2.4 Evidence and Climate Science

Any discussion about the judicial treatment of climate change science in the *Smith v Fonterra* litigation should start with the premise that the courts' decisions have all concerned interlocutory applications to strike out proceedings. In the strike-out context, the conventional position (uncontested as a matter of law) is that the plaintiff is entitled to have the courts 'assume the pleaded material facts are true save for those that are entirely speculative and without foundation'.[168] The defendants' tactical decision to pursue strike-out applications meant that, at that stage of the litigation, Smith could focus on matters of principle without the onerous task of having to prove his factual allegations in evidence. At this stage of the litigation, Smith was able to focus argument on matters of principle without the onerous task of having to establish his factual allegations in evidence. Therefore, the Supreme Court decision to allow the claims to go to trial contains no findings on core evidentiary issues where Smith and the defendants are at odds, where tort claims in other jurisdictions have faltered. These include critical tenets of Smith's claim, such as the existence and strength of any causal link between the activities of the defendants and climate change-related harm claimed by Smith.

However—and in this regard, following a trend in climate change litigation both in New Zealand and other jurisdictions[169]—each of the three judgments in the High Court, Court of Appeal and Supreme Court included introductory

---

166 The Court referred in particular to *Ellis v R (Continuance)* [2022] NZSC 114, (2022) 1 NZLR 239 [121]–[125] (Glazebrook J), [181] (Winkelmann CJ) and [261]–[273] (Williams J).

167 *Smith v Fonterra (Supreme Court)* (n 1) [189].

168 *Smith v Fonterra (Court of Appeal)* (n 27) [38], cited with apparent approval at *Smith v Fonterra (Supreme Court)* (n 1) at [74].

169 In New Zealand, see e.g. *Thomson v Minister for Climate Change Issues* [2017] NZHC 733, (2018) 2 NZLR 160 [8]–[18]; *Hauraki Coromandel Climate Action Inc v Thames-Coromandel District Council* [2021] NZRMA 22 [3]–[7]. In other jurisdictions, see eg *Milieudefensie et al v Shell Plc* (Court of Appeal The Hague, 12 November 2024) [3.2]–[3.13]; *Verein KlimaSeniorinnen Schweiz and Others v Switzerland* (ECtHR, 9 April 2024) [64]–[120].

sections dealing with high-level findings from climate science. A five-paragraph section of Wylie J's High Court judgment recorded key points from IPCC reports, including the statement that 'unless substantial and rapid reductions in anthropogenic emissions occur, temperature rise will exceed 2°C and likely be substantially more'.[170] Writing for the Court of Appeal, French J's overview of climate science was comparatively briefer. In the first of four concise paragraphs, French J observed:[171]

> Climate change is commonly described as the biggest challenge facing humanity in modern times. Its causes and its effects are now widely recognised, with scientists predicting that if greenhouse gas emissions keep increasing, the planet will eventually reach a point of no return.

The remaining three paragraphs relating to foundational climate science were all qualified with the descriptors 'contends' or 'alleged'. For example:[172]

> Mr Smith **alleges** in the statement of claim that the release of greenhouse gases by the respondents is human activity that has contributed and will continue to contribute to dangerous anthropogenic interference with the climate system and to the adverse effects of climate change.
>
> [...]
>
> It is also **alleged** that poor and minority communities will be disproportionately burdened by the adverse effects of climate change. (emphasis added)

The Supreme Court addressed matters of climate science in significantly more detail than either the High Court or Court of Appeal. Included in its judgment under the heading 'Climate Change' is a 14-paragraph summary of points which 'may be taken as common ground or indisputable'.[173] Unlike the Court of Appeal, the Supreme Court's high-level overview of climate science includes no caveats or qualifying terms, nor indications that these matters might be revisited at trial. This includes statements such as:[174]

> Climate change threatens human well-being and planetary health...the window of opportunity to ensure a liveable and sustainable future for all is rapidly closing. The choices made, and actions implemented, in this decade will have impacts both now and for thousands of years.

---

170 *Smith v Fonterra (High Court)* (n 27) [28(e)].

171 *Smith v Fonterra (Court of Appeal)* (n 27) [2].

172 ibid [4].

173 *Smith v Fonterra (Supreme Court)* (n 1) [13]–[26].

174 ibid [14].

The Court accepted as either common ground or indisputable that 'between 3.3 and 3.6 billion people "live in contexts that are highly vulnerable to climate change". Vulnerable communities, which historically have contributed the least to the problem, are being "disproportionately affected" by climate change'.[175] The Court also accepted that 'the probability of "abrupt and/or irreversible changes" increases with higher global warming levels, as does the probability of low-likelihood outcomes that have potentially "very large adverse impacts"'.[176]

The treatment of general matters of climate science coming from the country's highest court can be expected to influence not only lower courts in this matter, but also other claims on climate matters. However, as noted earlier, there remain fundamental questions of fact which will be fiercely contested, and proceedings before the High Court can be expected to involve the most comprehensive array of expert witnesses and competing expert evidence of any climate change case in New Zealand legal history.

It is difficult to predict the outcome of that anticipated contest of evidence beyond noting that considerable challenges await both the plaintiff and the defendants at trial. The trial judge or judges also face the monumental task of navigating through a maze of complex evidence—judicial adjudication on contested factual matters will require an advanced level of sophistication and expertise.

3. REFLECTIONS

3.1 Judicial Competence—The 'Political Question'

In jurisdictions around the world, the issue of the appropriateness of judicial assessment and determination of claims involving elements of climate change policy—in United States jurisprudence and legal commentary dubbed the 'political question doctrine'[177]—has been an obstacle for many climate change-related tort claims.[178] In *Smith v Fonterra*, the Court of Appeal bluntly

---

175 ibid [18] (footnotes omitted).

176 ibid [22].

177 See e.g. Amelia Thorpe, 'Tort-Based Climate Change Litigation and the Political Question Doctrine' (2008) 24 *J Land Use & Environmental L* 79.

178 See e.g. *Native Village of Kivalina v ExxonMobil Corp*, 696 F.3d 849, 854 (9th Cir 2012) [7]; *Minister for the Environment v Sharma* (2022) 400 ALR 203 [15], [17]. See also the discussion (in an extrajudicial context) by Justice Susan Glazebrook in 'The Role of Judges in Climate Governance and Discourse' (paper for Asia-Pacific Judicial Conference on Climate Change, November 2020) cited in submissions to the Supreme Court for the first to fifth defendants: DR Kalderimis and others (n 127) n 13.

expressed its view: 'Climate change provides a striking example of a polycentric issue that is not amenable to judicial resolution'.[179] Five months before Smith's hearing before the Supreme Court, Australian Federal Court judge Allsop CJ observed that 'the posited duty throws up for consideration at the point of breach matters that are core policy questions unsuitable in their nature and character for judicial determination' in *Sharma* (a decision overturning a first instance finding that the Australian Minister for the Environment owed a duty of care to Australian children when assessing an application to extract coal from a coal mine).[180]

The Supreme Court's decision to allow Smith's claim to proceed to trial appears to be a major departure from the approach taken in other courts on the issue of whether complex questions of polycentric climate policy are suitable for determination in judicial fora. It *is* a departure—a significant one—but it by no means follows that the Supreme Court has enabled the construction of a parallel judicial framework for control of emissions. This is contrary to the assertion of counsel for the first to fifth defendants, described as 'creation of a parallel—and inconsistent—regulatory regime'[181] and characterised by the Court of Appeal as 'ineffective and inefficient, and likely to be socially unjust'.[182]

It is helpful to distinguish between the substantive assessment of the intended tort claims and the question of remedies. As noted prior, Smith's relief sought a range of remedies. As recognised in oral argument before the Supreme Court,[183] it is conceivable that a trial court could reach a position that the activities of the defendants (or a subset of them) amount to tortious breaches, but, sensitive to separation of power principles, confine relief to a suspended injunction ('personally favour[ed]' by Smith's counsel),[184] or declaration. Both forms of relief would allow the legislative branch to intervene, maintaining a measure of 'constitutional balance' in which both judicial and legislative branches cooperate in legal and policy responses.[185]

179 *Smith v Fonterra (Court of Appeal)* (n 27) [26].
180 *Minister for the Environment v Sharma* (2022) 400 ALR 203 [7].
181 Synopsis of submissions for first to fifth respondents (n 127) [51].
182 *Smith v Fonterra (Court of Appeal)* (n 27) [35].
183 See eg, *Smith v Fonterra (Supreme Court)* (n 1) transcript 70.
184 ibid 68.
185 The term was offered by counsel for Smith: ibid [70].

3.2 *Tikanga* Māori

One of the features of the Supreme Court's decision in *Smith v Fonterra* that has attracted worldwide curiosity is the affirmation of the relevance of Māori *tikanga* to evolving tort law doctrine in the climate change context. This interest is understandable. For many lawyers steeped in English legal traditions, the prospect of infusing tort law with a set of unfamiliar values and principles may be confronting, even bewildering.

The Supreme Court's affirmation of the relevance of *tikanga* to the evolving common law on public nuisance, negligence and beyond is not especially remarkable, nor should it have been unexpected (at least, to New Zealand lawyers). As documented in the decision, New Zealand judges have long accepted the place of *tikanga* in a variety of legal contexts, including tort claims.[186] In this regard, the Supreme Court followed reasonable precedent from its own decisions, including *Takamore*,[187] *Trans-Tasman Resources,*[188] and *Ellis*[189] as well as decisions of other New Zealand courts.[190]

As acknowledged in the *Smith v Fonterra* decision, *tikanga* has potential relevance to climate change-focused tort claims involving Māori at different levels. To the extent that a special damage requirement remains an element of public nuisance claims, evidence on *tikanga*, including '*tikanga* conceptions of loss that are neither physical nor economic'[191] may become a routine aspect of pleadings and trial process.

Less certain but perhaps more intriguing is the potential for *tikanga* to more radically reshape tort doctrine, including (as addressed in submissions by Coates discussed previously)[192] providing impetus for judicial adoption of an 'alternative [tort] lens' with reduced focus on 'individualized relational proximity' and greater weight given to ecocentric conceptions of harm and proximity. This type of approach would align with the view of academics such as Maria Hook and others who, writing in a 2021 publication cited in submissions for Smith,[193] considered that 'there is a sound basis for extending a new tort to the environment to climate harm'.[194] Analysis of this potential development is

186 *Smith v Fonterra (Supreme Court)* (n 1) [177]–[189].

187 *Takamore* (n 64).

188 *Trans-Tasman Resources* (n 64).

189 *Ellis* (n 64).

190 See Law Commission Te Aka Matua o te Ture (n 65) 168–210.

191 *Smith v Fonterra (Supreme Court)* (n 1) [177]–[188].

192 *Smith v* Fonterra *(Supreme Court)* (n 1) transcript 87–88 (Natalie Coates).

193 Davey Salmon KC and others (n 40) n 227.

194 Maria Hook and others, 'Tort to the Environment: A Stretch Too Far or a Simple Step Forward?' (2021) 33 *J Environmental L* 195, 209.

beyond the scope of this chapter, but the author awaits the next chapter of these proceedings with anticipation, alongside existing and prospective litigants, defendants, jurists and legal commentators in New Zealand and further afield.

3.3 Is It Worth It?

As discussed in this chapter, while the Supreme Court's decision enabling Smith's claim to proceed to trial represents a major development in climate litigation in New Zealand, a positive substantive outcome is far from certain. Multiple procedural, doctrinal and evidential obstacles await Smith at trial, and very likely, in subsequent appeals. Even if Smith is able to secure rulings in his favour, injunctive relief may be elusive. Additionally, the spectre of legislative intervention will hover over the (likely years of) proceedings.[195]

The following questions arise: 'Is it worth it?' Would the energy, time and resources taken up in the next phase of the proceeding be better directed elsewhere? Emphatic responses from climate litigation sceptics such as Australian environmental and climate change lawyer Guy Dwyer are: 'No', and 'Yes', respectively. In Dwyer's words, 'Climate litigation is too slow, ad hoc, and piecemeal to serve as an effective tool for mitigating the causes of climate change. Any effective regulatory approach to climate change must be driven primarily by a top-down plan'.[196] Although primarily focused on judicial review rather than private law actions, New Zealand academic and former Prime Minister Geoffrey Palmer has commented in a similar vein that 'the scope for improving the quality of climate change stewardship via the courts is limited…Here and there the courts may give policy a nudge, but do not expect quantum leaps…'[197]

While these observations are worth considering, success, in litigation as in other spheres, can be measured in different ways. Securing a tightly constructed injunction requiring seven of New Zealand's largest emitters to curb their contribution to climate change would not be the only meaningful outcome of the *Smith v Fonterra* case. Another consequence of a merits trial testing Smith's claims largely independent of the final result is a greater measure of transparency concerning the defendants' corporate knowledge, actions and

---

[195] See Stephen Mills, 'Legislating to Stop Climate Claim Unwise' [5 March 2024] *Capital Letter*.

[196] Guy Dwyer, 'Climate Litigation: A Red Herring among Climate Mitigation Tools' in Benoit Mayer and Alexander Zahar (eds), *Debating Climate Law* (Cambridge University Press 2021) 133–34.

[197] Sir Geoffrey Palmer, 'Can Judges Make a Difference? The Scope for Judicial Decisions on Climate Change in New Zealand Domestic Law' (2018) 49 *Victoria University of Wellington L Rev* 191, 204.

inaction on climate change issues. In academic commentary, Salmon noted 'the ability to compel discovery allows scrutiny of parties' claimed positions in a way that would quickly collapse much of what has been said to cloud understanding of climate change and knowledge of it by major polluters'.[198]

The transparency enhancing function of discovery continues at trial, where both corporate representatives and expert witnesses present and are questioned on evidence under oath. Speaking extrajudicially, New Zealand Supreme Court judge Glazebrook J has acknowledged the limitations on courts' contribution to climate governance,[199] but also observed that 'the courts are also fora where evidence is presented, there is (usually) rational argument based on that evidence and a reasoned judgment follows. This may be missing from more general public and political discourse'.[200]

Dispassionate scrutiny of contested evidence about corporate responses and contributions to climate change-related harm as well as contested matters of materiality, causation, proximity and other aspects of climate science relevant to New Zealand would be useful outcomes of a trial process. Finally, confronting the potential for a 'no liability' outcome, a carefully analysed, clearly articulated judgment which identifies gaps or shortcomings in legal and political frameworks could be a critical contribution to both the design of climate governance and the strategies for non-government, corporate and public actors in this area. As Ewing and Kysar observe:[201]

> Rather than counseling against common law adjudication, therefore, the complexity and enormity of the climate change problem counsel in its favor, in order that baseline norms of responsibility—whatever their content—may be more clearly specified as public and private actors embark on what undoubtedly will be a centuries-long struggle to deal with greenhouse gas emissions and their impacts.

---

198 Davey Salmon (n 13) 151.

199 Susan Glazebrook, 'The Role of Judges in Climate Governance and Discourse' (Asia Pacific Judicial Conference on Climate Change: Adjudication in the Time of COVID-19, Virtual, November 2020) 19.

200 ibid [25].

201 Benjamin Ewing and Douglas A Kysar, 'Prods and Pleas: Limited Government in an Era of Unlimited Harm' (2011) 121 *Yale LJ* 350, 356.

5. *Centre for Oil Pollution Watch (COPW) v Nigeria National Petroleum Corporation (NNPC)*: locus standi and the emerging trends in climate litigation in Nigeria

Funmi Abioye

1. INTRODUCTION

Climate change and its effects are felt daily, all over the Earth. From the raging fires that burn across North America during the often referred to 'fire season', to the unprecedented weather patterns (cyclones, tsunamis, tornados) leading to floods experienced across continents, humankind is faced with one of its greatest challenges. Reports from scientists and expert bodies echo the devastating effects of climate change and warn that if left unmitigated, the Earth's climate will continue its spiral as global warming increases.[1] The Intergovernmental Panel on Climate Change (IPCC) in its Sixth Assessment Report observed that many of the current changes being experienced in relation to the Earth's climate are unprecedented.[2] That report also noted that many of the changes, especially in relation to rising sea and ocean levels, will remain irreversible for centuries to millennia,[3] and will progressively get worse. These changes have been noted to result in extreme weather patterns and more frequent

[1] 'The Climate Crisis – A Race We Can Win' (*United Nations*) <www.un.org/en/un75/climate-crisis-race-we-can-win> accessed 5 May 2025.

[2] Valérie Masson-Delmotte and others (eds), IPCC 2021: *Climate Change 2021: The Physical Science Basis. Contribution of Working Group I to the Sixth Assessment Report of the Intergovernmental Panel on Climate Change* (CUP 2021).

[3] ibid 21.

occurrences of 'natural disasters' throughout the world.[4] The global impacts of climate change include food and water insecurity, economic disruptions and displacements as a result of severe weather patterns, conflicts and even acts of terrorism.[5]

These expert reports have also indicated that Africa and other developing regions are most affected by and bear the brunt of climate change impacts even though they are not among the big contributors of greenhouse gas (GHG) emissions.[6] As a region, Africa is highly vulnerable to climate change and its impacts, both in temperature and extreme precipitation patterns,[7] while at the same time having limited economic and institutional capacities and flexibility to cope with and adapt to climate variability.[8] Nigeria is also experiencing these impacts in the nature of increasing incidences of climate change-related droughts, storms, flooding, deforestation, loss of biodiversity and many others.[9] This reality, coupled with the obtuse response of political leadership in African countries to climate change crises, has affected the efforts to ensure sustainable development and climate justice on the continent.[10] The existential threat posed by all of this to African countries makes an African response imperative.

---

[4] The increase in the number and severity of floods in parts of Europe and parts of Africa in 2024 was quite remarkable. Some of the worst floods in 2024 impacted countries like Spain and Germany in Europe, and Mozambique, Nigeria, Chad, Cameroon, Mali and Niger in Africa. These were all affected by severe floods, displacing over 500,000 persons, and had over 1,000 casualties. See e.g., Mark Maslin and others, 'Understanding the health impacts of the climate crisis' (2025) 12:1 *Future Healthcare J* <www.sciencedirect.com/science/article/pii/S2514664525000190> accessed 7 May 2025.

[5] Funmi Abioye, 'The judiciary and environmental protection in Africa in pursuit of sustainability: Creating an environmental rule of law jurisprudence in response to climate change imperatives' in Jean-Claude Ashukem and Semie M Sama (eds) *Human Rights and the Environment in Africa* (Routledge 2023) 199.

[6] Abioye (n 5); Masson-Delmotte and others (n 2) 24. According to Michel Boko and others, 'Africa' in Martin Parry and others (eds), *Climate Change 2007: Impacts, Adaptation and Vulnerability. Contribution of Working Group II to the Fourth Assessment Report of the Intergovernmental Panel on Climate Change* (CUP 2007) 433, 435, 'Africa is one of the most vulnerable continents to climate variability and change'.

[7] Masson-Delmotte and others (n 2).

[8] Abioye (n 5).

[9] Uzuazo Etemire, 'The Future of Climate Change Litigation in Nigeria: COPW V NNPC in the Spotlight' (2021) 2 *Carbon & Climate L Rev* 158.

[10] ibid 199.

Key among the many responses from the continent has been the enhanced environmental protection provisions seen in the constitutions of African states and the enhanced environmental protection and climate change mechanisms at the regional level. The African Charter on Human and Peoples Rights (the African Charter) is an example. In South Africa, Section 24 of the 1996 Constitution of the Republic of South Africa guarantees a right to an environment that is not harmful to the health or well-being of the people. It also guarantees a right 'to have the environment protected, for the benefit of present and future generations, through reasonable legislative and other measures that prevent pollution and ecological degradation, promote conservation, and secure ecologically sustainable development and use of natural resources while promoting justifiable economic and social development'.[11] This provision introduces the sustainable development balance that is crucial in our interactions with the environment and with climate change. In Kenya, the 2010 Constitution of the Republic of Kenya recognizes the right of 'everyone to a clean and healthy environment' in article 42, and articles 69, 70 and 72 provide an extensive framework for environmental rights protection in Kenya.[12]

Section 20 of the 1999 Constitution of the Federal Republic of Nigeria provides that the State shall 'protect and improve the environment, including water, air, lands, forests and wildlife of Nigeria'.[13] Interestingly, this provision is contained in Chapter II of the Constitution which is titled Fundamental Objectives and Directive Principles of State Policy.[14] These are objectives that the State sets out to achieve and as such do not give rise to the same protections that fundamental rights have as contained in Chapter IV of the same Constitution.[15] Thus, the 1999 Constitution does not provide a legal guarantee of socio-economic rights on its own, but rather reflects them as objectives of State policy, which are viewed as aspirational objectives of the State and

---

11 Constitution of the Republic of South Africa 1996, s 24.

12 See generally, Constitution of Kenya 2010. As a result of the constitutional protection granted to environmental rights, Kenya has developed an advance governance structure to protect this right. The Environmental and Land Court is an example of such.

13 Constitution of the Federal Republic of Nigeria 1999, s 20 (Constitution of Nigeria).

14 Constitution of Nigeria, ch II. Thus, the provisions of ss 13–24 in that chapter are recognized as non-justiciable in the courts, as they are not guaranteed as rights, but as desirable objectives of the State.

15 Constitution of Nigeria, ch IV (ss 33–46) deals with fundamental rights such as the right to life, right to personal liberty, freedom of expression, freedom of association and others.

not guaranteed rights that are justiciable in courts.[16] There is thus no outright constitutional protection given by these provisions under the Constitution, including Section 20, which views the protection and improvement of the environment in Nigeria as a State objective to be aspired to by the State. The courts have historically recognized that Chapter II provisions (including environmental objectives) are not justiciable under the Nigerian Constitution.[17]

Nigeria has the largest economy in Africa, in which the oil and gas industry plays a dominant role. This industry attracts the participation of various multinational and national corporations. The revenue from the oil industry accounts for about 90% of Nigeria's exports, making it one of Africa's main oil producers at about 1.5 million barrels of oil per day in 2023.[18] The oil industry however accounts for an overwhelming proportion of the nation's environmental challenges. Frequent oil spills in the Niger Delta region have led to environmental degradation as a result of the contamination of the air, waters and land, posing hazards to biodiversity, and to animal and human populations.[19] Gas flaring, a product of the process of oil exploration, remains a problem for the environment. Nigeria is currently listed as one of eight top gas flaring

---

[16] Eghosa Ekhator and Edward O Okumagba, 'Climate Change and Multinationals in Nigeria: A Case for Climate Justice' in Kim Bouwer and others (eds), *Climate Litigation and Justice in Africa* (Bristol University Press, 2024) 247, 265 <https://doi.org/10.2307/jj.9941112> accessed 15 May 2025.

[17] See generally, Ogugua VC Ikpeze, 'Non-justiciability of Chapter II of the Nigerian Constitution as an impediment to Economic Rights and Development' (2015) 5:18 Developing Country Studies 48; Halima D Kutigi, 'Towards Justiciability of Economic, Social, and Cultural Rights in Nigeria: A Role for Canadian-Nigerian Cooperation?' (2017) 4:1 *The Transnational Human Rights Rev* 1 <https://digitalcommons.osgoode.yorku.ca/thr/vol4/iss1/7> accessed 6 May 2025.

[18] Doris Dokua Sasu, 'Oil industry in Nigeria – statistics and facts' (*Statista*, 5 November 2024) <www.statista.com/topics/6914/oil-industry-in-nigeria/> accessed 20 June 2025; see also Uzuazo Etemire 'Global Perspectives on Corporate Climate Legal Tactics: Nigeria National Report' (BIICL, February 2024) <www.biicl.org/documents/12171_global_perspectives_on_corporate_climate_legal_tactics_-_nigeria_national_report_v1.pdf> accessed 20 June 2025.

[19] Ephraim Elenwo and Justine Akankali, 'Environmental Policies and Strategies in Nigeria Oil and Gas Industry: Gains, Challenges and Prospects' (2014) 5:14 *Nature Resources* 884; Ifeoma Ezenwa, 'The Right to a Safe and Clean Environment with regard to Oil Pollution and Climate Change in Nigeria' (*OAL*, 8 January 2025) <https://oal.law/the-right-to-a-safe-and-clean-environment-with-regard-to-oil-pollution-and-climate-change-in-nigeria/> accessed 22 June 2025.

countries in the world[20] that account for about 65% of the global gas flare and a significant amount of GHG emissions as a result. The environmental challenges associated with this industry highlight the need for a governance framework to ensure the protection of the environment.

As a member of the global and regional communities, Nigeria is a signatory to several rights-based international and regional instruments. Relevant amongst these are the International Covenant on Economic, Social and Cultural Rights (ICESCR), which provides protections for socio-economic and cultural rights, and the African Charter on Human and Peoples Rights (African Charter).[21] These have been ratified by and domesticated in Nigeria and are binding under Nigerian law. The African Charter recognizes and protects both political and socio-economic rights and imposes an obligation on member states to ensure the promotion and protection of these rights.[22] This obligation has been recognized by the Supreme Court of Nigeria in its decision recognizing that the African Charter Act is justiciable in Nigerian courts.[23] Of particular relevance to the discussion in this chapter is article 24 of the African Charter which provides that 'all peoples shall have the right to a generally satisfactory environment favourable to their development'. Other provisions of the African Charter impacting environmental rights are article 4 (right to life), article 16 (right to the best attainable state of physical and mental health) and article 22 (right to economic, social and cultural development).

Without a direct recognition of an environmental right in the Constitution of Nigeria, the legislature and the judiciary (through court decisions) have worked to set a standard in ensuring that Nigeria guarantees environmental quality that is conducive to human health and well-being, as well as being consistent with sustainable development criteria. This has resulted in the passing of specific environmental laws over time that provide a governance mechanism

---

20 Global Gas Flaring Data (*World Bank Group*) <www.worldbank.org/en/programs/gasflaringreduction/global-flaring-data> accessed 22 June 2025.

21 African Charter on Human and Peoples' Rights (adopted 27 June 1981, entered into force 21 October 1986) (1982) 21 ILM 58 (African Charter) was ratified on 22 June 1983 and domesticated by the *African Charter on Human and Peoples' Rights (Ratification and Enforcement) Act*, No 2 of 1983, which directly incorporates the African Charter into Nigerian law, and makes it enforceable by all authorities and persons exercising legislative, executive and judicial powers in Nigeria.

22 African Charter, art 1 states that 'Member States... shall recognize the rights, duties and freedoms enshrined this Charter and shall undertake to adopt legislative or other measures to give effect to them'.

23 See *General Sani Abacha & Others v Chief Gani Fawehinmi* [2000) 4 SCNJ 401.

for environmental protection.[24] Nigeria's latest legislative addition to environmental protection governance mechanisms is the 2021 Climate Change Act,[25] which sets out an ambitious framework for mainstreaming climate change actions and provides for a system of carbon budgeting to reach net zero emissions targets between 2050 and 2070. The Act also establishes the National Council on Climate Change as a governance body.[26]

The judiciary plays an important role in ensuring that the legislative and treaty-based commitments governing climate change and environmental protection are properly interpreted and applied with the aim of ensuring that the provisions of article 24 of the African Charter on the right to a generally satisfactory environment favourable to development of the people, and other relevant rights, are actualized. In *Gbemre v Shell Petroleum Development Company Nigeria Ltd and Others*[27] for instance, environmental rights were recognized as fundamental human rights for the first time in Nigeria. The Court held that 'the Constitutionally guaranteed fundamental rights to life and dignity of human person provided in Sections 33(1) and 34(1) of the Constitution… inevitably includes the right to clean poison-free, pollution-free and healthy environment'.[28]

Thus, the interpretative mandate of the courts helps set standards for all members of society to adhere to, particularly the executive and legislative branches. This applies to climate litigation cases. Due to the urgency of the climate change crisis, climate change litigation has become one of the crucial mechanisms used to achieve climate justice. This is particularly so due to the limitations inherent in other initiatives. Courts are being called upon to devise creative interpretative mechanisms to secure environmental protection. Judges are utilizing their discretionary powers to fashion effective remedies.[29] Judges have a key role to play as the rate of climate litigation increases globally, with

---

24 A number of regulatory bodies were created to govern different aspects of environmental protection, such as the National Environmental Standards & Regulations Enforcement Agency (created by National Environmental Standards and Regulations Enforcement Agency (Establishment) Act 2007 (NESREA Act)), which replaced the Federal Environmental Protection Agency (created by a similarly named Act of 2004); the National Oil Spill Detection and Response Agency (created by National Oil Spill Detection and Response Agency (Establishment) Act 2006 (NOSDRA Act)).

25 Climate Change Act 2021.

26 Climate Change Act 2021 (n 25).

27 (2005) AHRLR 151 (FHC).

28 ibid 2.

29 For an extensive discussion of creative interpretation required of the judiciary in a time of climate crises, see Abioye (n 5).

more people turning to the courts as an avenue to have their environmental rights aired and protected and to combat the climate crisis.[30] This has become a key purpose of the emergent transnational litigation,[31] even in Nigeria, where climate litigation is recognized as relatively new.[32]

Data from the Sabin Center for Climate Change Law shows that as of September 2024, over 2,796 climate litigation cases had been instituted globally, with 70% of those having been filed in the past decade following the adoption of the Paris Agreement.[33] Climate change litigation is increasingly seen as a means to achieve the protection and enforcement of environmental rights by the courts. It has been described as 'an attempt to control, order or influence the behaviour of others in relation to climate governance', and it is being used by a wide variety of litigants (governments, private actors and civil society) at multiple levels (local, regional, national and international).[34] It recognizes the power and mandate of the courts to interpret legislation and to determine and/or control adherence of all parties, states and private citizens to that legislation. It can be an effective complementary tool in efforts to tackle climate change.[35] A broader conception sees climate change litigation as any adjudicative activity that raises legal questions relating to climate change miti-

---

30 Michael Burger and Maria Antonia Tigre, *Global Climate Litigation Report: 2023 Status Review* (UNEP, 2023) <www.unep.org/resources/report/global-climate-litigation-report-2023-status-review> accessed 25 May 2025.

31 Pedi Obani and Eghosa Ekhator, 'Transnational Litigation and Climate Change in Nigeria' (Afronomics Law, 4 December 2021) <www.afronomicslaw.org/category/analysis/transnational-litigation-and-climate-change-nigeria> accessed 10 May 2025.

32 See generally Uzuazo Etemire, 'The Future of Climate Change Litigation in Nigeria: COPW v NNPC in the spotlight' (n 9); Edward Okumagba, 'Examining global court practices in reducing climate change impacts through litigation: Lessons for Nigeria' in Eghosa Ekhator and others (eds), *Implementing the Sustainable Development Goals in Nigeria: Barriers, prospects and strategies* (Routledge, 2021) 107.

33 Martini Igini, 'Explainer: Climate litigation – trends and impact' (*Earth.org*, 20 September 2024) <https://earth.org/explainer-climate-litigation-trends-and-impact/> accessed 25 May 2025.

34 Navroz K Dubash and others, 'National and Sub-national Policies and Institutions' in Priyadarshi R Shukla and others (eds), IPCC, 2022: *Climate Change 2022: Mitigation of Climate Change. Contribution of Working Group III to the Sixth Assessment Report of the Intergovernmental Panel on Climate Change* (CUP 2022) 1375 <https://www.ipcc.ch/report/ar6/wg3/downloads/report/IPCC_AR6_WGIII_FullReport.pdf>.

35 See Kim Bouwer, 'Lessons from a Distorted Metaphor: The Holy Grail of Climate Litigation' (2020) 9:2 *Transnational Environmental L* 347, 353, where the

gation or adaptation, or engages with some aspect of climate law or policy, whether central to or peripheral to the litigation.[36] This broader view of the scope and definition of climate litigation is proposed for a proper understanding of climate litigation in Africa, which may not mirror how climate litigation is mapped and defined in other jurisdictions.

This chapter focuses on access to justice in climate litigation in Nigeria. It highlights the role of the judiciary in ensuring the protection of the environment through a flexible and proactive approach to climate change litigation. It argues that the evolving jurisprudence from the Supreme Court of Nigeria signifies the Nigerian courts' acknowledgement and awareness of the urgency with which climate change issues must be addressed to encourage an environment where climate justice is upheld. Though relatively late in the day, this change is a positive step for climate litigation in Nigeria. The analysis in this chapter focuses on the groundbreaking case of *Centre for Oil Pollution Watch v Nigerian National Petroleum Corporation* (*COPW*),[37] highlighting the legal arguments before the Court, and the Court's decision and reasoning. It examines the continuing impact of this decision on national jurisprudence and what this signifies for the future of climate litigation in the region.

Section 2 gives a brief historical overview of the approach of the courts in climate litigation in Nigeria. It explains the approach of the judiciary in climate litigation cases and how procedural considerations have historically been allowed to act as a bar to climate litigation. Section 3 analyzes the *COPW* case, the remedies sought in the case, and the pertinent issue of locus standi raised before the Supreme Court of Nigeria. Section 4 adopts a post-*COPW* approach to evaluate the potential impact of the *COPW* decision on climate litigation and environmental protection in Nigeria. It considers whether the decision has led to a shift in jurisprudence and especially the adoption of a more liberalized view on locus standi in climate litigation cases. The final section shares some concluding thoughts on this case and its impacts.

2. CLIMATE LITIGATION IN NIGERIA AND THE ATTITUDE OF THE COURTS: AN OVERVIEW

In Nigeria, courts have historically adopted a strict and restrictive interpretation of locus standi, by requiring that a plaintiff have a direct interest in or be directly affected by the subject matter of a case. Where this direct link is not

---

author talks about the need for forms of governance, including effective strategic litigation, to support the climate response.

[36] Kim Bouwer and others, 'Africa, Climate Justice and the Role of the Courts' in Kim Bouwer and others (eds), *Climate Litigation and Justice in Africa* (n 16) 7.

[37] (2019) 5 NWLR (pt 1666) 518 (SC) (*COPW*).

established, courts have refused to hear such cases due to a lack of locus standi.[38] This rigid interpretation of locus standi has heightened the threshold that plaintiffs must surmount for their claims to be heard before the courts. This has been particularly limiting in the field of climate litigation and environmental protection and has been a bar to the protection of socio-economic rights in Nigeria, particularly environmental rights, for many years.[39] The argument of the applicants in *COPW* was that the Court needed to consider a re-interpretation or a reversal of the rule on locus standi as it had previously applied in climate action cases.

The *COPW* case is regarded as groundbreaking in the field of environmental law because the Court clarified and settled the long-contentious issue of locus standi in Nigerian law. A principle with common law heritage, locus standi is the legal right to institute proceedings in a court of law – the 'standing to sue'.[40] Referring to previous decisions, the Supreme Court of Nigeria defined locus standi as 'the legal capacity to institute or commence an action in a competent court of law or tribunal without let or hindrance from any person or body whatsoever'.[41] The rule of locus standi generally postulates that a person must have suffered a legal grievance, or have been wrongfully deprived of something or wrongfully refused something, or had their title to something wrongfully curtailed, in order to be able to bring a claim before the court.[42] This conception requires the breach of a direct and personal legal interest and it has historically led to what is considered a rigid application of the doctrine of locus standi in Nigeria. The result of this is that Nigerian courts have, over time, refused to entertain otherwise valid claims where the plaintiff(s) before the court are unable to show that they have suffered a specific and direct legal

38 See generally, *Olawoyin v AG Northern Nigeria* (1961) 1 SCNLR 5 (FHC); *Owodunmi v Reg Trustees of CCC* (2000) 10 NWLR 315; *Gamioba v Esezi* (1961) 2 SCNLR 237; *AG Eastern Nigeria v AG Federation* (1964) 2 All NLR 224; *Odeneye v Efunnuga* (1990) 7 NWLR 618 (SC); *Adesanya v President of FRN* (1981) 2 NCLR 358; *Oloriode v Oyebi* (1984) 1 SCNLR 390.

39 Alex Cyril Ekeke, 'Liberalization of the Rule on Locus Standi before Nigerian Courts: Lessons from India' (2022) 66:2 *J of African L* 339, 343.

40 A judge made rule enunciated in *Ex parte Sidebotham* (1880) 14 Ch D 458 as 'a person aggrieved must be a man who has suffered a legal grievance, a man against whom a decision has been pronounced which has wrongfully deprived him of something or wrongfully refused him something or wrongfully affected his title to something'.

41 *COPW* (n 37) 10; *Owodunni* (n 38) 565.

42 Elijah A Taiwo, 'Enforcement of fundamental rights and the standing rules under the Nigerian Constitution: A need for a more liberal provision' (2000) 9:2 AHRLJ 546, 552; *COPW* (n 37).

injury by reason of actual or threatened violation of their legal right or legally protected interests.[43]

The application of this restrictive approach to standing by the courts in Nigeria raised the bar on who could approach the courts based on the requirement that an applicant must show 'sufficient legal interest'. The Supreme Court has reiterated this narrow and rigid conception of locus standi over time. The first notable case in which locus standi was used to bar a suit is the case of *Olawoyin v Attorney-General of Northern Nigeria*.[44] This 1961 case related to a challenge to the validity of a law passed by the executive, prohibiting children from engaging in political activities. The appellant, a father, challenged this law on the basis that he had a desire to educate his children politically. The Court held that a party seeking to challenge a law must demonstrate that they have sustained or are immediately in danger of sustaining some direct injury as a result of its enforcement, not merely that they suffer in some indefinite way in common with the rest of the public. The trial judge dismissed the claim on the ground that no right of the plaintiff was alleged to have been infringed and that 'it would be contrary to principle to make the declaration asked for in vacuo'.[45]

This thinking of the Court was reiterated in subsequent cases. In *Senator Abraham Adesanya v President, Federal Republic of Nigeria (FRN)*,[46] the Supreme Court of Nigeria dismissed the appellant's appeal against a political appointment by the President and held that the appellant lacked the locus to bring the matter before the court. The appellant was a serving senator in the National Assembly at the time, and he challenged a presidential appointment to an executive position. The Court observed that it was accepted in common law countries that the claimant must have some 'justifiable interest' which may be affected by the action being challenged, or that he will 'suffer injury or damage because of the action'.[47] The Court acknowledged that this stance would 'deter meddlesome interlopers, professional litigants and the likes from suing in matters that do not directly concern them'.[48] These cases and subsequent

---

43 See *Olawoyin* (n 38); *Gamioba* (n 38) where the Court held that any person who invokes the judicial power of the court to declare a law or enactment invalid must be able to show not only that the statute is invalid, but that they have sustained or are immediately in danger of sustaining some direct injury as a result of its enforcement and not merely that they suffer in some indefinite way in common with people generally (emphasis added).

44 *Olawoyin* (n 38).

45 ibid.

46 *Adesanya* (n 38).

47 ibid, per Fatayi-Williams, CJN.

48 *Adesanya* (n 38).

ones solidified the interpretation that locus standi required a plaintiff to show that they had suffered a direct and personal injury as a result of the action being contested. Thus, Nigeria became known as a jurisdiction with a narrow and rigid conception of locus standi.[49]

This protectionist stance of the courts, aimed at avoiding a floodgate of litigation and preventing 'meddlesome interlopers', has been criticized as limiting access to justice and public interest litigation by NGOs, human rights activists and advocates on behalf of the poor and vulnerable.[50] It is seen as an obstacle to the enforcement of fundamental rights and a negation of a purposive interpretation of the Constitution.[51] The restrictive interpretation greatly limits the ability of aggrieved groups, peoples and especially indigenous communities, to take legal action in a bid to protect their rights in cases of degradation, pollution and other types of harm and damage to the environment as a result of the acts of the State, private enterprises or both. The failure of the Nigerian judiciary to support environmental protection cases on the basis of lack of locus standi has led to the resort to regional bodies and even other foreign jurisdictions in climate litigation cases emanating from Nigeria.[52]

In the *COPW* case, the Supreme Court of Nigeria clarified and settled the long-contentious issue of locus standi in environmental protection and climate litigation cases in Nigeria. The next section takes an in-depth look at this case, in which the Supreme Court recognized the need for a more liberal approach to locus standi. The court stated that there would be a 'great lacuna in the system of public law if a pressure group or even a single public-spirited taxpayer were prevented by outdated technical rules of locus standi from bringing

---

[49] See *Chief (Dr) Irene Thomas & 5 Others v The Most Reverend Timothy Omotayo Olufosoye* (1986) NGSC 14, 1 NWLR (pt 18) 669.

[50] Ekeke (n 39); Ngwu Godwin Emeka, Chioma Nnajiofor Blessing and Maureen Obiageli Ugwu Esq, 'The Role of the Judiciary in Interpreting the Doctrine of "Locus Standi" in Environmental Pollution in Nigeria' (2023) 2:10 *Law & Economy* 1 <www.paradigmpress.org/le/article/view/827> accessed 15 May 2025; Taiwo (n 42).

[51] See Adenekan Ademola JCA, 'Human rights and national development' in Ayo Ajomo and B Owasanoye (eds), *Individual Rights Under the 1989 Constitution* (Nigerian Institute of Advanced Legal Studies 1993) 28.

[52] See *Socio and Economic Rights Action Centre (SERAC) and the Center for Economic and Social Rights (CESR) v Federal Republic of Nigeria* (2001) Communication 155/96 ACHPR; *Socio-Economic Rights and Accountability Project v Nigeria*, ECW/CCJ/APP/08/09 Ecowas Community Court of Justice. These are cases that had to be taken to regional and sub-regional bodies as a result of a failure of lack of locus standi in Nigeria.

a matter to the attention of the Court to vindicate the rule of law and get an unlawful conduct stopped'.[53]

3. *COPW V NNPC*

3.1 Facts

The plaintiff (appellant), a non-governmental organization (NGO) incorporated in Nigeria, is set up as a private environmental watchdog. Its mission includes addressing environmental pollution, particularly crude oil spills and advocating for the rights of communities affected by such spills. In this particular public interest, litigation commenced by originating motion before the Federal High Court, Lagos, in February 2006, the appellant organization was described as carrying out the 'function of ensuring reinstatement, restoration and remediation of environments impaired by oil spillage/pollution particularly the un-owned environment'.[54] The appellant organization undertakes to ensure that environments that belong to no one in particular are kept clean and safe for human and aquatic consumption. Thus, the appellants carry out a public interest mandate and sought to activate that mandate in this case.

The defendant (respondent before the Supreme Court), the Nigerian National Petroleum Corporation (NNPC), is a State-owned corporation created by Act of Parliament[55] (the Act), and carrying on the business of prospecting, mining, exploring and storing of crude oil, other hydrocarbons and other by-products. The Act charges the respondent corporation with the duty of owning and managing crude oil installations, pipelines and rigs across the country, some of which pipelines run underground and undersea at various points.[56] To that effect, the respondent owns offices, oil installations, oil pipelines and rigs

---

[53] *COPW* (n 37) 21.

[54] *COPW* (n 37) 37.

[55] Established 1 April 1997 by the Nigerian National Petroleum Corporation Act 1997, ch 320 LFN 1990 (NNPCA).

[56] See NNPCA, ss 5(1)(a), (d) which provides that:

> Subject to the provisions of this Act, the Corporation shall be charged with the duty of
>
> (a) exploring and prospecting for, working, winning or otherwise acquiring, possessing and disposing of petroleum; . . .
>
> (d) providing and operating pipelines, tanker-ships or other facilities for the carriage or conveyance of crude oil, natural gas and their products and derivatives, water and any other liquids or other commodities related to the Corporation's operations.

in different parts of Nigeria, which it is tasked with running and managing as required by law.[57]

A number of environmental laws in force at the time controlled and guided the activities of the defendant. For example, the National Environmental Standards & Regulations Enforcement Agency (Establishment) Act, created an agency tasked with the function of enforcing 'compliance with international agreements, protocols, conventions and treaties on the environment, including climate change, biodiversity, conservation… pollution, sanitation, and such other environmental agreements as may from time to time come into force.[58] This applied to the operations of the respondent, and informed the argument of the appellant in this case that the respondent was under a duty to ensure proper remediation of the effects of the oil spill in question.

The appellant claimed that the respondent corporation was negligent in its running of the oil pipelines that it had laid down in and around the Acha community of Abia State, Nigeria. As a result of such negligence, one of the pipelines owned and managed by the respondent ruptured and spilled hydrocarbon oil into the only streams in the area serving the communities: the Ineh and Aku rivers. These rivers are the main sources of water for the surrounding communities. The appellant further claimed that not only was the respondent negligent in causing the spill, it was also negligent in managing the effects of the oil spill.[59] Such negligent conduct stemmed from the failures of the respondent to maintain its facilities such as the pipelines and rigs, and to have early warning systems and technology in place to alert to possible leakages along the pipelines, especially those buried undersea.

While the respondent stopped the spill and repaired the pipelines when it became aware of it, it failed to clean up or to adequately remediate the effects of the spilled oil.[60] This resulted in a total contamination of the streams and the soil, affecting the ecosystem, marine life and the forest systems.[61] It left

---

57 *COPW* (n 37) 27; see also the provisions of the Oil Pipeline Act, ch 07 LFN 2004 and the regulations thereunder.

58 NESREA Act (n 24), s 7; other relevant legislation include NOSDRA Act (n 24), ss 6(2)–(3), 7; Oil Pipeline Act (n 57); Oil Spill Recovery, Clean-up, Remediation and Damage Assessment Regulations 2011.

59 The appellants had noticed a strange, oily substance circulating on top of the streams and which quickly, within days, overflowed from the streams and surged into the adjoining lands, estuaries, creeks and mangroves. Upon inspection, it was determined that the respondent's oil pipeline, which had corroded due to lack of maintenance, had fractured and spewed its contents of hydrocarbon mineral oil into the surrounding streams and river of Ineh and Aku.

60 *COPW* (n 37) 38.

61 ibid.

the streams impure, soiled, contaminated and no longer able to serve their ordinary and natural use: they were no longer good for human consumption, aquatic life and sea birds. The failure to clean up the underbed of the streams meant that there was excessive crude hydrocarbon oil in bottom sediments of the streams that continued to contaminate the streams and continued the pollution and its effects long after the spill had stopped. This gravely affected the communities' ability to make use of the streams.

This situation persisted for years before the suit was filed. Scientific reports and expert evidence before the court showed that the effects of an oil spill on the ecosystem, marine life and the forest systems could last for several decades if proper and adequate remediation measures were not implemented.[62] Contact with or ingesting the contaminated water from the streams had human impacts, which were reported to manifest in health conditions like skin cancers, reproductive and respiratory health issues, and other psychological impacts (like depression and anxiety) among the communities using the streams.[63] The fact that they did not have an alternative source of water exacerbated these circumstances and increased the suffering of the people affected.

The appellant alleged that these issues infringed the rights of the affected communities, including their environmental rights, and sought remedies on behalf of these communities. The appellant requested the following remedies from the court: 1) a claim for the reinstatement, restoration and remediation of the environment, which had been impaired/contaminated in the Acha community in Abia State; 2) a claim for the provision of a safe and potable water supply as a substitute for the water provided by the streams, and 3) a claim for the provision of medical facilities for evaluation of those affected by the spill and for the treatment of the victims following the negative health effects of the spillage.

This suit is one of many of its kind that has been filed before the courts in Nigeria with a public interest and climate justice focus. This action was brought on behalf of vulnerable communities seeking the enforcement of their rights to a generally satisfactory environment. That right is protected under the African Charter, incorporated in Nigeria and now regarded as binding in Nigeria, including article 24 which grants the right to a generally satisfactory environment favourable to the development of the people.

The appellant averred that the respondent's history of negligent conduct led to substantial abuse and damage to the environment and the people affected and that the respondent had never been held accountable for the damage or the

---

62 ibid.

63 ibid.

failure to remedy that damage.[64] The respondent averred that the plaintiff had no standing to sue and sought an order striking out the lawsuit.

3.2 Lower Court Decisions

The Court of First Instance struck out the suit without going into the substance of the matter, on the grounds that the appellant lacked the locus standi to institute and maintain the action.[65] The appellant appealed to the Court of Appeal, and this appeal was dismissed on the same grounds. The appellant then appealed to the Supreme Court for a determination on the issue of the locus standi issue.

3.3 Supreme Court Decision

The issue on appeal before the Supreme Court was limited to the locus standi issue: whether the Justices of the Court of Appeal were right in dismissing the appellant's appeal for want of locus standi to maintain the suit.[66] This was argued in the broader context of two issues: the locus standi on environmental matters, and the civil rights and obligations of parties.

On the civil rights argument, counsel for the appellants argued that every Nigerian citizen had the right to see the law being enforced in public matters and public law.[67] He submitted that the civil rights of the appellant included seeing that the laws of the Federal Republic of Nigeria are enforced in matters of public interest. He further submitted that the failure of the respondents to properly and fully remediate the impacted and contaminated streams was in direct breach of the provisions of the Federal Environmental Protection Agency (FEPA) Act and the Oil Pipelines Act.[68] While these arguments were raised before the court, the focus of the Supreme Court was on the locus standi issue.

On the pertinent issue of locus standi, counsel for the appellant argued for a qualification or a relaxation of the restrictive requirement of direct injury or harm in analysis of plaintiff's standing in environmental cases. Counsel argued that such a change was warranted in environmental cases brought in the public interest, with the intention of vindicating the rule of law, even if the

64 *COPW* (n 37) 41.
65 ibid.
66 ibid 39.
67 *Adesanya* (n 38) [85].
68 FEPA, ibid; section 17(4) of the Oil Pipelines Act which makes oil mining licenses that are issued by the State subject to all public safety and prevention of pollution regulations in force at any point in time.

plaintiff had not suffered any direct injury.[69] The appellants argued that based on a previous decision of the Supreme Court in *Adesanya v President F.R.N*, the law on locus standi in public interest litigation in environmental matters had changed 'to the extent that pressure groups, NGOs and even public-spirited tax payers are cloaked with the locus to maintain a public interest action even though they may not have suffered any injury at all, let alone any particular injury above other members of the society from the subject matter of the suit'.[70] On this issue, the appellant noted that some of its members were actual indigenes of, or lived in, the Acha community and thus were directly affected by the contamination of the water source from the two streams. This was an attempt to create a sufficient, albeit tenuous, link upon which public interest litigation and a broad interpretation of locus standi could rest.[71] However, this fact still did not amount to direct and personal injury to the plaintiff that was a requirement of the prevailing rigid interpretation of locus standi.

In the case of *Abraham Adesanya v President F.R.N*, the Supreme Court had noted in obiter that a plaintiff who seeks the maintenance of a law for the public interest should have locus standi in the matter irrespective of whether he has any sufficient interest in the matter or has suffered any injury above every other member of the society.[72] The question this view begs is how, then, do we assess locus standi? What are we looking to determine if a certain party has 'sufficient interest' and is not just a 'meddlesome interloper', as the Court in *COPW* put it? Even though the Court in *Abraham Adesanya v President* went on to apply a rigid interpretation of locus standi to bar the appellant from continuing with the suit, this obiter resonates with a liberalized interpretation of locus standi that puts the need to see justice done above the particular interest of the plaintiff in public interest litigation cases.

The Supreme Court in the *COPW* case invited senior members of the legal profession to act as amici curiae (friends of the court), to assist the Court by submitting considered opinions and arguments on the issues before the Court, particularly on the contested issue of the locus standi of the appellants. Nine Senior Advocates of Nigeria (SANs)[73] and some senior State officials from the Office of the Attorney General made submissions to the Court. The Court

---

69 *COPW* (n 37) 45.

70 ibid 40; see also *Adesanya* (n 38).

71 *COPW* (n 37) 51, 53–54 where the court relied on the submissions of: Adegboyega Awomolo (SAN), AB Mahmoud (SAN) and Lucius E Nwosu (SAN), amici curiae to the Court.

72 *Adesanya* (n 38) 358, per Fatayi-Williams, CJN (then Chief Justice of Nigeria).

73 This is the prestigious title conferred on legal practitioners in Nigeria, who have distinguished themselves in the legal profession. It requires a minimum of ten

further consulted foreign law and regional law, including the African Charter of Human and Peoples Rights, to reach its decision. The Court recognized the fact that the principles of locus standi, though derived from the common law, have been liberalized by the English courts and other common law jurisdictions, such as Canada, Australia and India over the years, to allow for non-governmental organizations to have standing before courts in such matters.[74]

Nweze, JSC delivered the lead judgement. He referred to the Canadian case of *Thorson v AG of Canada*, in which liberal views on standing were expressed and balanced against the requirement of sufficiency of a plaintiff's interest in the subject matter of the suit.[75] The law on standing in environmental matters in Canada allows for both private and public interest standing. Public interest standing has been recognized in cases where the courts have allowed environmental advocacy groups to represent broader environmental interests[76] and in environmental class action cases.[77]

Justice Nweze also considered and referred to jurisprudence from India, where the courts, guided by the 'need to do justice generally', have liberalized the traditional rule on locus standi in environmental protection and climate change cases.[78] This philosophy of the 'need to do justice' has guided the courts in India in their expansive thinking when it comes to climate change

---

years of active service at the bar and is equivalent to the title of Kings Counsel in the UK.

74 *COPW* (n 37) 18. See the lead judgement by Nweze, JSC where he cited the English cases of *R v Inland Revenue Commissioners Ex p National Federation of Self Employed and Small Businesses Ltd* [1981] UKHL TC_55_133, [1982] AC 617, 639; *R v Secretary of State for Foreign Affairs ex p The World Development Movement Ltd* [1994] EWHC Admin 1, [1995] 1 All ER 611, 620, where a non-governmental organization was recognized as having locus standi to bring an application on the bases of broader public interests, and representing people in developing countries.

75 1974 CanLII 6 (SCC), [1975] 1 SCR 138 where the Court held that the 'issue of sufficiency of interest was the foundation upon which the decisions where reached'. Canada has adopted a liberal approach to locus standi, especially in public interest litigations, where the courts have allowed environmental advocacy groups to represent broader environmental interests.

76 See for example, *Zoocheck Canada Inc v Alberta (Minister of Agriculture and Forestry)*, 2019 ABCA 208, where the Court acknowledged that public interest standing was recognized and allowed in Canada, and it was a discretionary decision of the Court.

77 *Dewey v Corner Brook Pulp and Paper Limited*, 2025 NLCA 8.

78 See generally, *Maharaj Singh v State of Uttar Pradesh* & Others, AIR 1976 SC 2602; *SP Gupta v Union of India & Anr,* AIR 1982 SC 149, 189.

litigation cases. They have adopted creative and proactive thinking to protect the environment, as previously argued by this author elsewhere.[79]

English law was also consulted, with the learned justice noting that a more expansive and liberal stance is taken in the United Kingdom when locus standi issues are raised before the courts.[80] The ideology informing these bold steps by the English courts is the understanding of the need to provide access to justice and to allow more members of society to have access to the mechanisms of justice.[81]

In *COPW v NNPC*, the Supreme Court distinguished locus standi considerations from the analysis of the justiciability of a claim. The Court held that 'when a party's standing to sue is in issue, the question is whether the person whose standing is in issue is the proper person to request an adjudication of a particular issue, and not whether the issue is itself justiciable'.[82] Drawing this fine line, the Court corrected a misinterpretation that issues of lack of locus automatically mean that the subject matter of the action is not justiciable. Referring to its decision in a previous case, the Court reiterated the fact that the traditional interpretation of locus standi should not be allowed to deny access to the courts and that a rigid adherence to 'the ubiquitous principle inherent in locus standi' may no longer stand.[83] The Court's reasoning was based on notions of access to justice, especially in environmental matters, and on the enforcement of the rule of law. Its view was that refusing a pressure group, or even a single public-spirited taxpayer, access to the court on the basis of the technical rules of locus standi would lead to a grave lacuna in the system of justice and would hinder enforcement of the rule of law.[84] The Court held that the accountability owed by officials and departments of government to a court of justice under the Constitution is based on the lawfulness (or otherwise) of their actions.[85] It stands to reason that in upholding such accountability and the rule of law, it was important for the Court to facilitate the means whereby members of the public could be part of the process, by liberalizing the rules on access to the courts.

Aka'ahs, JSC noted in a concurring judgement that this issue was particularly important in environmental matters and that the time had come to relax the application of the rule of locus standi in environmental public interest litigation cases in Nigeria. He stated that this was due to the fact that the

---

79 See Abioye (n 5) 210–211.

80 *COPW* (n 37) 18.

81 ibid.

82 ibid 16.

83 *COPW* (n 37) 102; *Ladejobi v Oguntayo* (2004) 18 NWLR (pt 904) 153, 149.

84 *COPW* (n 37) 67.

85 ibid.

environment was a common good of all, and it was the duty of the government to protect the environment for the benefit of all and against the actions of those damaging the environment, even if against itself.[86] The Court noted that the increasing concern about climate change, depletion of the ozone layer, waste management, flooding, global warming and other environmental hazards, warranted the adoption of stronger measures to protect and safeguard the environment. This included the liberalization of the rule of locus standi to afford access to those on a mission to protect the environment.[87]

In a unanimous decision of the Supreme Court of Nigeria, the seven justices of the court held that the appellant had standing to sue the respondent. The Court emphasized that public-spirited individuals and non-governmental organizations can bring an action in courts against relevant public authorities and private entities to demand their compliance with relevant laws and to ensure the remediation, restoration and protection of the environment.[88] The Court, referring to the Stockholm Convention, recognized the sustainable development imperatives and held that the national resources of the earth 'must be protected and conserved for the benefit of present and future generations through careful planning and management as appropriate'.[89] Building on the current human rights trend in climate litigation, the Court linked environmental protection rights to the right to life protected in Section 33 of the Constitution, as an unsafe environment would no doubt have an impact on the quality of life.[90] In granting the appeal, the Supreme Court remitted the case to the Federal High Court for trial on the merits based on the substantive claims brought by the appellants. It is unclear at the time of writing if this matter has proceeded to the trial court, as there are currently no publicly available records.

Nigerian courts have slowly been developing more creative ways of interpreting the law, and interpreting the obligations under the African Charter, and other internationally binding treaties, with the aim of providing environmental justice to affected peoples and communities.[91] The Supreme Court decision on locus standi in this case is another one of these creative developments and marks a radical change in the law governing locus standi. It goes a long way in the development of national and regional jurisprudence and agrees with the transnational trend in ensuring that affected persons and groups in

---

86 ibid 60, see the observation of Aka'ahs JSC at p 95 of the decision.

87 ibid.

88 *COPW* (n 37) 75, per Justice Onnoghen, CJN.

89 ibid 106; Stockholm Convention on Persistent Organic Pollutants, https://www.pops.int, accessed 5 July 2025.

90 ibid, see variously p 79, 82, 89, 103.

91 See *Adesanya* (n 38); *Gbemre* (n 27); and now *COPW* (n 37).

environmental and climate change actions can access the courts to claim their rights, whether directly or through representation. The Supreme Court, by holding that 'any person (including NGOs) who can show bona fide interest in seeking the performance and enforcement of statutory functions and public laws, should be regarded as having locus standing before the court, particularly about public health and the protection of the environment',[92] brought Nigeria in line with the expanded transnational scope of access to justice in environmental protection and climate change cases. This decision also highlights the burden on the State to comply with its international and national obligations, and to protect and vindicate all rights entrenched in international mechanisms to which Nigeria is a signatory, including article 24 of the African Charter. This portends to have a huge impact on environmental protection and climate justice litigation not only in Nigeria, but in the region as well.

4. IMPACTS IN NIGERIA OF *COPW V NNPC*

The judiciary in Nigeria has acknowledged the need to liberalize its locus standi requirements in public interest and environmental litigation cases. The Court in the *COPW* case, recognizing the urgency and imperative nature of environmental protection and climate change issues, moved away from the rigid requirement of what constitutes 'sufficient interest'. For this reason, *COPW v NNPC* has been hailed as groundbreaking, revolutionary, a landmark case, and of great import in moving Nigeria towards a liberal conception of locus standi. Unfortunately, it does not appear that this momentum has been maintained by the Supreme Court, and other courts as well, in Nigeria.[93] One thing that the *COPW* decision did not do was to reverse previous decisions of the court on locus standi in their entirety, and to change the law permanently as the courts in the UK, Canada and India have done. Rather, the Supreme Court in the *COPW* case restricted its liberalization of locus standi rules to public interest litigation and environmental action cases.

The failure to overturn its previous decisions on locus standi, and limiting the more liberalized interpretation to public interest litigation and environmental action cases, potentially gives rise to a bifurcation of the law on locus standi in Nigeria. This will lead to a situation wherein a liberal application is allowed in public interest (environmental and perhaps, constitutional) actions, but a rigid application is applied in other types of actions. Some of the recent decisions of the Supreme Court highlight this. In a 2025 decision of the

92 *COPW* (n 37) 99.

93 *Okonjo-Iweala v Fawehinmi* (2025) 7 NWLR (pt 1988) 1.

Supreme Court on locus standi in *Okonjo-Iweala v Fawehinmi*,[94] the Court adopted the rigid interpretation, when it denied the plaintiff, a Senior Advocate of Nigeria and prominent member of society, the right to continue with his lawsuit. The plaintiff approached the court asking for a determination that the payment of salaries in foreign currency to certain members of the executive was illegal and amounted to an abuse of power by the President. The matter was compounded by the fact that the salaries paid in foreign currencies ended up being above the range prescribed by law.[95] The Court reiterated the reasoning underlying a liberal interpretation of locus standi, and recognized the fact that it was an attempt at a flexible and more purposive dispensation that granted locus to citizens on constitutional matters affecting the preservation of rights. However, it applied a rigid interpretation of locus standi on these facts.[96] The Supreme Court held that the plaintiff had no locus standi to institute proceedings, and that when it came to specific actions of the executive, a plaintiff must show personal injury above and beyond other members of society to have locus to institute such proceeding. In its words:

> If an Act made by the National Assembly or any Law made by the State legislature is unconstitutionally assaulted or is in itself unconstitutional every Nigerian should have access to court to vindicate the rule of law and arrest any breach of the law and the Constitution. ***But*** when the claim is directed towards an executive action over any subject-matter, the plaintiff's locus standi will inure if he shows a direct and personal effect of the action complained of on his own right as a Nigerian.[97]

A puzzling feature of this decision, as emphasized by Adegboruwa, is that the executive actions complained about are bound by law and such law comes from a constitutional mandate.[98] Yet the status of the plaintiff as a taxpayer, demanding fiscal responsibility, was deemed insufficient to constitute a

---

94 *Okonjo-Iweala* (n 93).

95 Pursuant to Certain Political, Public and Judicial Office Holders (Salaries and Allowances, etc) Act No 6 2002.

96 See opinion piece by Ebun-Olu Adegboruwa, 'Still on Okonjo-Iweala Vs Fawehinmi: Demystifying locus standi – Part 2' (*The Guardian*, 28 May 2025) <https://guardian.ng/opinion/still-on-okonjo-iweala-vs-fawehinmi-demystifying-locus-standi-part-2/> accessed 25 May 2025.

97 Okonjo-Iweala (n 93) 41 [B]–[E], quoted by Adegboruwa, 'Demystifying locus standi – Part 2' (n 96) (emphasis added).

98 See Ebun-Olu Adegboruwa, 'Still on Okonjo-Iweala vs Fawehinmi: Demystifying locus standi – Part 3' (*The Guardian*, 29 May 2025) <https://guardian.ng/opinion/columnists/still-on-okonjo-iweala-vs-fawehinmi-demystifying-locus-standi-part-3/> accessed 29 May 2025.

personal and direct interest or injury to grant him standing in the action.[99] The Court also distinguished this case from public interest litigation and held that the plaintiff had not deposed so. This suggests that, based on the *COPW* case, the Court might have taken a different approach if the matter had been argued as a public interest litigation case. That was not the case, with the result that the Court recognized the more liberal approach to standing in public interest cases but deliberately distinguished the matter before it on a spurious basis, in order to avoid applying that liberal approach.

5. CONCLUSION

The *COPW* case is a notable and landmark decision of the judiciary in public interest litigation cases and particularly climate litigation cases. It is a win for climate justice action in Nigeria. It is indicative of the recognition by the courts of the need to be part of the efforts to combat climate change and its effects, by being proactive and forward looking. The decision shows the courts taking on the challenge of ensuring access to justice in climate litigation cases on behalf of communities, societies and groups in their battle to ensure the protection of their environments. However, the failure of the Supreme Court in *COPW v NNPC* to reverse itself fully has left room for the rigid interpretation of locus standi to continue to be applied in other cases.[100] There is a need for the Court to extend its liberal approach to standing in climate ligation, to other causes of action, where locus standi is in issue. This approach would be consistent with the principles that informed the *COPW v NNPC* decision. While inconsistency in the law remains, this case nevertheless offers a solid precedent for a liberal approach to the concept of locus standi in environmental and climate change cases.

99 Adegboruwa, 'Demystifying locus standi – Part 3' (n 98).
100 See eg *Okonjo-Iweala* (n 93).

6. Human rights as procedural due process in climate mitigation policies: *Verein KlimaSeniorinnen Schweiz v. Switzerland*

Johannes Reich

1. SKETCHING A JUDICIAL FRAMEWORK GOVERNING THE NEXUS BETWEEN CLIMATE CHANGE MITIGATION AND HUMAN RIGHTS FOR EUROPE

On 9 April 2024, the European Court of Human Rights (ECtHR) in Strasbourg (France), in a Grand Chamber judgment, in the matter of *Verein KlimaSeniorinnen Schweiz and Others v. Switzerland* (hereafter referred to as *'KlimaSeniorinnen'*), found Switzerland to be in violation of the right to respect for private and family life as well as the right of access to a court, both enshrined in the European Convention on Human Rights (ECHR).[1] Switzerland, the ECtHR held, failed to comply with its 'positive obligations' due to the insufficient actions taken to mitigate climate change.[2] The judgment was delivered in conjunction with two other decisions, both pertaining to the nexus between human rights and climate change mitigation, namely *Duarte*

1 European Convention for the Protection of Human Rights and Fundamental Freedoms (Rome, 4 Nov. 1950), 312 E.T.S. 5, as amended by Protocols Nos. 11, 14 and 15 supplemented by Protocols Nos. 1, 4, 6, 7, 12, 13 and 16 on 1 June 2010.

2 European Court of Human Rights (sitting as a Grand Chamber), *Verein KlimaSeniorinnen Schweiz and Others v. Switzerland*, Application no. 53600/20, 9 April 2024, paras. 574 and 640. The ECtHR's judgments and decisions are available at: https://hudoc.echr.coe.int/.

Agostinho and Others v. Portugal and 32 Others[3] and *Carême v. France*.[4] In contrast to *KlimaSeniorinnen*, the Court declared both *Duarte* and *Carême* inadmissible, albeit on different grounds. In *Duarte* the applicants had failed to exhaust domestic remedies,[5] while in *Carême* the Court determined that the applicant had relinquished his victim status upon his departure from the municipality of Grande-Synthe near Dunkirk (France), which is considered to be especially vulnerable to the impacts of climate change, including flooding.[6] While judgments of the ECtHR are merely binding *inter partes*,[7] the underlying interpretation of the ECHR by the Court affects all of the contracting parties (principle of *res interpretata*).[8] Against this backdrop, *KlimaSeniorinnen* sketched a framework governing the nexus between human rights enshrined in the ECHR and climate change mitigation obligations within the jurisdiction of the ECtHR, that is, the *Council of Europe* (CoE) with its 46 member states having a combined population of about 675 million.[9]

---

3 European Court of Human Rights (sitting as a Grand Chamber), *Duarte Agostinho and Others v. Portugal and 32 Others*, Application no. 39371/20, 9 April 2024.

4 European Court of Human Rights (sitting as a Grand Chamber), *Carême v. France*, Application no. 7189/21, 9 April 2024.

5 ECtHR, *Duarte Agostinho* (note 3), para. 227; see ECHR (note 1), article 35 section 1 (according to which '[t]he Court may only deal with the matter after all domestic remedies have been exhausted'.).

6 ECtHR, *Carême* (note 4), paras. 15, 83 and 88; see ECHR (note 1), article 34 sentence 1 (according to which '[t]he Court may receive applications from any person, non-governmental organisation or group of individuals claiming to be the victim of a violation by one of the High Contracting Parties of the rights set forth in the Convention or the protocols thereto'.).

7 ECHR (note 1), article 46 section 1 (according to which '[t]he High Contracting Parties undertake to abide by the final judgment of the Court in any case to which they are parties'.).

8 Angelika Nussberger, *The European Court of Human Rights* (OUP 2020), 173; see also Oddný Mjöll Arnardóttir, 'Res Interpretata, Erga Omnes Effect and the Role of the Margin of Appreciation in Giving Domestic Effect to the Judgments of the European Court of Human Rights' (2017) *European Journal of International Law* 819, 823–4 (pointing out that '[t]he principle of *res interpretata* rests on the simple truth that despite the fact that ECHR law contains no doctrine of binding precedent, once the ECtHR has pronounced on an issue, it is to be expected that the Convention will be interpreted and applied in the same manner if the Court is confronted with the same issue again in a different state'.).

9 The Council of Europe (CoE) with the ECtHR as 'its "jewel"' (Nussberger (note 8) 38) based in Strasbourg is, of course, distinct from the European Union (EU) and its Luxembourg-based European Court of Justice (ECJ). All 27 member states of the EU are, at the same time, members of the CoE. States such as

KlimaSeniorinnen falls into the category of *strategic climate change litigation.*[10] The application relies on human rights to compel governments to implement specific emission reduction policies in order to mitigate climate change.[11] The judgment of the ECtHR was preceded by a finding by an administrative agency and two judgments of domestic courts – the Federal Administrative Court (appellate court)[12] and the Federal Supreme Court (apex court)[13] – both declining to consider the applicants' motion on its merits. This chapter will highlight these different judicial approaches to climate change litigation. Sections 3 and 4 outline the domestic legal and factual context in which the case has arisen. Section 5 analyses the strategy behind the lawsuit. This chap-

Albania, Armenia, Azerbaijan, Iceland, Moldova, Norway, Serbia, Switzerland, Türkiye, Ukraine and the United Kingdom are members of the CoE but not of the EU. Canada, the Holy See, Japan, Mexico and the United States are Observer States to the CoE. Israel is an Observer to the CoE's Parliamentary Assembly. On 16 March 2024, the CoE's Committee of Ministers adopted Resolution CM/Res(2022)2 on the cessation of the membership of the Russian Federation to the CoE. The CoE's Committee of Ministers further decided to suspend all relations with Belarus owing to the country's active participation in the aggression of the Russian Federation against Ukraine on 17 March 2022 (CM/Del/Dec (2022)1429/2.5). While having joined the CoE on 6 May 1963, Switzerland is not a member of the European Union (EU), despite a densely knit web of bilateral treaties. See the official register of sectoral treaties with the EU (in French) at: https://www.fedlex.admin.ch/fr/sector-specific-agreements/EU-acts-register.

[10] See, in more detail, Johannes Reich, 'Strategische Prozessführung im Kontext eines hegemonialen politischen Systems' (2025) *Jahrbuch des öffentlichen Rechts der Gegenwart* 439, [DOI:10.1628/joer-2025-0020].

[11] Jacqueline Peel & Hari M. Osofsky, *Climate Change Litigation: Regulatory Pathways to Cleaner Energy* (Cambridge University Press 2015), 4–9 (distinguishing various categories of climate change litigation).

[12] [Swiss] Federal Administrative Court, decision BVGer A-2992/2017 (27 November 2018); all decisions of the Court are available at: https://www.bvger.ch/en/jurisprudence/judgments-database; the judgment in question (in German) is available at: https://bvger.weblaw.ch/pdf/A-2992-2017_2018-11-27_8c2c4cc9-5956-4efb-a4a2-0471e19a4108.pdf.

[13] [Swiss] Federal Supreme Court [official designation: *Bundesgericht* (German), *Tribunal fédéral* (French), *Tribunale federale* (Italian); literal translation: Federal Court], decision BGE 146 I 145 (5 May 2020); all decisions of the Court are available at: www.bger.ch; for a detailed assessment of the judgment (in German) see Johannes Reich, 'Bundesgericht, I. öffentlich-rechtliche Abteilung, 1C_37/2019, 5. Mai 2020 [Verein KlimaSeniorinnen Schweiz et al. gegen Eidgenössisches Departement für Umwelt, Verkehr, Energie und Kommunikation (UVEK)]' (2020) *Schweizerisches Zentralblatt für Staats- und Verwaltungsrecht* [ZBl] 489, 497–507, available at: https://t.uzh.ch/1q6.

ter then gives an overview of previous court decisions in section 6. A detailed analysis of the ECtHR *KlimaSeniorinnen* judgment is given in section 7, and its aftermath is analysed in sections 8 and 9. The chapter concludes with observations in section 10 on the possible legal and political consequences of the case.

2. SWITZERLAND'S AMBIVALENT GHG EMISSIONS RECORD: VAST DIFFERENCE BETWEEN DOMESTIC AND CONSUMPTION-BASED EMISSIONS

In 2023, Switzerland's greenhouse gas (GHG) emissions amounted to 40.85 million tonnes of CO_2 equivalents (territorial emissions).[14] Based on the emissions generated on Switzerland's territory, the country was responsible for 0.09 percent of global CO_2 emissions in 2023.[15] The country's share of global cumulative emissions of CO_2 since 1750 (historical emissions) stands at 0.17 percent.[16] Switzerland's GHG emissions decreased by 26.1 percent between 1990 and 2023, while its permanent resident population grew by 32.1 percent.[17] This corresponds to a reduction of 44 percent per capita.[18] In the sector 'households and services' ('buildings'), a significant reduction of 46 percent was achieved during that time, while the 'industry' sector reduced its GHG emissions by 32 percent.[19] GHG emissions from transport account for 33.6 percent of total territorial emissions but merely dropped by a dismal 8 percent between 1990 and 2023.[20] In line with the system boundaries set out in the

---

14 *Federal Office for the Environment,* Kenngroessen zur Entwicklung der Treibhausgasemissionen in der Schweiz 1990–2023 [as of April 2025], 2025, 9, 12, 41, 61, 63, available (in German) at: https://www.bafu.admin.ch/dam/bafu/de/dokumente/klima/fachinfo-daten/kenngroessen_thg_emissionen_schweiz.pdf.download.pdf/Kenngr%C3%B6ssen_2025_DE.pdf.

15 Our World in Data, Switzerland: What share of global CO2 emissions are emitted by the country?, available at: https://ourworldindata.org/co2/country/switzerland#what-share-of-global-co2-emissions-are-emitted-by-the-country.

16 Hannah Ritchie, Pablo Rosado, & Max Roser in: Our World in Data, Switzerland: What share of global cumulative CO_2 has the country emitted? (base line: year of 1750), available at: https://ourworldindata.org/grapher/share-of-cumulative-co2?time=latest&country=~CHE.

17 Federal Office for the Environment (note 14), 11–12.

18 Federal Office for the Environment (note 14), 39.

19 Federal Office for the Environment (note 14), 39.

20 Federal Office for the Environment (note 14), 39.

Paris Agreement, these figures do not take into account emissions from international aviation and shipping.[21]

In comparison, Switzerland's record on GHG emissions is *ambivalent at best*. The country's per capita GHG emissions of 4.6 tonnes of CO_2 equivalents per year[22] are above global average but relatively moderate when considered in the context of Switzerland's status as one of the most advanced and highly developed free market economies. Taking into account all consumption-based emissions generated on Swiss territory and abroad, including imported goods and international air travel, however, Switzerland's GHG footprint stands at 11.3 tonnes per capita and year, well above the global average of about 6 tonnes.[23] On this basis, Switzerland *ranks among the highest global emitters per capita* – surpassing countries such as Japan (10.6 tonnes) and Germany (10.4 tonnes), though considerably lagging behind the United States (18.1 tonnes), Australia (17.9 tonnes) or Canada (15.2 tonnes).[24] There are three main factors that contribute to the significant difference between domestic and consumption-based emissions. The first is the fact that domestic electricity production is largely CO_2 free.[25] The second is the absence of virtually any heavy industry.[26] The third is that international air traffic is not included in territorial emissions due to the defined system boundaries.[27] Taking international air traffic into account would increase territorial emissions by around 13 percent.[28]

---

21 See Federal Office for the Environment (note 14), 3.

22 Federal Office for the Environment (note 14), 12.

23 Federal Council, *Environment Switzerland 2022* (2022) 73, available at: https://www.bafu.admin.ch/en/publication?id=GjWv4YDOaKui.

24 Federal Council (note 23), 73.

25 Federal Council (note 23), 73.

26 Federal Council (note 23), 73.

27 On these system boundaries see Federal Office for the Environment (note 14), 3.

28 See Federal Office for the Environment, Greenhouse Gas Emissions from Aviation (15 April 2025), available at: https://www.atlas.bfs.admin.ch/maps/13/de/17804_229_228_227/27579.html.

3. AN UNFAVOURABLE CONSTITUTIONAL ENVIRONMENT FOR CLIMATE CHANGE LITIGATION: WEAK CONSTITUTIONAL REVIEW AND DIRECT DEMOCRACY

Switzerland is a federal state characterised by 'extreme decentralization',[29] comprising the federal government (Federation), the 26 constituent units (cantons) and 2,131 municipalities.[30] Federal Parliament is assigned a more prominent role than that of the executive and the judiciary. The Federal Constitution[31] explicitly states that the Federal Parliament, subject to the rights of the People and the Cantons, 'is the supreme authority of the Confederation'.[32] This elevated position of Parliament also affects the powers of the judiciary. Although the United States Supreme Court set out the constitutional blueprint for the Federal Supreme Court in the 1870s, the Federal Constitution requires all courts to adhere to federal statutes enacted by Parliament, as well as to international law. This applies even if such domestic legislation or treaties conflict with the Federal Constitution. Therefore, it is the responsibility of Federal Parliament rather than the courts to determine, in the process of legislation (with regard to domestic law) or ratification (as far as treaties are concerned), respectively, whether a federal statute or a treaty under consideration is compatible with the Federal Constitution. Once Parliament has answered this question in the affirmative, the courts ought to abstain, in principle, from re-examining the Federal Parliament's decision. Furthermore, neither the Federal Constitution nor federal legislation provide for a mechanism by which Federal Statutes (federal acts) or Federal Ordinances may be subjected to abstract constitutional review.

Switzerland's ratification of the ECHR on 28 November 1974 has somewhat alleviated this system of *weak constitutional review* of Federal Statutes.[33]

---

[29] Lori Thorlakson, 'Comparing Federal Institutions: Power and Representation in Six Federations' (2003) 26(2) *West European Politics*, 1, 18.

[30] See Swiss Federal Office of Statistics, Les 2131 communes de la Suisse au 1.1.2024, available at: https://www.atlas.bfs.admin.ch/maps/13/de/17804_229_228_227/27579.html.

[31] Federal Constitution of the Swiss Confederation of 18 April 1999 [unofficial English translation available at: www.admin.ch/opc/en/classified-compilation/19995395/index.html] (Fed. Const.).

[32] Fed. Const. (note 31), article 148 section 1 (official French version): 'L'Assemblée fédérale est l'autorité suprême de la Confédération, sous réserve des droits du peuple et des cantons'.

[33] Fed. Const (note 31), article 190 (official French version): 'Le Tribunal fédéral et les autres autorités sont tenus d'appliquer les lois fédérales et le droit

In accordance with its mandate, the ECtHR monitors Switzerland's compliance with the ECHR in response to an application, including when an alleged violation of ECHR rights stems from a federal statute. In order to prevent the ECtHR from ruling against Switzerland, the Federal Supreme Court has declared that the Convention takes precedence in the event of a conflict between a right guaranteed by the ECHR and federal law, including not only federal statutes but the Federal Constitution.[34] The Federal Supreme Court has generally adopted a pragmatic approach in order to avoid conflicts between international and domestic law. However, in a judgment handed down in 2012, the Court argued in fundamental terms on the basis of the principle of *pacta sunt servanda* under international law and presented the political authorities with the choice of either fully complying with the obligations arising from the ECHR or withdrawing from the Convention.[35]

In addition to a comparatively weak constitutional review, the Swiss political and legal system is characterised by a strong emphasis on *direct democracy*. All amendments to the Federal Constitution are subject to a mandatory referendum. In a similar vein, each Federal Statute passed and each treaty ratified by Federal Parliament is subject to an optional referendum. Such a referendum is held if 50,000 signatures of Swiss citizens with the right to vote are collected within 100 days of the enactment's official publication. As a result, at the federal level alone, 682 popular votes were held between 1848 and 2023.[36] While only about 7 percent of the legislative acts eligible for an optional referendum are in fact challenged in a referendum at the ballot box,[37] about 42 percent of those bills passed by the Federal Parliament that were indeed challenged by an optional referendum failed to win a majority at the ballot box.[38] The

international'; for a more comprehensive analysis see Johannes Reich, 'Verhältnis von Demokratie und Rechtsstaatlichkeit', in Oliver Diggelmann, Maya Hertig Randall & Benjamin Schindler (eds), *Verfassungsrecht der Schweiz – Droit constitutionnel suisse* (Schulthess 2020), 333, 349–55 (DOI: doi.org/10.5167/uzh-184637); see also Johannes Reich, 'Switzerland', in Richard Albert et al. (eds.), *2024 Global Review of Constitutional Law* (2025), 308–312, 311 (available at: https://papers.ssrn.com/sol3/papers.cfm?abstract_id=5912623), terming the ECHR 'a shadow- or supra-constitution from a Swiss perspective'.

34 See in more detail Johannes Reich, 'Switzerland', in Richard Albert et al. (eds.), *2018 Global Review of Constitutional Law* (2019), 298–302, 299 (available at: https://ssrn.com/abstract=3471638).

35 Reich (note 34), 299 (with reference to the decision BGE 139 I 16).

36 Adrian Vatter, *Das politische System der Schweiz* (5th ed, Nomos 2024), 358.

37 Vatter (note 36) 290.

38 Pascal Sciarini & Anke Tresch, 'Votations populaires', in Yannis Papadopoulos et al (eds), *Handbuch der Schweizer Politik – Manuel de la politique suisse* (7th ed, 2022), 559 and 565.

higher the rate of approval of a bill in parliament, the lower the empirical probability that an optional referendum will be called, while the level of parliamentary consensus has no impact at all on the outcome of a popular vote on such an optional referendum.[39] The high degree of stress and uncertainty that the threat of an optional referendum induces in the legislative process makes building broad coalitions around reforms at the cost of *mere incrementalism in policy-making* a prudent political strategy.[40]

In essence, Switzerland's constitutional order is distinguished by the pervasive influence of the political system.[41] The prevailing tenets of republicanism have led to the predominance of direct democratic instruments in conflict resolution, superseding the utilisation of court judgments (see also section 8). The characteristics of the legal and political system have been identified as factors which limit the possibility of initiating legal action on the grounds of climate-related issues.

4. FEDERAL CLIMATE CHANGE LEGISLATION: INCREMENTALISM, TRIAL AND ERROR, AND OMISSIONS[42]

International treaties entered into by the Federation[43] become part of Swiss federal law upon ratification ('monism').[44] Switzerland ratified the 1992 United Nations Framework Convention on Climate Change (UNFCCC),[45] the

39 Alexander Trechsel & Pascal Sciarini, 'Démocratie directe en Suisse: l'élite politique victime des droits populaires?' (1996) 2 *Swiss Political Science Review* 201, 220–25.

40 Yannis Papadopoulos, 'How does direct democracy matter? The impact of referendum votes on politics and policymaking' (2001) 24 *West European Politics* 35, 42–3.

41 See Reich (note 10), 450–51.

42 For an in-depth account see Johannes Reich, 'Federalism and Mitigating Climate Change: The Merits of Flexibility, Experimentalism, and Dissonance' (2021) *Transnational Environmental Law* 263, 270–76.

43 Fed. Const (note 31), article 54 section 1 (official French version): 'Les affaires étrangères relèvent de la compétence de la Confédération'; see e.g. Federal Supreme Court (note 13), decision 96 I 737 (23 December 1970), section 3.

44 See, e.g. Federal Supreme Court (note 13), decision 94 I 669 (22 November 1968), section 2.

45 United Nations Framework Convention on Climate Change, New York (USA), 9 May 1992, in force 21 March 1994, available at: https://unfccc.int/resource/docs/convkp/conveng.pdf. (ratified by Switzerland on 10 December 1993, No. 0.814.01, available at: https://www.admin.ch/opc/fr/classified-compilation/19920113/index.html; authentic French version).

1997 Kyoto Protocol including the 2012 Doha Amendment,[46] and the 2015 Paris Agreement.[47] With regard to the latter, Switzerland pledged 'to reduce its greenhouse gas emissions by 50 percent by 2030 compared to 1990 levels'.[48] Between 1990 and 2020, domestic GHG emissions decreased by 19 percent and stood at 4 tonnes of carbon dioxide (CO_2) equivalents per capita in 2020.[49] Switzerland therefore missed, albeit narrowly, its national target of minus 20 percent for 2020,[50] whereas it remained in compliance with its commitment under the Kyoto Protocol for the period 2013–2020 due to emission reductions by way of projects initiated outside of Switzerland.[51]

Targets deriving from Switzerland's pledges under international law and measures to reduce GHG emissions attributable to the use of fossil fuels as energy sources are stipulated in the Federal Act on the Reduction of CO_2 Emissions (Federal CO_2 Act).[52] Owing to direct democracy, incrementalism and an approach of trial and error have been the prevailing characteristics

46 Kyoto Protocol on the United Nations Framework Convention on Climate Change, 5 October 2011, available at: http://unfccc.int/resource/docs/convkp/kpeng.pdf (ratified by Switzerland on 9 July 2013, No. 0.814.011, available at: https://www.fedlex.admin.ch/eli/cc/2004/802/fr; authentic French version).

47 Paris Agreement, Paris (France), 12 Dec. 2015, in force 4 November 2016, available at: https://unfccc.int/sites/default/files/english_paris_agreement.pdf. (ratified by Switzerland on 6 October 2017, No. 0.814.012. Available at: https://www.admin.ch/opc/fr/classified-compilation/20162916/index.html; authentic French version).

48 'Switzerland's intended nationally determined contribution (INDC) and clarifying information' (27 February 2015), p. 1. Available at: https://www4.unfccc.int/sites/ndcstaging/PublishedDocuments/Switzerland%20First/15%2002%2027_INDC%20Contribution%20of%20Switzerland.pdf.

49 Federal Council (note 23), 68.

50 Federal Council (note 23), 68.

51 Federal Office for the Environment FOEN, 2020 target achievement review (for the years 2013 to 2020) at: https://www.bafu.admin.ch/bafu/en/home/topics/climate/info-specialists/emission-reduction/target-achievement-review/2020-target.html.

52 Federal Act on the Reduction of CO_2 Emissions, 23 December 2011, No. 641.71 (Federal CO2 Act); official titles: *Bundesgesetz über die Reduktion der CO_2-Emissionen* (German)/*Loi fédérale sur la réduction des émissions de CO_2* (French)/*Legge federale sulla riduzione delle emissioni di CO_2* (Italian). Available at: https://www.admin.ch/opc/en/classified-compilation/20091310/index.html (non-official English translation).

of Switzerland's domestic climate policy.[53] The 'CO_2 levy', an incentive tax ('Pigouvian tax') provided for in the Federal CO_2 Act, has been the centrepiece of Swiss climate policy since 1 May 2000.[54] The levy is collected on the production, extraction and import of 'thermal fuels' (heating oil, natural gas, coal etc.) as opposed to 'motor fuels' (petrol, diesel, kerosene, etc.).[55] Both businesses and private individuals pay the CO_2 levy when purchasing energy sources such as heating oil, natural gas, coal and other fossil fuels. The amount of the levy is determined by the carbon content of each energy source. Only partially in line with the concept of an incentive tax, which ought to be fiscally neutral, approximately two-thirds of the revenue from the CO_2 levy are redistributed to both the resident population and the companies headquartered in Switzerland, evenly divided according to recipients.[56]

On 25 September 2020, Federal Parliament decided to increase the ceiling of the existing CO_2 levy and to introduce a new levy on motor fuels and commercial passenger flights.[57] At the time, it was estimated that the new levy would increase the price of gasoline by up to CHF 0.12 (EUR 0.11/USD 0.13) per litre and result in a rise of prices for commercial flights of between CHF 30 to 120 (EUR 28 to 111/USD 33 to 130) per ticket as of 2025.[58]

After a fierce campaign, supported by the oil importers' association, the homeowners' association and conservative parties, among others, on 13 June 2021, the revision of the Federal CO_2 Act was narrowly *defeated in a referendum*, with 51.59 percent of the votes cast.[59] In order to avoid a regulatory gap, since the overall 20 percent emissions reduction target was only set until 2020, Federal Parliament extended the target by a further 1.5 percent per year first until the end of 2021, then until the end of 2024.[60] These temporary revisions of the Federal CO_2 Act took effect on 1 January 2021 and 2022, respectively.[61] On 18 June 2023, in yet another popular vote, the 'Federal Act on

---

53 Johannes Reich & Nina Boss, 'Switzerland' (2023) 33:1 *Yearbook of International Environmental Law* 149, 149–53.

54 Reich (note 42) 273.

55 See Federal CO_2 Act (note 52), article 29 section 1 in connection with article 2 sections 1 and 2.

56 See Federal CO_2 Act (note 52), article 36.

57 See Reich (note 42) 273 (with further references).

58 Flora Hausammann & Johannes Reich, 'Switzerland' (2022) 32:1 *Yearbook of International Environmental Law* 145, 146–7 (offering a more detailed account).

59 Reich (note 42) 273–4 (with further references).

60 Hausammann & Reich (note 58) 147.

61 See Official Collection of Federal Law, RO 2020 1269 (in French), at: https://www.fedlex.admin.ch/eli/oc/2020/243/fr, and RO 2022 262 (in French), at: https://www.fedlex.admin.ch/eli/oc/2022/262/fr.

Climate Protection Targets, Innovation, and Strengthening Energy Security (Federal Act on Climate Protection)', providing for both a roadmap to 'net zero' by 2050 and subsidies for technical innovation,[62] was *approved* with a majority of 59.1 percent.[63] The Federal Act on Climate Protection, with its emission reduction targets for the period from 2025 to 2050,[64] and the revised Federal CO_2 Act, adopted by the Federal Parliament on 15 March 2024,[65] both took effect on 1 January 2025. Against this backdrop, from 1 January 2021 to 31 December 2024 Switzerland lacked domestic regulations setting *interim* targets and pathways to meet the commitment made under the Paris Agreement to 'reduce its GHG emissions by 50 percent (…) by 2030'.[66] Nevertheless, from 1 January 2021 to 31 December 2024, domestic regulations were implemented that at least set *annual* GHG emission reduction targets.[67] Notwithstanding this, all *instruments* of Switzerland's climate change law, in particular the CO_2 levy, remained in force and were legally binding for the entire period from 1 January 2021 to 31 December 2024. The same is true for the article stating the purpose of the Federal CO_2 Act to reduce GHG emissions, 'particularly CO_2 emissions from the use of fossil fuels, with the aim of keeping the global temperature increase below 2°C'.[68]

Two major reasons explain the omissions with regard to specific categories of targets and reduction pathways: direct democracy and not being a member of the EU. Direct democracy adds a high dose of unpredictability and uncertainty to public policy making, often resulting in a trial and error approach to legislation. Moreover, as Switzerland is not a member of the EU, it was not bound by the EU's climate law with its interim targets.[69]

---

62 For a detailed account see Reich & Boss (note 53) 149–53.

63 See Federal Chancellery, Votation no. 663: Résultats dans les cantons, available at: https://www.bk.admin.ch/ch/f/pore/va/20230618/can663.html.

64 See Systematic Collection of Federal Law, RS 814.310 (in French), at: https://www.fedlex.admin.ch/eli/cc/2023/655/fr.

65 See Official Collection of Federal Law, RO 2024 376 (in French), at: https://www.fedlex.admin.ch/eli/oc/2024/376/fr.

66 See note 48.

67 See notes 60 and 61.

68 Federal CO_2 Act [in force from 1 January 2013 until 31 December 2024] (note 52), article 1 section 1 (transl.).

69 See e.g. Parliament and Council Regulation 2021/1119 of 30 June 2021 Establishing the Framework for Achieving Climate Neutrality and Amending Regulations (EC) No. 401/2009 and (EU) 2018/1999, 2021 OJ L243/1, article 4 (on intermediate Union climate targets) and article 8 (with regard to 'the pathway for the reduction of net emissions at Union level').

5. LITIGATION STRATEGY: MAKING CLIMATE CHANGE MITIGATION A HUMAN RIGHTS OBLIGATION

5.1 Overcoming Obstacles Relating to Standing: 'Association' and Particular Vulnerability to Heatwaves due to Old Age and Female Gender

KlimaSeniorinnen is a case rooted in a deliberate decision by civil society actors to transform mitigating climate change from a policy under international and domestic administrative law into an obligation under both constitutional and international human rights law. On 24 June 2015, the District Court of The Hague decided on an application brought by *Stichting Urgenda* (Urgenda Foundation) on behalf of several hundred private citizens and ordered the State of the Netherlands to reduce its GHG emissions by at least 25 percent by 2020 measured against 1990 levels.[70] Greenpeace Switzerland, a foundation under Swiss law and the national branch of an environmental non-governmental organisation,[71] sought legal advice with two legal counsels on how to initiate similar proceedings against Swiss federal authorities.[72]

Unlike the Dutch Civil Code,[73] Swiss federal law does not grant standing to private organisations to initiate proceedings to protect public interests. Only a handful of provisions of federal statutory law grant non-profit organisations standing to seek a remedy against administrative acts in selected areas of environmental law and only after meeting certain criteria and in pursuance of

---

[70] Rechtbank Den Haag [The Hague District Court], *Stichting Urgenda [Urgenda Foundation] v. The State of the Netherlands (Ministry of Economic Affairs and Climate Policy)*, Case No. ECLI:NL:RBDHA:2015:7145, 24 June 2015, available at: https://uitspraken.rechtspraak.nl/inziendocument?id=ECLI:NL:RBDHA:2015:7145.

[71] See Commercial Register of the Canton of Zurich: Entry 'Greenpeace Schweiz' (UID: CHE-106.028.233), available at: https://zh.chregister.ch/cr-portal/auszug/auszug.xhtml?uid=CHE-106.028.233.

[72] See Cordelia Ch. Bähr, Ursula Brunner, Kristin Casper & Sandra H. Lustig, 'KlimaSeniorinnen: Lessons from the Swiss Senior Women's Case for Future Climate Litigation' (2018) 9 *Journal of Human Rights and the Environment* 194, 201 (noting that Greenpeace Switzerland requested the two authors of the paper who are mentioned first 'to advise whether it would be possible to bring a similar case' [similar to *Urgenda* that is]).

[73] See Dutch Civil Code, article 3:305a; official title: *'Burgerlijk Wetboek'*, available at: https://www.admin.ch/opc/en/classified-compilation/19830267/index.html (non-official English translation).

specific public interests.[74] Mechanisms for collective redress generally remain scarce in Switzerland's procedural laws.[75] Above all, statutory law fails to offer class actions or any meaningful measure serving as a functional equivalent of a class action.[76] Several claimants may thus merely act jointly as plaintiffs, to the extent that their rights result from similar circumstances or legal grounds.[77] Such simple joinder of parties, however, neither limits a claimant's potential liability for court costs and party compensation, nor is it an appropriate measure to seek compensation for damage being both low in value and scattered among numerous aggrieved parties.[78]

Longstanding case law from Switzerland's highest courts, nonetheless, allows organisations with full legal capacity to file actions in their own name on behalf of their members. The case law requires that the organisation's bylaws authorise it to safeguard the relevant interests and that the majority or a large number of its members would individually have standing to seek a remedy. This is known as a 'group action'.[79] An organisation invoking sub-

---

[74] See *inter alia*, Federal Act on the Protection of the Environment, 7 Oct. 1983, No. 814.01, article 55 (granting standing to certain environmental non-governmental organisations regarding rulings on planning, construction or modification of installations for which an environmental impact assessment is required), available at: https://www.admin.ch/opc/en/classified-compilation/19830267/index.html (non-official English translation); Federal Act on the Protection of Nature and Cultural Heritage, 1 July 1966. No. 451, article 12 section 1b, available at: https://www.admin.ch/opc/en/classified-compilation/19660144/index.html (non-official English translation); Federal Act on Forest, 4 October 1991. No. 921, article 46 section 3, available at: https://www.admin.ch/opc/en/classified-compilation/19910255/index.html (non-official English translation).

[75] See Federal Council, *Exercice collectif des droits en Suisse: état des lieux et perspectives* [Report on Collective Redress in Switzerland] (2013), 15–42, available at: https://www.bj.admin.ch/bj/fr/home/publiservice/publikationen/berichte-gutachten/2013-07-03.html.

[76] See Samuel P. Baumgartner, 'Class actions and group litigation in Switzerland' (2006) 27 *Northwestern Journal of International Law & Business*, pp. 301–350, at 309–16.

[77] See Swiss Civil Procedure Code, 19 Dec. 2008, No. 272, article 71 section 1, available at: https://www.admin.ch/opc/en/classified-compilation/20061121/index.html (non-official English translation).

[78] For a recently dismissed equivalent of a 'class action' seeking compensation from the Volkswagen Company on account of its software designed to deceive regulators measuring emissions from diesel powered cars, see Federal Supreme Court (note 13), decision 4A_483/2018 (8 February 2018), available at: www.bger.ch.

[79] See e.g. Federal Supreme Court (note 13), decision 142 II 80 (9 September 2015), section 1.4.2.

stantive rights on behalf of its members in this way is typically established as an 'association' (German: *Verein*; French: *association*; Italian: *associazione*) under the Swiss Civil Code.[80] Founding such an entity merely requires two or more natural or legal persons to consent to establishing an 'association' and to agree on the respective bylaws (articles of association) defining the association's purpose, resources and structure.[81]

Subject to statements in the bylaws to the contrary, an association's liability is limited to its own assets, thereby excluding any personal liability of its members.[82] Bringing an action by an association on behalf of its members thus significantly reduces the financial risks of litigation that individual claimants would otherwise face. This explains why the statutory purpose of *Verein KlimaSeniorinnen Schweiz* (German) (or *Association des Aînées pour la protection du climat Suisse* in French and *Associazione delle anziane per il clima* in Italian)[83] is narrowly confined to lodging 'legal actions at the federal level in order to safeguard the interests of its members' and to manage associated media relations.[84] The association's bylaws limit the association's liability to its own assets,[85] which basically derives from private donations, voluntary member fees and grants by Greenpeace Switzerland.[86]

Under domestic law, organisations such as *KlimaSeniorinnen* are required to demonstrate that the majority of their members would have standing in their individual capacities regarding the matter in dispute, when bringing a case before a court.[87] Standing (*locus standi*) requires an individual claimant to be affected by the matter in dispute in a specific and particular manner that distinguishes them from the general public. Moreover, they must demonstrate a distinct, significant and close connection to the issue at hand.[88] Claimants are also under an

---

80 Swiss Civil Code, 10 Dec. 1907, No. 210, articles 60–79, available at: https://www.admin.ch/opc/en/classified-compilation/19070042/index.html (non-official English translation).

81 See Swiss Civil Code (note 80), article 60.

82 See Swiss Civil Code (note 80), article 75a.

83 German, French and Italian are all official languages of the Swiss Confederation; see Fed. Const. (note 31), article 70 section 1 sentence 1 (defining 'German, French and Italian' as the Federation's official languages).

84 Articles of Association, article 3, available at: https://www.klimaseniorinnen.ch/wp-content/uploads/2016/08/2023_Statuten_KlimaSeniorinnen-Schweiz_V2.pdf.

85 Articles of Association (note 84), article 8 section 3.

86 Articles of Association (note 84), article 8 sections 1 and 2, article 4 section 3.

87 See the text in section 2.2 accompanying notes 70–73.

88 See, e.g., Federal Act on Administrative Procedure (Administrative Procedure Act; APA), 20 December 1968, No. 172.021, article 48 section 1b & c,

obligation to establish that their interest in seeking a remedy is not only 'worthy of protection', but also has practical value in improving their specific legal or factual situation.[89] Anthropogenic climate change, by its very nature, affects everyone, making it difficult to prove that an individual or a group is uniquely affected, rather than sharing a generalised grievance with society as a whole.[90]

On account of their members being female and on average 73 years old, *Verein KlimaSeniorinnen Schweiz* argued that its members are significantly and adversely affected by heatwaves occurring both more frequently and severely due to omissions by the Swiss federal authorities to reduce the country's GHG emissions.[91] Multiple lines of evidence in fact not only indicate a consistent and almost linear relationship between cumulative emissions of GHGs, above all CO_2, and projected global temperature change,[92] but underscore that the risk of heatwaves exceeding the temperature level being harmful to human health increases disproportionately as a direct effect of the rise in

available at: https://www.admin.ch/opc/en/classified-compilation/19680294/index.html (non-official English translation); and Federal Act on the Federal Supreme Court (Federal Supreme Court Act, FCA), 17 June 2005, No. 173.110, article 89 section 1b & c, available at: https://www.admin.ch/opc/fr/classified-compilation/20010204/index.html.

[89] See note 99 and, e.g. Federal Supreme Court (note 13), decision 140 II 214 (2 April 2014), section 2.1 (stating that complainants are required to 'have a specific relation to the matter in dispute . . . and derive a practical benefit from the annulment or amendment of the contested decision'; transl.).

[90] See, in more detail, Johannes Reich, ''Litigating the commons': KlimaSeniorinnen als gerichtliche Durchsetzung der Verfassung der globalen Allmende' (2025) *Klima und Recht* [KlimR] section 3.

[91] *Verein KlimaSeniorinnen Schweiz*, Begehren um Einstellung von Unterlassungen im Klimaschutz im Sinne von Art. 25a VwVG sowie Art. 6 Ziff. 1 und 13 EMRK [Request to halt omissions in climate protection pursuant to APA, article 25a and ECHR, article 6 and article 13 section 1], 25 November 2016, available at: https://www.klimaseniorinnen.ch/wp-content/uploads/2016/11/161124-Gesuch-um-Erlass-anfechtbarer-Verfuegung_final.pdf (German original), https://www.klimaseniorinnen.ch/wp-content/uploads/2016/10/161024_summary_swiss-climate-case_def.pdf (English summary) and https://www.klimaseniorinnen.ch/wp-content/uploads/2016/10/161024_synthese_action-en-justice_def.pdf (French summary) (thereinafter referred to as '*Verein KlimaSeniorinnen Schweiz*, Initial Request'), 47–54.

[92] Intergovernmental Panel on Climate Change [IPCC], *Climate Change 2014. Synthesis Report* (2015), 5 (according to which CO_2 accounts for around two-thirds of total GHG-emissions) and 8, available at: https://www.ipcc.ch/site/assets/uploads/2018/05/SYR_AR5_FINAL_full_wcover.pdf; and *Federal Office for the Environment* (note 14), 6 (stating that CO_2 accounted for 78,5 percent of all of Switzerland's GHG emissions).

average global temperature.[93] According to recent findings, the heatwaves of the summers of 2003 and 2015 resulted in an excess mortality of 5 to 7 percent in Swiss cities, with age being one of the main determining factors of heat vulnerability: 77 percent of the excess deaths were of people aged 75 years and older.[94] However, the evidence as to the specific vulnerability of women within this age group, as compared to men, appears less conclusive.[95]

5.2 Object of Complaint: Overcoming Prohibition of Direct Constitutional Review by Challenging Administrative Inaction Rather Than Legislation

The Swiss Federal Constitution explicitly bars abstract review of acts of both Federal Parliament (the National Council and the Council of States, together referred to as the 'Federal Assembly') and the Federal Council (the executive branch of federal government), by stating that '[a]cts of the Federal Assembly or the Federal Council may not be challenged in the Federal Supreme Court.[96] Unlike, for example, the 2021 *Neubauer* case before the German Federal Constitutional Court,[97] where the applicants argued that

---

[93] Erich M. Fischer & Reto Knutti, 'Anthropogenic Contribution to Global Occurrence of Heavy-Precipitation and High-Temperature Extremes' (2015) *Nature Climate Change* 560–64.

[94] Ana M. Vicedo-Cabrera, Martina S. Ragettli, Christian Schindler &Martin Röösli, 'Excess mortality during the warm summer of 2015 in Switzerland' (2016) *Swiss Medical Weekly* 146: w14379, 6 (DOI: https://doi.org/10.4414/smw.2016.14379); see also Martina S. Ragettli, Ana M. Vicedo-Cabrera, Christian Schindler & Martin Röösli, 'Exploring the association between heat and mortality in Switzerland between 1995 and 2013' (2017) *Environmental Research* 703–9.

[95] See Vicedo-Cabrera et al. (note 94), 6 (indicating a slightly, albeit hardly statistically significant, higher excess mortality of men); but see Jean-Marie Robine, Siu Lan K. Cheung, Sophie Le Roy, Herman Van Oyen, Clare Griffiths, Jean-Pierre Michel & François Richard Herrmann, 'Death toll exceeded 70,000 in Europe during the summer of 2003' (2008) *Comptes Rendus Biologies* 171, 174 (finding that the 'female deaths represent 65% of the total number of the excess deaths' during 2003 in different European countries).

[96] '[A]cts of the Federal Assembly or the Federal Council may not be challenged in the Federal Supreme Court' Fed. Const (note 31), article 189 section 4 sentence 1 (official French version): 'Les actes de l'Assemblée fédérale et du Conseil fédéral ne peuvent pas être portés devant le Tribunal fédéral'.

[97] See German Federal Constitutional Court, *Neubauer et al. v. Federal Republic of Germany*, Application No. 1 BvR 2656/18, 1 BvR 78/20, 1 BvR 96/20, and 1 BvR 288/20, 24 March 2021, available (in the Court's English translation) at: https://www.bundesverfassungsgericht.de/SharedDocs/Entscheidungen/EN/2021/03/rs20210324_1bvr265618en.html.

the climate targets in the Federal Climate Protection Act (KSG) violated their fundamental rights under the German Basic Law (Grundgesetz, GG), the *Verein KlimaSeniorinnen Schweiz* were barred from making the claim that the Federal CO_2 Act, any of its specific provisions, or other parts of federal law, failed to meet the requirements of specific guarantees of either the Federal Constitution or international law, including the ECHR. Faced with this obstacle, *Verein KlimaSeniorinnen Schweiz* had to find a different legal route. They argued that the Federal Council's failure to act – by not revising the existing federal climate legislation or by weakly implementing existing statutory provisions – amounted to an 'informal administrative act' by omission.[98]

According to the Federal Administrative Procedure Act judicial review is, in principle, only granted against 'administrative acts' (French: *décisions*; German: *Verfügungen*), i.e. formal decisions (rulings) issued by administrative agencies on the basis of federal public law.[99] Deviating from the traditional principles of administrative procedural law and owing to the rights to a fair trial and to an effective remedy enshrined in the ECHR,[100] the Federal Administrative Procedure Act has provided for remedy against '*informal* administrative acts' (French: *actes matériels*; German: *Realakte*) since 2007.[101] Such 'informal administrative acts' are not formally notified in writing with a statement of reasons and information on legal remedies, but they have nonetheless the potential to infringe the rights of private individuals. According to the Federal Supreme Court's case law, not only a campaign by a federal agency to raise awareness of sexually transmitted diseases such as HIV/AIDS, but an administrative agency shutting down a specific rubbish collection point or the local police physically hindering a journalist to board a bus to prevent him from travelling to Davos to report on the World Economic Forum (WEF), all amount to such 'informal administrative acts'.[102] Petitioners must, however, not only demonstrate that the act in

---

98 *Verein KlimaSeniorinnen Schweiz*, Initial Request (note 91), 7–9, 100–110.

99 See APA (note 88), article 5 and article 44.

100 ECHR (note 1), article 6 and article 13; see also Fed. Const (note 31), article 29a.

101 See APA (note 88), article 25a.

102 See *inter alia* Federal Supreme Court (note 13), decision 143 I 336 (12 April 2017), sections 4.1–4.4 (finding that the closure of a specific rubbish collection point in a small municipality in the Swiss Alps amounts to an 'informal administrative act'), decision 130 I 369 (7 July 2004), section 6 (stating that a journalist who was physically prevented by local police from travelling to Davos to report on the World Economic Forum has a right to effective remedy), and decision 144

question infringed upon their 'rights or obligations' with 'a certain degree of severity', but that they were affected by the act in a specific and distinct manner compared to the general public.[103] If these conditions are met, the competent authority must issue a formal administrative act retroactively regarding the previous informal administrative act. This administrative act can then be appealed to the Federal Administrative Court.[104] In a nutshell, the Federal Administrative Procedure Act aims to grant judicial review in the absence of a formal ruling by an administrative agency by providing for a conditional right for an administrative act to be issued on previous 'informal administrative acts'.

6. DOMESTIC PROCEEDINGS: DECLINING TO CONSIDER THE CASE ON ITS MERITS

6.1 Grounds of Complaint: Constitutional Principles and International Human Rights

In the domestic proceedings, *Verein KlimaSeniorinnen Schweiz* argued that the Federal Council failed to take the necessary steps to revise the allegedly too-lenient federal climate legislation, in particular the Federal CO_2 Act.[105] These ostensible omissions, according to the complaint, not only amounted to violations of the constitutional principles of precaution and environmental sustainability[106] but to infringements on the right to life and the right to respect

II 233 (15 June 2018), section 7.1 (holding that a rather explicit campaign by the Federal Office of Public Health, entitled 'Love Life', intended to prevent the spread of sexually transmitted diseases, constituted an 'informal administrative act'); on the latter decision see Johannes Reich, 'A Bigger Bang for a Buck'. Staatliche Warnungen und Empfehlungen zwischen Grundrechtsschutz, Kindeswohl und Aufmerksamkeitsökonomie, in Ruth Arnet et al. (eds), *Der Mensch als Mass* (Schulthess 2019), 185–99 available at: https://doi.org/10.5167/uzh-170160.

[103] See e.g. Federal Supreme Court (note 13), decision 144 II 233 (note 102), sections 7.1, 7.3.2 and 8.4.

[104] See APA (note 88), article 5 and article 44 in connection with Federal Act on the Federal Administrative Court (Administrative Court Act, ACA), 17 June 2005, No. 173.32, article 31, available at: https://www.admin.ch/opc/fr/classified-compilation/20010206/index.html.

[105] See *Verein KlimaSeniorinnen Schweiz*, Initial Request (note 91), 6–9.

[106] Fed. Const (note 31), article 73 and article 74 section 2 sentence 1.

for private and family life guaranteed by both the Federal Constitution[107] and the ECHR.[108]

6.2 Federal Administration: Insisting on the Difference Between Legislation and the Administrative Process

Verein KlimaSeniorinnen Schweiz as well as four of its members as individual applicants filed their motion on 25 November 2016 with the Federal Council, the government department responsible for climate, energy, and the environment (i.e. the Federal Department of the Environment, Transport, Energy and Communications, DETEC), and two administrative agencies that report to DETEC (i.e. the Federal Office for the Environment and the Federal Office for Energy).[109] In its decision of 28 April 2017, DETEC stated that the aim of the petition was for the federal authorities to enact legislation.[110] The purpose of the administrative process, however, DETEC reasoned, was to protect the rights of individuals, as the provision on 'informal administrative acts' explicitly emphasises.[111] DETEC therefore refused to consider the merits of the application, giving a statement of reasons of fewer than three pages.[112] DETEC held that the conditions set out in the Federal Administrative Procedure Act for issuing an administrative act regarding the alleged 'informal administrative act'[113] committed by the federal authorities – namely, failing to revise existing

[107] Fed. Const (note 31), article 10 section 1 and article 13 sentence 1.

[108] ECHR (note 1), article 2 and article 8; see *Verein KlimaSeniorinnen Schweiz*, Initial Request (note 91), 23–5, 61–91.

[109] See *Verein KlimaSeniorinnen Schweiz*, Initial Request (note 91), 1, 10–11.

[110] Eidgenössische Departement für Umwelt, Verkehr, Energie und Kommunikation (UVEK) [Federal Department of the Environment, Transport, Energy and Communications (DETEC)], Verfügung [Administrative Decision], 25 April 2017 [hereinafter referred to as 'DETEC, Administrative Decision'], 13–4, available at: https://ainees-climat.ch/wp-content/uploads//2019/01/Verfu%CC%88gung_UVEK_KlimaSeniorinnen.pdf (in German) and https://www.klimaseniorinnen.ch/wp-content/uploads/2017/11/Verfuegung_UVEK_Abschnitt_C_English.pdf (English summary).

[111] DETEC, Administrative Decision (note 111), 14; on 'informal administrative acts' see the text in section 5.2 between notes 109 and 116.

[112] See DETEC, Administrative Decision (note 111), 13–15 (reasoning), 16 (decision).

[113] On these criteria see DETEC, Administrative Decision (note 111), section 5.2.

federal climate legislation and implementing the statutory provisions too leniently – were not met.[114]

6.3 Federal Administrative Court: Climate Change Affects Everyone, Yet No One in Particular

After *Verein KlimaSeniorinnen Schweiz* and four of its members filed an appeal against the ruling of the DETEC before the Federal Administrative Court, the Court handed down its judgment on 27 November 2018.[115] The Court emphasised that, in order to avoid giving rise to an *actio popularis*, judicial review of 'informal administrative acts' must be subject to the condition that they affect the rights or obligations of the applicants in a specific and particular manner.[116] Against this backdrop, the appellate court noted that the 'effects of climate change on humans, animals and plants are (…) of a general nature'.[117] Although the Federal Administrative Court conceded that 'not everyone' was affected by the consequences of climate change 'to the same extent',[118] a 'brief overview of possible effects of climate change' was sufficient for the Court to conclude 'that the group of women over 75 years of age' akin to the members of *Verein KlimaSeniorinnen Schweiz* were 'not particularly affected by the effects of climate change'. Even though this conclusion appears to be inconclusive (see section 6.5), the Federal Administrative Court relied on it to dismiss the appeal and uphold DETEC's decision.[119]

---

[114] See DETEC, Administrative Decision (note 111), 16.

[115] See Federal Administrative Court, decision A-2992/2017 (note 12).

[116] See Federal Administrative Court, decision A-2992/2017 (note 12), section 6; on these criteria see the text in the section 5.2 between notes 109 and 116.

[117] Federal Administrative Court, decision A-2992/2017 (note 12), section 7.4.2 (transl.;*'Die Auswirkungen des Klimawandels auf Menschen, Tiere und Pflanzen sind mithin allgemeiner Art, auch wenn nicht alle gleichermassen betroffen sind.'*); for a critical assessment of this approach see Johannes Reich, 'Abwendung der Klimakatastrophe durch Gerichte?' (2019) Schweizerisches Zentralblatt für Staats- und Verwaltungsrecht 413, available at: https://t.uzh.ch/1pZ.

[118] Federal Administrative Court, decision A-2992/2017 (note 12), section 7.4.2 (transl.; German original: *'… auch wenn nicht alle gleichermassen betroffen sind.'*).

[119] See Federal Administrative Court, decision A-2992/2017 (note 12), 24; for an assessment see Meret Rehmann, 'Bundesverwaltungsgericht, Abteilung I, Urteil A-2992/2017 vom 27. November 2018 Verein KlimaSeniorinnen Schweiz et al. gegen Eidgenössisches Departement für Umwelt, Verkehr, Energie und Kommunikation UVEK, Verfügung über Realakte im Zusammenhang mit dem Klimaschutz' [Case Note], (2019) *Aktuelle Juristische Praxis* 653, 655–9.

6.4 Federal Supreme Court: (Mis-)Reading Science to Conclude That 'There is Still Some Time' to Avoid a '> 1.5/2.0°C-World'

On 21 January 2019, *Verein KlimaSeniorinnen Schweiz* and four of its members filed an appeal in matters of public law against the Federal Administrative Court with the Federal Supreme Court.[120] In its judgment of 5 May 2020, Switzerland's apex court first emphasised that the Federal Constitution offered a multitude of instruments to initiate and shape federal legislation, ranging from the right to elect members to the two chambers of the Federal Parliament to the right to initiate a popular initiative to amend the Federal Constitution.[121] Turning to the question whether the alleged omissions by federal authorities indeed amounted to an 'informal administrative act', the Federal Supreme Court based its reasoning on the Intergovernmental Panel on Climate Change's (IPCC) 'Global Warming of 1.5°C'.[122] According to the Court's reading of the IPCC's special report, an 'increase in the global average' in excess of 1.5°C as defined by the Paris Agreement would only occur 'around the year 2040'.[123]

This reasoning led the Court to conclude 'there is still some time to prevent global warming' to rise beyond the threshold defined by the Paris Agreement.[124] This led the Court to conclude that the applicants' rights weren't affected to the extent required by the Federal Administrative Procedure Act for them to have the right to the issuance of an administrative act on the alleged omissions by federal authorities that would have opened the avenue to judicial review.[125] Although such a reading of climate science is highly questionable (see section 6.5),[126] the Court nevertheless inferred from it that the alleged omissions by the federal authorities would 'at the present time' fail to impair

---

120 Federal Supreme Court, decision 146 I 145 (note 13).

121 Federal Supreme Court, decision 146 I 145 (note 13), section 4.3.

122 Intergovernmental Panel on Climate Change [IPCC], Global Warming of 1.5°C (2018), available at: https://www.ipcc.ch /sr15/.; see Federal Supreme Court, decision 146 I 145 (note 13), section 5.3; on 'informal administrative acts' see section 5.2 between notes 109 and 116.

123 Federal Supreme Court, decision 146 I 145 (note 13), section 5.3 (transl.; German original: '*Der Weltklimarat … kommt … zum Schluss, die globale Erwärmung werde ungefähr um das Jahr 2040 … 1,5 Grad Celsius erreichen …*').

124 Federal Supreme Court, decision 146 I 145 (note 13), section 5.3 (transl.; German original: '*Es wird davon ausgegangen, dass für die Abwendung einer diesen Wert übersteigenden Erderwärmung noch ein gewisser Zeitraum zur Verfügung steht*'.

125 On these criteria set out, see Federal Supreme Court, decision 146 I 145 (note 13), section 5.2.

126 For a critical and more detailed assessment of the Federal Supreme Court's reasoning see Reich (note 13) 497–507.

both the complainants' right to life and the right to respect for private and family life to the extent required by the Administrative Procedure Act[127] in order to vindicate legal remedy.[128] For these reasons, the encroachment of individual rights provoked by the alleged 'informal administrative acts' (omissions) failed to reach the threshold set forth by federal statutory law according to the Court that would have opened the avenue to judicial review on the merits.[129] Consequently, the Federal Supreme Court upheld the decisions by the Federal Administrative Court holding that the DETEC was right not to consider the application by *Verein KlimaSeniorinnen Schweiz* and four of its members in their individual capacities.[130]

6.5 Conclusion: Non-Interference in the Political Sphere as a 'Passive Virtue' of a Court in a Direct Democracy?

On behalf of the Federal Council and the Federal Administration, DETEC thus refused to issue a formal administrative act on the application. The two courts involved in the domestic proceedings both upheld DETEC's finding and declined to consider the application on its merits with regard to both *Verein KlimaSeniorinnen Schweiz* and the four individual applicants. Neither of the judgments handed down by the two courts is, however, entirely convincing. The Federal Administrative Court's rather blunt argument, that because climate change affects and will affect everyone, no one will be *particularly* affected by its consequences (see section 6.3), defies logic.[131] In the same vein, all people on Earth are exposed to *some* harmful ultraviolet light, but those who reside near the poles under a depleted ozone layer are *particularly* affected by such radiation. The complainants also carefully explained why the health effects of climate change vary considerably according to age

127 See APA (note 88), article 25a.

128 Federal Supreme Court, decision 146 I 145 (note 13), section 5.3 (transl.; German original: *'Unter den genannten Umständen erscheint das Recht auf Leben der Beschwerdeführerinnen (...)durch die gerügten Unterlassungen im heutigen Zeitpunkt nicht in einem Ausmass bedroht, dass von einem hinreichenden Berührtsein in eigenen Rechten im Sinne von Art. 25a VwVG* [APA] *gesprochen werden könnte (...). Dasselbe gilt für ihr Privat- und Familienleben sowie ihre Wohnung (...).').*

129 On these criteria set out, see Federal Supreme Court, decision 146 I 145 (note 13), section 5.2; Federal Supreme Court, decision 146 I 145 (note 13), section 5.3; on these criteria see the text in the section 5.2 between notes 109 and 116.

130 See Federal Supreme Court, decision 146 I 145 (note 13), section 6.2.

131 Reich (note 104) 413–4.

and gender. The Federal Administrative Court did not address this in any substantive way.

Similarly, the Federal Supreme Court's reading of the physical basis of climate change, which culminates in the determination that there remains 'some time' to avert a rise in the global average temperature above the 1.5°C threshold, is not persuasive either (see section 6.4).[132] To begin with, given that the average global temperature increased by 0.9°C between 1864 and 2016, while the temperature in Switzerland rose by 2.0°C,[133] it appears inconsistent from a human rights perspective to identify the global rather than the local surface temperature as the criterion to assess whether an infringement of the applicants' *individual* rights occurred. What is more, the Court's reading fails to take into account that, due to the high heat capacity of the Earth system, an average of 10.2 years elapses between the emission of CO_2 (cause) and its maximum effect in terms of the resulting global warming (effect), with a 90 percent probability range of 6.6 to 30.7 years.[134] The Federal Supreme Court's reasoning does not appear to have taken into account the fact that reductions in CO_2 emissions only take full effect after a delay of around a decade.

In view of these shortcomings, it is insightful to consider the political process running alongside the administrative and judicial proceedings. The initial motion was filed on 25 November 2016 (see section 6.2), whereas the judgment of the Federal Supreme Court, given on 5 May 2020, marked the end of the domestic proceedings (see section 6.4). Between 31 August and 30 November 2016, the Federal Council held a public consultation on amending the Federal CO_2 Act.[135] From December 2018 until September 2020, the legislation was subject to fierce debate in both houses of Federal Parliament until a compromise that found the support of a majority in both houses was reached on 25 September 2020.[136] Within the 100-day timeframe, the

132 For a critical and more detailed assessment of the Federal Supreme Court's reasoning see Reich (note 13) 497–507.

133 Federal Office for the Environment, Switzerland's Seventh National Communication and Third Biennial Report under the UNFCCC. Fourth National Communication under the Kyoto Protocol to the UNFCCC (2018) 11, available at: www.bafu.admin.ch/en/climate-reporting.

134 On this insight see Katharine L. Ricke & Ken Caldeira, 'Maximum warming occurs about one decade after a carbon dioxide emission' (2014) *Environmental Research Letters* 9 124002, 3.

135 See Federal Council, '*Message du 1er décembre 2017 relatif à la révision totale de la loi sur le CO2 pour la période postérieure à 2020*' [Report of 1 December 2017 on the total revision of the CO2 Act for the period after 2020], Federal Gazette 2018 229–371, 230, 268.

136 See Federal Gazette 2020 7607 (decision of 25 September 2020).

requisite 50,000 signatures of Swiss citizens[137] were collected to initiate a referendum on the bill.[138] Following wide-ranging debates, the bill failed to garner majority support at the ballot box on 13 June 2021, albeit by a narrow margin (see section 4).[139] In addition, various civil society organisations collected signatures for the popular initiative 'For a healthy climate (Glacier Initiative)' between 30 April 2019 and 30 October 2020.[140] The initiative, aiming to make the net zero target by 2050 and phasing-out of fossil fuels from 2050 onwards part of the Federal Constitution, resulted in the creation of the Federal Climate Protection Act, which was approved in a referendum on 18 June 2023.[141]

In view of this political backdrop, had the Federal Administrative Court or the Federal Supreme Court directed DETEC to address the merits of the motion brought by *Verein KlimaSeniorinnen Schweiz* and four individual claimants, the courts would have been bound, at a subsequent stage of the proceedings, to indirectly rule on whether the legislative projects – which were the subject of intense debate in Parliament and among the general public at the time – complied with the requirements of both the Federal Constitution and the ECHR. In the absence of the authority to evaluate federal legislation in an abstract manner (see section 5.2), it is conceivable that the courts would have been subject to intense and pervasive condemnation from the political branches of government in such a scenario. From this standpoint, the rulings of the Federal Administrative Court and the Federal Supreme Court were also aimed at insulating the institutions from such pressures, and thus amount to illustrations of *passive virtues*.[142]

---

137 See Fed. Const. (note 31), article 141 section 1a.

138 See Federal Gazette 2021 462.

139 Reich (note 42) 273–4 (with further references).

140 See Federal Gazette 2019 3075.

141 See Reich & Boss (note 53) 153.

142 See Alexander Bickel, *The Least Dangerous Branch: The Supreme Court at the Bar of Politics* (2nd edition, Yale University Press 1962), 113–98 (coining the phrase 'passive virtues' with regard to the United States Supreme Court).

7. EUROPEAN COURT OF HUMAN RIGHTS: ESTABLISHING A HUMAN RIGHTS CLAIM TO DUE PROCESS IN DOMESTIC CLIMATE CHANGE MITIGATION POLICIES

7.1 Landmark Ruling without Prescriptive Remedies and No Individuals as 'Victims'

On 26 November 2020, *Verein KlimaSeniorinnen Schweiz* and four of its members filed an application against Switzerland with the ECtHR.[143] The number of third-party interventions – a staggering 23 in all, eight of them by governments of Council of Europe member states[144] – testifies to the importance attached to the proceedings. In its 260-page judgment of 9 April 2024, the ECtHR found Switzerland, the respondent state, to be in violation of both article 8 (right to respect for private and family life) and article 6 (right to access to court and to a fair trial) of the ECHR.[145] The Court reached its finding, according to which Switzerland failed to comply with the positive obligations deriving from ECHR, article 8, 'encompassing a right for individuals to effective protection by the State authorities from serious adverse effects of climate change on their life, health, well-being and quality of life',[146] with a majority of 16 votes to one.[147] With regard to Switzerland's failure to comply with the right to access to court and to a fair trial as enshrined in ECHR, article 6, the Court's decision was unanimous.[148] Against this backdrop, the Court deemed it 'not necessary' to examine whether the right to life according to ECHR, article 2 was applicable.[149]

Two features of the judgment stand out. First, the ECtHR granted the association *Verein KlimaSeniorinnen Schweiz locus standi* (standing) to bring forward its claims under ECHR, articles 2 and 8, while *refusing* to grant the

143 See ECtHR, *KlimaSeniorinnen* (note 2), para. 1.

144 See ECtHR, *KlimaSeniorinnen* (note 2), paras. 366–409 (the governments of Austria, Ireland, Italy, Latvia, Norway, Portugal, Romania and Slovakia all intervened as did, among others, the United Nations High Commissioner for Human Rights, the United Nations Special Rapporteurs on toxics and human rights, on human rights and the environment, and the Independent Expert on the enjoyment of all human rights by older persons).

145 See ECtHR, *KlimaSeniorinnen* (note 2), paras. 574 and 640.

146 ECtHR, *KlimaSeniorinnen* (note 2), para. 519.

147 ECtHR, *KlimaSeniorinnen* (note 2), holding no. 7.

148 ECtHR, *KlimaSeniorinnen* (note 2), holding no. 11.

149 ECtHR, *KlimaSeniorinnen* (note 2), holding no. 6.

four individual applicants victim status with regard to the same provisions.[150] Moreover, the Court, despite ruling against Switzerland, deemed itself 'unable to be detailed or prescriptive as regards any measures to be implemented in order to effectively comply with the (…) judgment'.[151] Amid 'the complexity and the nature of the issues involved', the ECtHR instead left it 'to the Committee of Ministers to supervise (…) the adoption of measures aimed at ensuring that the domestic authorities comply with Convention requirements, as clarified in the (…) judgment'.[152] In short, the Court granted standing to an association, but not to individuals, in a human rights case, and was unable to identify concrete remedies. The Committee of Ministers of the CoE, a body composed of the Ministers for Foreign Affairs of the 46 member states or their Permanent Representatives in Strasbourg, is, according to the ECHR, responsible 'for the supervision of the execution of a final judgment' of the ECtHR.[153]

7.2 Locus Standi and Victim Status: The 'Special Features of Climate Change' as Grounds to Establish Standing for Representative Associations with Relevant Expertise

The ECtHR may, according to ECHR, article 34, 'receive applications from any person, nongovernmental organization or group of individuals claiming to be the victim of a violation by one of the High Contracting Parties of the rights set forth in the Convention or the Protocols thereto'.[154] When determining whether *Verein KlimaSeniorinnen Schweiz* as an association and/or the four individual claimants, all members of *Verein KlimaSeniorinnen Schweiz*, had standing with regard to the violation of ECHR, articles 2 and 8, the Court time and again reiterated the 'specific features' of climate change and climate-change litigation.[155] Vaguely echoing a similar observation by the Federal Administrative Court (see sections 5.2 and 5.4), the ECtHR held 'the number of persons affected' by the consequences of anthropogenically-induced climate change to be 'indefinite'.[156]

These observations led the Court to conclude that in order for an *individual* to qualify as a 'victim' with regard to 'complaints concerning harm or risk of harm resulting from alleged failures by the State to combat climate change',

150 ECtHR, *KlimaSeniorinnen* (note 2), paras. 526 and 535.
151 ECtHR, *KlimaSeniorinnen* (note 2), para. 657.
152 ECtHR, *KlimaSeniorinnen* (note 2), para. 657.
153 ECHR (note 1), article 46 section 3.
154 ECHR (note 1), article 34 sentence 1.
155 See, *inter alia*, ECtHR, *KlimaSeniorinnen* (note 2), paras. 501 and 502, but also 538, 547 and 554.
156 ECtHR, *KlimaSeniorinnen* (note 2), para. 479.

the applicant in question not only 'must be subject to a high intensity of exposure to the adverse effects of climate change' but there also 'must be a pressing need to ensure the applicant's individual protection, owing to the absence or inadequacy of any reasonable measures to reduce harm'.[157] The four individual applicants *failed* to meet this 'high threshold',[158] according to the Court, as the critical medical condition the applicants linked to heatwaves (see section 5.2) could 'be alleviated by the adaptation measures available in Switzerland or by means of reasonable measures of personal adaptation'.[159]

With respect to *Verein KlimaSeniorinnen Schweiz*'s standing, the Court pointed to its own case law granting associations standing, the collective and intergenerational dimensions of climate change, the practical considerations to bring a case as an organisation (on this see also section 5.1), the widespread ratification of the Aarhus Convention on Access to Information, Public Participation in Decision-making and Access to Justice in Environmental Matters[160] by most members states of the CoE, and an 'evolution in contemporary society' with regard to 'the importance of associations to litigate issues of climate change on behalf of affected persons'.[161] Against this backdrop, the ECtHR devised the following 'test for the standing of associations in climate-change litigation under the Convention'.[162] In order to enjoy standing, an association must be:

a) lawfully established in the jurisdiction concerned or have standing to act there;
b) able to demonstrate that it pursues a dedicated purpose in accordance with its statutory objectives in the defence of the human rights of its members or other affected individuals within the jurisdiction concerned, whether limited to or including collective action for the protection of those rights against the threats arising from climate change; and
c) able to demonstrate that it can be regarded as genuinely qualified and representative to act on behalf of members or other affected individuals within the jurisdiction who are subject to specific threats or adverse effects of climate change on their lives, health or well-being as protected under the Convention.[163]

157 ECtHR, *KlimaSeniorinnen* (note 2), para. 487.
158 ECtHR, *KlimaSeniorinnen* (note 2), para. 533.
159 ECtHR, *KlimaSeniorinnen* (note 2), para. 533.
160 Convention sur l'accès à l'information, la participation du public au processus décisionnel et l'accès à la justice en matière d'environnement (Convention d'Aarhus), Aarhus (Denmark), 25 June 1998, in force 1 June 2014; ratified by Switzerland on 27 September 2013; No. 0.814.07. Available (in its authentic French version) at: https://www.fedlex.admin.ch/eli/cc/2014/235/fr.
161 See ECtHR, *KlimaSeniorinnen* (note 2), paras. 476, 489, 494 and 497.
162 ECtHR, *KlimaSeniorinnen* (note 2), para. 501.
163 ECtHR, *KlimaSeniorinnen* (note 2), para. 502.

The Court held that *Verein KlimaSeniorinnen Schweiz* met all of the noted criteria and was thus granted *locus standi* to bring its claims under ECHR, articles 6 and 8.[164]

7.3 Violation of the Positive Obligations Deriving from the Right to Privacy and Family Life

With regard to ECHR, article 8 ('Everyone has the right to respect for his private and family life, his home and his correspondence'), the Court's previous environmental case law referred to situations in which harm (for example, toxic waste, pollution) inflicted on applicants may be traced directly to a *specific source*, such as an industrial steelworks complex or landfill, located within the jurisdiction of a member state of the CoE.[165] The ECtHR found states that consistently failed to limit hazardous emissions of sources to the detriment of the local population to be in violation of 'positive obligations' derived from the ECHR, especially articles 2 and 8.[166] The Court acknowledged that in these previous cases there was a direct link 'between a source of harm and those affected by the harm', and the measures necessary to alleviate the harm were 'identifiable and available to be applied at the source of the harm'.[167]

This is different when compared to the consequences of anthropogenic climate change.[168] CO_2 accounts for two-thirds of all GHGs emitted.[169] Multiple lines of evidence indicate a causal and 'almost linear relationship between cumulative CO_2 emissions and projected global temperature change'.[170] Each tonne of CO_2 emitted into the atmosphere anywhere on Earth at any given time thus had, has, and will have an almost identical effect on the average global temperature. Climate change induced by increased atmospheric CO_2 concentration 'remains largely irreversible for 1,000 years after emissions

[164] See ECtHR, *KlimaSeniorinnen* (note 2), para. 524.

[165] Johannes Reich, 'KlimaSeniorinnen and the Choice Between Imperfect Options: Incorporating International Climate Change Law to Maintain the ECHR's Relevance Amid the Climate Crisis' in Maxim Bönnemann & Maria Antonia Tigre (eds.), *The Transformation of European Climate Litigation* (Verfassungsbooks 2024), 41, 44.

[166] See ECtHR, *KlimaSeniorinnen* (note 2), paras. 538–40; for an instructive example see ECtHR, *Cordella and Others v. Italy*, Application nos. 54414/13 and 54264/15, 24 January 2019.

[167] See ECtHR, *KlimaSeniorinnen* (note 2), para. 415.

[168] On the following see Reich (note 168) 44.

[169] IPCC (note 92) 5; Federal Office for the Environment (note 14) 6.

[170] IPCC (note 92) 8.

stop'.[171] The rise in the global average temperature is therefore, as the ECtHR acknowledged,[172] essentially determined by the *cumulative* level of all GHG emissions accrued over centuries, to the effect that '[m]ost aspects of climate change will persist for many centuries even if emissions of CO_2 are stopped'.[173]

The ECtHR therefore drew 'some inspiration from the principles set out in the Court's existing case law',[174] while stressing 'the specificity of climate-change disputes, in comparison with classic environmental cases'.[175] After reviewing the relevant scientific evidence at length, the Court concluded that 'anthropogenic climate change' not only 'exists' but 'that it poses a serious current and future threat to the enjoyment of human rights guaranteed under the Convention', and that member states are 'capable of taking measures to effectively address' the challenge.[176] On that basis, the Court stressed that a sufficiently severe risk to the rights protected under ECHR, article 8 is enough to create a positive obligation for governments to take measures with regard to potential harmful activities.[177]

Based on the principle of 'common but differentiated responsibilities and respective capabilities' derived from the UNFCCC,[178] the ECtHR held that Switzerland was under an obligation to take 'reasonable measures' even if these would not prevent the harm in question from occurring.[179] Taking into account the broad choice of instruments to mitigate anthropogenically induced climate change and the margin of appreciation member states enjoy under the Convention, the Court sketched a five-pronged test to assess whether domestic climate mitigation policies comply with the positive obligations set forth by

171 Susan Solomon, Gian-Kasper Plattner, Reto Knutti & Pierre Friedlingstein, 'Irreversible Climate Change due to Carbon Dioxide Emissions' (2009) 106:6 *Proceedings of the National Academy of Sciences* 1704.

172 See ECtHR, *KlimaSeniorinnen* (note 2), paras. 416–17, 425, 439.

173 IPCC, 'Summary for Policymakers. Climate Change 2013: The Physical Science Basis. Contribution of Working Group I to the Fifth Assessment Report of the Intergovernmental Panel on Climate Change' 27, available at: https://www.ipcc.ch/site/assets/uploads/2018/02/WG1AR5_SPM_FINAL.pdf.

174 ECtHR, *KlimaSeniorinnen* (note 2), para. 422.

175 ECtHR, *KlimaSeniorinnen* (note 2), para. 423 and paras. 414–21.

176 ECtHR, *KlimaSeniorinnen* (note 2), para. 436.

177 See ECtHR, *KlimaSeniorinnen* (note 2), paras. 418–20, 425, 435, 437.

178 UNFCCC (note 45), article 3 clause 1.

179 ECtHR, *KlimaSeniorinnen* (note 2), paras. 441–2, 444, 571 (without elaborating further what exactly the principle and its different parts – 'common' as opposed to 'differentiated' – entail).

ECHR, article 8.[180] According to this test (and in addition to providing relevant information to the public),[181] states are required to:

a) adopt general measures specifying a target timeline for achieving carbon neutrality and the overall remaining carbon budget for the same time frame, or another equivalent method of quantification of future GHG emissions, in line with the overarching goal for national and/or global climate-change mitigation commitments;
b) set out intermediate GHG emissions reduction targets and pathways (by sector or other relevant methodologies) that are deemed capable, in principle, of meeting the overall national GHG reduction goals within the relevant time frames undertaken in national policies;
c) provide evidence showing whether they have duly complied, or are in the process of complying, with the relevant GHG reduction targets (…);
d) keep the relevant GHG reduction targets updated with due diligence, and based on the best available evidence; and
e) act in good time and in an appropriate and consistent manner when devising and implementing the relevant legislation and measures.[182]

The Court also stressed the duty that is incumbent on CoE member states to allow for *transparency* regarding 'the relevant regulations and measures to tackle climate change' as well as ensuring that the necessary *procedures* are in place to consider 'the views of the public' on 'the relevant regulations and measures, or the absence thereof'.[183]

The Court found that Switzerland failed to meet these criteria in various ways. There was, the Court concluded, an absence of relevant legislation after the unsuccessful revision of the Federal CO_2 Act (see section 4).[184] Regardless of such legislative 'lacunae',[185] the Court noted that Switzerland proved unable to meet its (rather modest) previous reduction target (see section 4) and abstained from explicitly quantifying its national GHG emissions limitations, 'through a carbon budget or otherwise'.[186] For these reasons, the Court found Switzerland to be in violation of ECHR, article 8.[187] In view of its finding that Switzerland failed to comply with ECHR, article 8, the Court deemed it

180 See ECtHR, *KlimaSeniorinnen* (note 2), paras. 541–8.
181 See ECtHR, *KlimaSeniorinnen* (note 2), para. 554.
182 ECtHR, *KlimaSeniorinnen* (note 2), para. 550.
183 ECtHR, *KlimaSeniorinnen* (note 2), para. 554.
184 ECtHR, *KlimaSeniorinnen* (note 2), paras. 560–62, 573.
185 ECtHR, *KlimaSeniorinnen* (note 2), paras. 562 and 573.
186 See ECtHR, *KlimaSeniorinnen* (note 2), paras. 558–9 and 573.
187 ECtHR, *KlimaSeniorinnen* (note 2), para. 574.

'unnecessary' to determine whether the threshold of an infringement of the right to life under ECHR, article 2, had been met.[188]

Accordingly, the Court derived from ECHR, article 8, in connection with the 2015 Paris Agreement, above all a duty to initiate comprehensive government planning arrangements to mitigate the risks posed by the effects of anthropogenic climate change to the private and family lives of individuals within the scope of the ECHR. From this perspective, ECHR, article 8 imposes a specific *duty of care* (*due diligence*) on CoE member states with regard to adapting to and mitigating climate change. While the Court set out the process of the governmental action to be taken in relatively high detail (e.g. drawing up a carbon budget, setting long-term targets, defining interim targets etc.),[189] the judgment, despite its considerable length, largely remains silent on the specific *substance* of the targets that states must achieve to comply with ECHR, article 8, but instead emphasises the importance of transparency and the obligation to provide for procedural avenues through which the views of those affected by the relevant matter can be taken into account.[190]

This *discrepancy between substance and procedure* mirrors the idea that also underpins the 'Nationally Determined Contributions' (NDCs)[191] in the Paris Agreement, namely that a transparent process, once initiated and organised around intermediate goals, *generates its own logic*, and is therefore bound to yield results with regard to reducing GHG emissions, despite the ambiguous nature of the obligations shaping the desired outcomes. From this perspective, what might be identified as due diligence obligations derived from ECHR, article 8, turn out to be, first and foremost, procedural obligations in the sense that it conveys a right to compel CoE member states *to establish, maintain and continually adapt structured governmental and administrative planning and procedures to mitigate climate change*. Or, to put it more succinctly, the ECHR thus enshrines *a right to an adequate and due procedure* in the domain of climate change law.[192]

---

188 See ECtHR, *KlimaSeniorinnen* (note 2), para. 536.

189 ECtHR, *KlimaSeniorinnen* (note 2), para. 550.

190 ECtHR, *KlimaSeniorinnen* (note 2), para. 554.

191 Cf Paris Agreement (note 47), article 4 section 2.

192 This procedural approach was confirmed in the ECtHR's second merits judgment on the nexus between human rights and climate change in the case of *Greenpeace Nordic and Others v. Norway*, handed down on 28 October 2025. The case concerned the approval process for fossil fuel activities in the Barents Sea. In that regard, the Court held that in the context of fossil fuel projects, the State bears a procedural obligation under Article 8, ECHR to conduct an 'adequate, timely and comprehensive environmental impact assessment in good faith and based on the best available science' including a quantification of the anticipated

7.4 Violation of the Right of Access to Court and to a Fair Trial

ECHR, article 6(1), in its first sentence, provides that 'everyone is entitled to a fair and public hearing within a reasonable time by an independent and impartial tribunal established by law' in 'the determination of his civil rights and obligations'. With regard to this right of access to court and to a fair trial, the Court noted that 'the applicants' action instituted at the domestic level largely concerned requests for legislative and regulatory action'.[193] Such 'policy decisions' being 'subject to the relevant democratic processes' would thus fall outside the scope of ECHR, article 6.[194] However, *Verein KlimaSeniorinnen Schweiz* and the individual applicants also sought to compel the domestic authorities to adequately implement *existing* mitigation measures (see sections 5.2, 6.1 and 6.2).[195] The Court further reiterated (see section 7.3) 'the important role of associations in defending specific causes in the sphere of environmental protection'.[196] Although the claims brought forward were subject to public-law proceedings at the domestic level, the Court still qualified the rights in question as 'civil rights' in the meaning of ECHR, article 6(1), in line with previous case-law as the outcome of the proceedings would interfere with the property and lifestyles of the applicants.[197] Given the violation of ECHR, article 8 (see section 7.3), the Court, as a consequence, stressed that, based on ECHR, article 6, a requirement exists for 'an adequate corrective action for the alleged failures and omissions on the part of the authorities in the field of climate change' under such

---

GHG emissions, which also must account for the downstream emissions occurring from the burning of the extracted fossil fuels. Further than that, 'at the level of the public authorities, there must be an assessment of whether the activity is compatible with their obligations under national and international law to take effective measures against the adverse effects of climate change' and 'informed public consultation must take place at a time when all options are still open and when pollution can realistically be prevented at source', see European Court of Human Rights (sitting as a Chamber), *Greenpeace Nordic and Others v. Norway*, Application no. 34068/21, 28 October 2025, paras 317–9; see also on this "proceduralisation" Johannes Reich, 'Urteil des Europäischen Gerichtshofs für Menschenrechte in Sachen Verein KlimaSeniorinnen Schweiz – ein Fanal "gerichtlichen Aktivismus"' (2026) Juristische Blätter [JBl] (forthcoming) p. 22.

193 ECtHR, *KlimaSeniorinnen* (note 2), para. 615.

194 ECtHR, *KlimaSeniorinnen* (note 2), paras. 615–16.

195 See also ECtHR, *KlimaSeniorinnen* (note 2), paras. 614 and 621.

196 ECtHR, *KlimaSeniorinnen* (note 2), para. 622.

197 See ECtHR, *KlimaSeniorinnen* (note 2), paras. 597–602, 621–2.

circumstances.[198] The Court therefore held that ECHR, article 6 applied to the application brought by *Verein KlimaSeniorinnen Schweiz* but not to the ones brought by the individual applicants as the latter lacked standing with respect to ECHR, article 8.[199] As neither the Federal Administrative Court (see section 6.3) nor the Federal Supreme Court (see section 6.4) reviewed the case on its merits, the Court, stressing 'the key role which domestic courts have played and will play in climate-change litigation',[200] found Switzerland in violation of ECHR, article 6.[201]

8. FEDERAL COUNCIL AND FEDERAL ASSEMBLY: TIMING AND DIRECT DEMOCRACY AS REASONS TO LEVEL HARSH CRITICISM AT THE ECTHR'S JUDGMENT

In the wake of *KlimaSeniorinnen*, both chambers of Switzerland's Federal Parliament (Federal Assembly) – i.e. the Council of State and the National Council – adopted declarations stating that the judgment exceeded 'the limits of dynamic interpretation', and that the ECtHR exposes itself 'to the reproach of exercising inappropriate and inadmissible judicial activism' and thus weakens not only its own legitimacy but that 'of the effective protection of human rights in Europe' as a whole.[202] The Federal Council, as (according to the Federal Constitution) the federal authority representing Switzerland in foreign affairs,[203] criticised the 'wide interpretation of the ECHR by the ECtHR' and voiced its opinion that, in view of the revised Federal CO_2 Act, in effect since

198 ECtHR, *KlimaSeniorinnen* (note 2), para. 621; see also ibid., para 644 (holding that ECHR, article 13, i.e. the right to an effective remedy, is 'being absorbed by the more stringent requirements' of ECHR, article 6).

199 ECtHR, *KlimaSeniorinnen* (note 2), paras. 623–5.

200 See ECtHR, *KlimaSeniorinnen* (note 2), para. 639.

201 See ECtHR, *KlimaSeniorinnen* (note 2), paras. 629–40.

202 See Déclaration du Conseil des Etats. Arrêt de la CEDH 'Verein KlimaSeniorinnen Schweiz et autres c. Suisse' (23 May 2024), No 24.053 (transl.), available at: https://www.parlament.ch/fr/ratsbetrieb/suche-curia-vista/geschaeft?AffairId=20240053; and Déclaration du Conseil national. Arrêt de la CEDH 'Verein KlimaSeniorinnen Schweiz et autres c. Suisse' (29 May 2024), No 24.054 (transl.), available at: https://www.parlament.ch/fr/ratsbetrieb/suche-curia-vista/geschaeft?AffairId=20240054.

203 Fed. Const (note 31), article 184 section 1 (official French version): 'Le Conseil fédéral (…) représente la Suisse à l'étranger'.

1 January 2025 (see section 4), Switzerland in the meantime complies with 'the climate policy requirements of the judgment'.[204]

The critical stance taken by the legislative and executive branches of the Swiss federal government is partly due to the fact that the ECtHR chose 14 February 2024 as the cutoff date for its assessment of Switzerland's climate policy.[205] This was after the referendum of 18 June 2023 on the Federal Climate Protection Act, which remedied some of the legislative 'lacunae' identified by the ECtHR (see sections 4 and 7.3). On 15 March 2024, the revised Federal CO_2 Act was adopted by the Federal Parliament. Both federal statutes entered into force on 1 January 2025 (see section 4).

Another, more fundamental reason for the harsh reaction may lie in the *specific concept of separation of powers underpinning the Swiss Federal Constitution* (see section 3), mirrored not only in the cautious approach employed by both the Federal Supreme Court and the Federal Administrative Court in the relevant proceedings (see sections 6.2–6.3), but in the trial-and-error approach to legislation inherent in the slow-working mechanism of direct democracy (see section 3). On 14 February 2024, the omissions in Swiss climate policy had not yet been remedied, but such remedies were well under way, not in spite of but *because* of direct democracy, since the Federal Climate Protection Act is rooted in a popular initiative.[206] Among other things, the ECtHR, while taking note of the Climate Protection Act not yet in force, took issue with the lacuna persisting for the time period between 2025 and 2030 left unregulated in both the unrevised CO_2 Act and the new Climate Protection Act, as well as with the fact that concrete measures under the CO_2 Act were to be taken "in good time", according to the new Climate Protection Act. These issues were, to some extent, addressed by the revised CO_2 Act, adopted by Federal Parliament shortly after the cutoff date. The ECtHR, however, had nothing to say about (direct) democracy in its judgment, except for the undisputed but rather shallow statement that 'democracy cannot be reduced to the will of the majority of the electorate and elected representatives, in disregard of the requirements of the rule of law'.[207] Rather, the Court stressed 'the key role which domestic courts have played and will play in climate-change liti-

204 Federal Council, 'Le Conseil fédéral clarifie sa position sur le verdict de la Cour européenne des droits de l'homme concernant la protction du climat' (transl.), available (in French) at: https://www.admin.ch/gov/fr/accueil/documentation/communiques.msg-id-102244.html.

205 See ECtHR, *KlimaSeniorinnen* (note 2), para. 556: '[T]he Court's assessment is limited to examining the domestic legislation as it stands on the date of adoption of the present judgment, namely 14 February 2024 (…)'.

206 See Reich & Boss (note 53) 153.

207 ECtHR, *KlimaSeniorinnen* (note 2), para. 412.

gation' (see section 7.4).[208] For better or worse, this is unlikely to apply to Switzerland. Alexis de Tocqueville famously observed that there is hardly a political question in the United States that does not 'sooner or later, turn into a legal one to be decided by the courts'.[209] The opposite is true for Switzerland: there is hardly a major legal issue that does not, sooner or later, become a political issue to be decided by the people at the ballot box.[210]

9. COMMITTEE OF MINISTERS OF THE COUNCIL OF EUROPE: SUPERVISION AS A PROCESS

Following their 1521st meeting of 4–6 March 2024, the Committee of Ministers of the CoE, responsible 'for the supervision of the execution of a final judgment' of the ECtHR,[211] published a brief decision on *KlimaSeniorinnen*.[212] The Committee noted, with reference to the Federal Act on Climate Protection and the revised Federal CO_2 Act (see section 4) 'that certain issues have been resolved on the federal level, such as the filling of legislative lacunae' and requested further information on the methodology of the relevant 'carbon budget' (see section 7.3).[213] Furthermore, it invited the domestic authorities to 'update the Committee on the evolution of domestic case law, regarding both the standing of associations to bring climate change-related cases and on courts' assessments of the merits of such cases' in view of the direct application of the ECHR in Switzerland ('monism'; see section 4).[214] The Committee made it clear that monitoring the execution of *KlimaSeniorinnen* will be an ongoing process.[215]

In a further review in September 2025, the Committee 'noted that Switzerland, by a method of its choice, has quantified the future greenhouse gas emissions

---

208 See ECtHR, *KlimaSeniorinnen* (note 2), para. 639.

209 Alexis de Tocqueville, 'De la démocratie en Amérique', vol I (first published 1835) in Jean-Paul Mayer (ed), *Alexis de Tocqueville: Oeuvres complètes, vol I* (Gallimard 1961), 281.

210 Johannes Reich, 'Information als Voraussetzung des Verwaltungshandelns', in Frank Schorkopf (ed), *Bestimmungsfaktoren 'guter Verwaltung'*, Veröffentlichungen der Vereinigung der Deutschen Staatsrechtslehrer vol 83 (De Gruyter 2024) 317, 356–7 (DOI: https://doi.org/10.0000/9783111293288).

211 ECHR (note 1), article 46 section 3.

212 Committee of Ministers (Deputies), H46-30 *Verein KlimaSeniorinnen Schweiz and Others v. Switzerland* (Application No. 53600/20), Supervision of the execution of the European Court's judgments, Decision, CM/Del/Dec(2025)1521/H46-30.

213 Committee of Ministers (note 215) para. 5.

214 Committee of Ministers (note 215) para. 8.

215 See Committee of Ministers (note 215) para. 9.

corresponding to the general measures and the target timeline put in place'.[216] This was an 'implicit carbon budget' derived from the interim targets enshrined in the Swiss CO_2 Act, the Climate Protection Act and Switzerland's NDC, which Switzerland communicated to the Committee as 'in conformity with its margin for manoeuvre recognised by the Court'.[217] This approach was, however, criticised by the Swiss National Human Rights Institution (NHRI) as 'a simple inventory of planned GHG emissions, not a budget of permissible emissions',[218] as well as by the 34 NGOs and the applicant association *Verein KlimaSeniorinnen Schweiz*,[219] in their respective communications to the Committee. The Committee decided to resume the supervision in a further meeting, while also 'invit[ing] Switzerland to consider the opportunity of establishing an independent national body suited to the national political structure to monitor its climate policy (. . .)'.[220] By recognizing an 'implicit carbon budget' grounded in existing legislation, the Committee effectively attenuates the obligations deriving from the initial judgment. Moreover, recent studies demonstrate that Switzerland's 'planned CO_2 emissions reductions are insufficient to remain within its fair share of the global [remaining carbon budget] under almost all allocation methods considered'.[221]

---

216 Committee of Ministers (Deputies), H46-37 *Verein KlimaSeniorinnen Schweiz and Others v. Switzerland* (Application No. 53600/20), Supervision of the execution of the European Court's judgments, Decision, CM/Del/Dec(2025)1537/H46-37, para. 5.

217 See 1537th meeting (September 2025) Rule 8.2a Communication from Switzerland in the case of *Verein KlimaSeniorinnen Schweiz and Others v. Switzerland* (Application No. 53600/20) [french only] p. 6–8, quote translated by the author.

218 1537th meeting (September 2025) Rule 9.2 Communication from an NHRI (Swiss human rights institution), H46-37 *Verein KlimaSeniorinnen Schweiz and Others v. Switzerland* (Application No. 53600/20) para. 14.

219 See 1537th meeting (September 2025) Rule 9.2 Communication from NGOs (Greenpeace International, Climate Litigation Network and 32 others) in the case of *Verein KlimaSeniorinnen Schweiz and Others v. Switzerland* (Application No. 53600/20); 1537th meeting (September 2025) Rule 9.2 Communication from an NGO (Verein KlimaSeniorinnen Schweiz) in the case of *Verein KlimaSeniorinnen Schweiz and Others v. Switzerland* (Application No. 53600/20).

220 Committee of Ministers (note 219) paras. 5 and 8.

221 See Konstantin Weber, Cyril Brunner & Reto Knutti, 'Tracking country-level mitigation progress using NGHGI-consistent carbon budgets' (2026) *nature communications* 17:1494, 7.

10. CONCLUSION: A RIGHT TO PROCEDURAL DUE PROCESS IN CLIMATE MITIGATION POLICIES IN LIEU OF SUBSTANTIVE PROTECTION

'[E]very dispute is normatively justiciable', *Aharon Barak*, the President of the Supreme Court of Israel from 1995 to 2006 and a prolific legal scholar, once boldly stated.[222] *KlimaSeniorinnen* may be considered a case in point. Despite its own assessment that 'the nature of climate change and its various adverse effects and future risks, the number of persons affected, in different ways and to varying degrees, is *indefinite*',[223] the Court was still in a position to unequivocally conclude in a very *definite* manner that Switzerland's climate mitigation legislation, as of 14 February 2024 (see section 8), failed to comply with the positive obligations derived from the right to private and family life according to ECHR, article 8 (see section 7.3). The seemingly commanding ease with which the Court identified *judicially manageable standards* to determine a common minimum standard derived from the ECHR's right to private and family life with regard to domestic climate change mitigation policies is anything but a matter of course. In its ruling, the Court was extremely cautious when it came to deriving substantive requirements from ECHR, article 8, which member states must not fall short of in their efforts to mitigate climate change. The Court's approach rather mirrors the concept of the Paris Agreement's 'Nationally Determined Contributions',[224] according to which initiating a transparent process to reduce GHG emissions is bound to generate its own logic and would thus yield substantive outcomes (see section 7.4). Against the backdrop of the principle of 'common but differentiated responsibilities and respective capabilities in light of different national circumstances',[225] it is proving difficult to define a GHG emissions threshold that a specific country is not permitted to exceed.[226] Despite all of this, climate change mitigation, it seems, proved to be *infinitely complex, albeit justiciable* for the ECtHR.

---

222 Aharon Barak, *The Judge in a Democracy* (Princeton University Press 2006) 179 and Aharon Barak, 'A Judge on Judging: The Role of a Supreme Court in a Democracy' (2002) 116 *Harvard Law Review* 19, 98: 'In my opinion, every dispute is normatively justiciable.'; see also Ran Hirschl, 'The Socio-Political Origins of Israel's Juristocracy' (2009) *Constellations* 476, 480: 'Barak's maxim [is] that "(…) anything and everything is justiciable" (…)'.

223 ECtHR, *KlimaSeniorinnen* (note 2), para. 479 (emphasis added).

224 Cf Paris Agreement (note 47), article 4 section 2.

225 Paris Agreement (note 47), article 2 section 2.

226 Cf in a similar vein ECtHR, *KlimaSeniorinnen* (note 2), Separate Opinion by Judge *Tim Eicke*, para. 14.

Considering its five-pronged test,[227] the key to identifying a common minimum standard is for the Court to reinterpret the *substantive* right to private and family life as a right to compel CoE member states *to establish, maintain and continually adapt structured governmental and administrative planning and procedures to mitigate climate change* (see section 7.3). At the same time, this requires states to exercise due diligence in their policies to address climate change (see section 7.3). While courts will generally lack relevant expertise to review such targets in substance (the Court's own refusal to specify concrete remedies speaks volumes in this respect[228]), they are well placed to monitor compliance with *procedural* requirements. Whether such *procedural due process in climate mitigation policies* ensures effective protection of the physical integrity of persons within the scope of the ECHR is, of course, another matter. Remarkably, these procedural requirements are broadly similar to those already established by EU law in particular.[229] This begs the question of what distinctive reasoning courts can offer when they resort to deriving fine-grained and essentially pre-existing legal obligations from vague human rights guarantees.

In the light of this transformation of a substantive right into a procedural obligation, establishing distinct criteria for the *locus standi* of associations (see section 7.2), being strictly limited to issues of climate change policy,[230] seems consistent in that the focus has shifted from protecting the rights of individual human beings to ensuring methodical and rational administrative and legislative procedures. This approach, if followed through at the domestic level, strengthens the relevance of the judicial branch in tackling the climate crisis – something which obviously chimes with the Court's intentions.[231]

'The justiciability of everything', however, comes at a price. Judicial applications to compel states to reduce their GHG emissions run the risk of being transformed "into a hybrid form of weak public interest litigation, akin to supervisory complaints in administrative law".[232] Furthermore, courts that not only proclaim to be authorised to rule on applications brought by associations whose members need not be victims but find themselves unable to specify concrete remedies may lose sight of their *raison d'être*: the victims.

---

[227] ECtHR, *KlimaSeniorinnen* (note 2), para. 550, and section 7.3.

[228] Cf ECtHR, *KlimaSeniorinnen* (note 2), para. 657; see section 7.1.

[229] On these requirements see, e.g. Regulation 2021/1119 (note 69).

[230] European Court of Human Rights (sitting as a Grand Chamber), *Cannavacciuolo and Others v. Italy*, Application no. 39371/20, 30 January 2025, paras. 220–21.

[231] Cf ECtHR, *KlimaSeniorinnen* (note 2), para. 639 (stressing 'the key role which domestic courts have played and will play in climate-change litigation'); see also section 7.4.

[232] See Reich (note 168) 47–8 (labelling this an 'administrative turn').

7. A constitutional duty to deploy climate finance: *PSB et al v Brazil (Climate Fund)*

Gabriel Mantelli, Matheus Henrique Cano, Isabela Soares Bicalho, and Marie-Louise Siemons

1. INTRODUCTION

According to the Edgar EU Database for Global Atmospheric Research, Brazil ranked sixth among the world's top total greenhouse gas (GHG) emitters in 2023, behind only China, the United States, India, the EU27 and Russia.[1] In contrast to other top-six global emitters, where GHGs primarily originate from energy production and industrial activity, Brazil's emissions are largely attributable to agriculture and land-use change—most notably deforestation in the Amazon rainforest, the majority of which lies within its territory. According to 2023 official data, 46% of the country's emissions came from land-use change and forestry, while agriculture contributed an additional 28%.[2] Although Brazil has made significant strides in combating deforestation and expanding renewable energy sources, its economy remains heavily dependent on oil extraction and mining. Laws and policies to reduce emissions from deforestation, agriculture and other land-use changes, and additional strategies to reduce emissions from other sectors are crucial to enable Brazil

[1] European Commission, Joint Research Centre, 'GHG emissions of all world countries' (JRC138862, 2024); Emissions per capita placed Brazil 67 out of 198 countries in 2021, according to Climatewatch, 'Brazil' (*Climatewatch*) <www.climatewatchdata.org/countries/BRA?end_year=2021&start_year=1990> accessed 11 June 2025.

[2] Observatório do Clima, 'Análise das Emissões de Gases de Efeito Estufa e suas Implicações para as Metas Climáticas do Brasil 1970–2023' (SEEG 2024) <https://seeg.eco.br/relatorios> accessed 8 January 2025.

to meet its economy-wide target of reducing net GHG emissions by 59 to 67% below 2005 levels by 2035.[3]

The World Bank has estimated the costs for Brazil to meet its previous Nationally Determined Contribution (NDC) (between 2016 and 2030) at around US$1.4 trillion.[4] Considering the critical role of finance to its climate commitments, Brazil established a Climate Fund in 2009 (Climate Fund or Fund) as a vital tool to support national initiatives to reduce GHG emissions, as well as to promote adaptation to the inevitable consequences of climate change. Considering the disproportionate challenges climate change poses to certain vulnerable communities, the Climate Fund had also been designed as a crucial instrument for promoting climate justice by directing resources to projects that especially meet the needs of vulnerable groups.

Following the 2018 election of far-right candidate Jair Bolsonaro, the federal government effectively paralyzed the operations of the Climate Fund. With available resources significantly underutilized, the Fund lost the ability to finance key climate mitigation and adaptation projects.[5] In 2020, the Brazil Socialist Party (PSB for its Portuguese acronym) and three other progressive opposition political parties filed a constitutional challenge before the Brazilian Supreme Court (Supreme Court or STF for its Portuguese acronym). In *PSB et al v Brazil on Climate Fund* (*Climate Fund* case),[6] the political parties argued that the federal government's failure to allocate available resources from the Climate Fund violated the right to an ecologically balanced environment (the Brazilian articulation of the right to a healthy environment), as guaranteed by the 1988 Constitution of the Federative Republic of Brazil (Brazilian Constitution or Constitution). As remedies, the political parties requested: a) a judicial order for the federal government to reactivate the operations of the Climate Fund; b) a judicial order for the Ministry of the Environment to develop an annual plan for allocation of available resources; and c) a judicial

---

3 Brazil 2024 NDC, UNFCCC.

4 The World Bank, Brazil Climate Finance Project <https://documents1.worldbank.org/curated/en/099800011302228261/pdf/P1788880aa8da5060b6c205df5e2d32150.pdf> accessed 23 May 2025.

5 Kathryn Hochstetler, 'Climate institutions in Brazil: three decades of building and dismantling climate capacity' (2021) 30:51 *Environmental Politics* 49.

6 Supremo Tribunal Federal (STF), *Arguição de Descumprimento de Preceito Fundamental (ADPF) 708 DF*, Relator Min Roberto Barroso, judgment of 4 July 2022, Diário da Justiça Eletrônico (DJE) 11 July 2022 (Brazil) (Climate Fund case).

order for the government to refrain from any future actions that would inhibit the use of the fund's resources.[7]

The July 2022 decision on the *Climate Fund* case was groundbreaking in multiple respects. For the first time the Brazilian apex court recognized a stable climate as part of the right to an ecologically balanced environment, enshrined in the Brazilian Constitution. Also, for the first time, the Supreme Court clearly established a state duty to combat climate change through both mitigation and adaptation measures, which includes implementing existing climate laws in Brazil. Furthermore, the Supreme Court elevated the Paris Agreement to supranational status, equivalent to the level previously reserved to international human rights agreements in the Brazilian legal framework.[8] Finally, the *Climate Fund* case is the first one worldwide to establish that a government has a constitutional obligation to disburse existing funds to climate actions when they have been legally allocated by the legislature. Considering the growing awareness of the importance of climate finance to meet the goals of the Paris Agreement, this is an important precedent that can inform efforts in other jurisdictions.

The *Climate Fund* case was part of a suite of strategic climate lawsuits, most of them initiated by a group of political parties, that challenged rollbacks of existing climate change laws and policies as a constitutional violation.[9] These cases formed part of the Supreme Court's 'green agenda'—a set of environmental and climate-related cases prioritized in 2022.[10] This chapter examines three noteworthy aspects of this landmark decision: 1) the special standing granted to political parties in Brazil to directly challenge the constitutionality of laws, policies and their (non-)implementation before the Supreme Court; 2) the Court's procedural innovation in convening a broad public hearing on climate-related constitutional claims; and 3) the decision's unique emphasis

---

7 Leticia Albuquerque, Gabrielle Tabares Fagundez and Roger Fabre, 'Climate Emergency and Human Rights: The Case of Climate Fund in Brazil and International Law Enforcement' (2022) 19:1 *Brazilian Journal of International Law* 127.

8 Angelica Rutherford and Flavianne Fernanda Bitencourt Nobrega, 'The Paris Agreement as a Human Rights Treaty: The Ruling in *PSB et al v Brazil (on Climate Fund)*' (2022) 3:2 Jus Corpus L J 456.

9 Other cases included *PSB et al v Brazil (on Amazon Fund)*, online <https://climatecasechart.com/non-us-case/psb-et-al-v-brazil/> and *PSB et al v Brazil (on deforestation and human rights)*, online <https://climatecasechart.com/non-us-case/brazilian-socialist-party-and-others-v-brazil/> accessed 16 September 2024.

10 'STF vota Pauta Verde' (*Instituto Clima e Sociedade*, 11 April 2022) <https://climaesociedade.org/stf-vota-pauta-verde/> accessed 04 October 2024.

on the obligation to operationalize and disburse climate finance as part of the state's duty to ensure a stable climate and a healthy environment. This chapter also briefly notes the STF's recognition of the Paris Agreement as a supraconstitutional norm and its application of the non-regression principle, with fuller analysis of these issues found elsewhere.[11]

We argue that this case illustrates how progressive political parties, supported by civil society, successfully used strategic climate litigation to challenge the Bolsonaro administration's dismantling of environmental policies and bring climate change, and climate finance, to the forefront of national debates. Their efforts responded to significant environmental harm, including record emissions from deforestation and rising forest fires in the Amazon.[12] We also argue that this was only possible due to a Supreme Court that has proven to be sensitive to the importance of upholding environmental rights.

The remainder of this chapter is structured as follows. Section 2 provides an overview of climate law and climate litigation in Brazil, to contextualize the analysis. Section 3 introduces the Climate Fund, highlighting its pivotal role in supporting initiatives to mitigate the impacts of climate change and promote climate adaptation and climate justice. Section 4 analyzes the *Climate Fund* case, highlighting three crucial aspects: the role of political parties as plaintiffs in a constitutional climate lawsuit against the government; the Supreme Court's inclusive process, which integrated diverse stakeholder interventions into the proceedings; and the landmark ruling establishing a state obligation to disburse dedicated climate funds—the first decision of its kind globally. Section 5 concludes with thoughts about the future of climate litigation in Brazil.

2. CLIMATE CHANGE LAW AND CLIMATE LITIGATION IN BRAZIL

Albeit still a relatively new field, climate litigation is rapidly evolving in Brazil.[13] Brazilian databases indicate that there are currently 120 climate

---

11 Maria Antonia Tigre and Joana Setzer, 'Human Rights and Climate Change for Climate Litigation in Brazil and Beyond: An Analysis of the Climate Fund Decision' (2023) 54:4 *Geo J Int'l L* 593.

12 Danielle de Andrade Moreira and others, 'Rights-Based Climate Litigation in Brazil: An Assessment of Constitutional Cases before the Brazilian Supreme Court' (2024) 16:1 *J of Human Rights Practice* 47.

13 Alessandra Lehmen, 'Advancing Strategic Climate Litigation in Brazil' (2021) 22:8 *German L J* 1471.

lawsuits in the judicial system.[14] That makes Brazil the fourth ranked jurisdiction in number of climate lawsuits. While reducing emissions from deforestation and land-use change has been central to Brazil's climate litigation,[15] lawsuits have also addressed the energy transition, fossil fuel regulation, climate risk and adaptation, environmental licensing and public investment in climate policy.[16] Another strand of litigation focuses on civil liability, seeking to hold individuals and companies accountable for GHG emissions and climate-related environmental harm.[17] Though less common, climate justice claims are increasingly highlighting the disproportionate impacts of climate change on vulnerable groups, including Indigenous peoples and traditional communities.[18] Finally, systemic cases—such as the *Climate Fund* case—challenge the effectiveness of climate laws, policies and governance, underscoring the judiciary's crucial role in advancing climate action and addressing governmental omissions.

As the number of cases continues to grow, it has become clear that one of the main strategies in Brazilian climate litigation, particularly when it comes to challenging climate laws and policies, has been to frame the protection of the climate system as a constitutional right, protected as part of the right to an ecologically balanced environment enshrined in Article 225 of the 1988 Brazilian Constitution.[19] Unlike liberal constitutional models such as those of the United

---

14 Danielle de Andrade Moreira and others, 'Brazilian Climate Litigation Bulletin 2024' (Law, Environment and Justice in the Anthropocene Research Group (JUMA/PUC-Rio) 2024) <https://juma.jur.puc-rio.br/pesquisas-litigancia-climatica> accessed 8 January 2025.

15 Maria Antonia Tigre and others, 'Just Transition Litigation in Latin America: An Initial Categorization of Climate Litigation Cases Amid the Energy Transition' (Sabin Center for Climate Change Law 2023) <https://ssrn.com/abstract=4317186> accessed 8 January 2025.

16 Danielle de Andrade Moreira, Carolina de Figueiredo Garrido and Maria Eduarda Segovia Barbosa Neves, 'Climate litigation and environmental licensing: Considering the climate dimension in the light of international climate treaties' (2022) 19:1 *International Law and Climate Litigation* 62.

17 Jacqueline Peel and Jolene Lin, 'Transnational Climate Litigation: The contribution of the Global South' (2019) 113:4 *AJIL* 679.

18 Tatiana Castelo Branco, 'Climate litigation sovereignties and gender: An intersectional analysis of women's quest for climate justice' (Centro Soberania e Clima 2024) <https://soberaniaeclima.org.br/wp-content/uploads/2024/10/Artigo-Tatiana-Castelo-Branco-INGLES-03.pdf?utm_source=chatgpt.com> accessed 10 January 2025.

19 Fernanda de Salles Cavedon-Capdeville and others, 'An Ecocentric Perspective on Climate Litigation: Lessons from Latin America' (2024) 16:1 *Journal of Human Rights Practice* 89.

States, the United Kingdom and Canada—which primarily protect civil and political rights and impose negative obligations on the state—the Constitution recognizes a broad spectrum of fundamental rights, including social, economic, cultural and environmental rights. It requires the state not only to respect individual liberties but also to actively implement laws and policies to realize these rights. This reflects a constitutional vision in which the state is a central actor in promoting social justice and reducing inequality, rather than merely refraining from interference in private life and individual rights.[20]

Many laws and policies in Brazil are enacted to fulfill this obligation to uphold and enforce fundamental rights enshrined in the Constitution. Environmental—and climate—laws and policies are grounded in Article 225, which reads:

> All persons are entitled to an ecologically balanced environment, which is an asset for common use by the people and essential to a healthy quality of life. Both the government and the community have a duty to defend and preserve it for present and future generations.

Article 225 thus enshrines the Brazilian version of the right to a healthy environment at the constitutional level, framed as a collective good, imposing a shared responsibility on both the government (public power) and society (the community) to protect and preserve it for present and future generations.[21]

The Brazilian National Environmental Policy (PNMA for its Portuguese acronym),[22] established in 1981, sets out a comprehensive framework of guidelines and instruments designed to protect the environment and ensure the sustainable use of natural resources. It became the cornerstone of Brazil's environmental governance after the Constitution was enacted, providing the broad regulatory structure to uphold the right to a healthy environment by guiding environmental actions across multiple sectors and levels of government.[23]

---

[20] José Afonso da Silva. Curso de Direito Constitucional Positivo (Revista dos Tribunai 1984).

[21] While the government holds primary responsibility for environmental protection, the legal system also imposes duties on citizens and private entities—requiring those who exploit natural resources to restore the environment and subjecting environmentally harmful activities to penalties. Danielle de Andrade Moreira and Stela Luz Andreatta Herschmann, 'The Awakening of Climate Litigation in Brazil: Strategies Based on the Existing Legal Toolkit' (2021) 59 *Revista Direito, Estado e Sociedade* 172.

[22] L 6.938/81.

[23] Hironobu Sano, 'The Brazilian National Environmental Policy: The Challenge of Plural Environmental Governance' (2012) 55:1 *Society for International Development* 119.

Building on this foundation, the National Policy on Climate Change (PNMC for its Portuguese acronym)[24] was introduced in 2009, with a more targeted focus of addressing climate change. PNMC now forms the backbone of Brazilian climate governance. Together, the PNMA and the PNMC aim to integrate environmental and climate change considerations across various spheres of government and promote sustainable development in Brazil.

The PNMC translates the broader overarching principles of the PNMA into concrete, climate-focused strategies, with its two primary objectives to reduce GHG emissions and build resilience to climate impacts (climate adaptation). In other words, while the PNMA establishes the fundamental principles of environmental stewardship in Brazil's legal system, the PNMC operationalizes these principles to guide laws and policies to respond to the pressing issue of climate change. The PNMC was drafted as a direct response to the emissions reduction pledges Brazil made at the United Nations Framework Convention on Climate Change's (UNFCCC) fifteenth session of the Conference of the Parties (COP15) in 2009.[25] Brazil's climate laws and policies are therefore closely linked to the country's obligations under international climate law. This reflects Brazil's evolving environmental agenda and efforts to align with global sustainability goals generally, and international climate change goals specifically.[26]

The National Climate Change System (SNMC for its Portuguese acronym) is responsible for implementing the PNMC, coordinating federal, state and municipal actions while actively involving civil society. This structure seeks to ensure effective, regionally responsive climate policies. The PNMC operates as a 'policy mix', combining economic instruments and command-and-control measures to advance Brazil's mitigation and adaptation goals.[27] Its tools include sectoral action plans—such as those targeting deforestation and fire in

---

[24] L 12.187/09.

[25] Eduardo Viola and Matías Franchini, 'Climate Politics in Brazil: Public Awareness, Social Transformations and Emissions Reduction' in Ian Bailey and Hugh Compston (eds), *Feeling the Heat: The Politics of Climate Policy in Rapidly Industrializing Countries* (Palgrave Macmillan UK 2012) 175.

[26] Danielly Molleta, Breno Simonini and Gustavo Luedemann, 'Esforços Nacionais sobre as Mudanças Climáticas' in Adriana Moura (ed), *Governança Ambiental no Brasil: Instituições, Atores e Políticas Públicas* (Ipea 2016) ch 11.

[27] André de Castro dos Santos, 'Política Nacional sobre Mudança do Clima no Brasil: Uma Avaliação de Instrumentos e de Efetividade' (2021) 6 Revista de Políticas Públicas da UFPE 1.

the Amazon and Cerrado[28]—and the Climate Fund, one of the government's key mechanisms for achieving PNMC objectives.

It is worth noting the crucial role civil society plays in Brazil's climate governance, contributing to both policy debate and formulation. The PNMC institutionalizes public consultations and forums to engage diverse actors—including local communities, NGOs and the private sector and their diverse views, experience and expertise—thereby enhancing the legitimacy and effectiveness of climate actions through inclusive participation.[29]

Despite its strengths, Brazil's climate governance faces significant challenges, including the need for greater ambition, better integration of environmental and economic policies, limited financial resources and institutional weaknesses in implementing bodies.[30] Given Brazil's obligations under the Paris Agreement, climate litigation has become—and is likely to remain—a key tool for advancing more ambitious and effective climate laws and policies, particularly in securing finance for mitigation and adaptation. This issue of climate finance is at the heart of the *Climate Fund* case, discussed next.

3. THE BRAZILIAN CLIMATE FUND

The Climate Fund was established in November 2009, ahead of Brazil's participation in COP 15 in Copenhagen, and is an integral component of the PNMC.[31] As one of Brazil's main financial instruments for climate action, it supports mitigation and adaptation projects by channeling revenues—particularly from oil and gas exploration—toward PNMC initiatives.[32]

The Climate Fund is institutionally housed at the Ministry of the Environment and operates through two main branches. The first, focused on refundable financing (e.g., low-interest loans), is managed by the Brazilian Development

---

28 The Cerrado, Brazil's second-largest biome, is a tropical savanna spanning at least 10 states and covering approximately 21% of the national territory.

29 Regiane Borsato, 'Governança Ambiental e as Certificações: Participação e Influência do Movimento Ambientalista nos Instrumentos Privados de Gestão Ambiental' (PhD thesis, Universidade Federal do Paraná 2012).

30 Mariana Maia, Lizandra Serafim and Elia Alves, 'Implementation of a Climate Policy Monitoring System in Brazil: Political Challenges for Coordination on Multilevel Governance Arrangements' (2024) 18:2 *Brazilian Political Science Review* 1.

31 Instituto de Estudos Socioeconômicos, 'Fundo Nacional sobre Mudança do Clima: governança, recursos, gestão e desafios' (2022) <https://www.inesc.org.br/wp-content/uploads/2022/11/Fundo-clima_Inesc.pdf> accessed 27 December 2024.

32 The Fund was established by L 12.114/09 art 6.

Bank (BNDES for its Portuguese acronym), a federal institution that funds economic development. It supports sustainable development through subprograms on urban mobility, renewable energy, solid waste, native forest preservation, carbon management and more. The second branch, overseen by the Ministry of the Environment, provides non-refundable financing (e.g., grants) to mitigation and adaptation initiatives by NGOs, state and municipal governments, and other entities. This dual financing structure reflects the Fund's overarching goal of reducing emissions while promoting climate adaptation and sustainable development. It allows the Climate Fund to act strategically by supporting a wide range of initiatives, from advancing the energy transition to helping vulnerable communities to adapt to the impacts of climate change.

One important aspect of the Climate Fund is how it obtains its financial resources. The Climate Fund sources of finance include: a) up to 60% of the 10% share of special participation from oil exploration[33] allocated to the Ministry of Environment; b) budget allocations specified in the federal Annual Budget Law; c) resources from agreements, adjustments, contracts and conventions with public administration entities; d) donations from national and international entities, whether public or private; e) loans from national or international financial institutions; f) reversion of unused balances; g) resources from interest and amortization of financing; h) earnings from the investment of the Climate Fund's resources; and i) other sources.[34] According to the federal government, until 2023 the special participation revenues from oil exploration was the primary source of resources for the Climate Fund. Since 2024, this role has been taken over by the issuance of sustainable public bonds by the National Treasury.[35]

The Climate Fund was expected to provide a steady and substantial flow of resources for implementing the PNMC, with revenues—such as oil royalties—legally earmarked for climate action. However, even before its complete suspension under the Bolsonaro administration, the Fund consistently fell short of its intended financing levels. Between 2010 and 2022, R$4.36 billion

---

[33] 'Special participation' is a 'quarterly tax due by field concessionaires for the exploration and production of oil and gas of high volume or high profit margin fields', Inesc (n 31).

[34] Inesc (n 31).

[35] Ministério do Planejamento e Orçamento, 'Fundo Nacional sobre Mudança do Clima' (*Governo Federal*, 6 September 2024) <www.gov.br/planejamento/pt-br/acesso-a-informacao/participacao-social/conselhos-e-orgaos-colegiados/cmap/politicas/2023/avaliacao-conjunta-cmas-cmag/fundo-nacional-sobre-mudanca-do-clima> accessed 24 December 2024.

was deposited,[36] yet only R$564 million was allocated to specific projects,[37] revealing chronic underutilization of available resources.[38] Several factors contributed to the Fund's low disbursement levels. First, from 2013 to 2018, transfers from oil royalties were suspended due to political disputes among federal, state and municipal governments over revenue sharing, leading to a legal amendment that redirected earmarked funds away from the Fund—later reversed. Second, even when funds were available, chronic underutilization persisted due to: a) repeated budget freezes (contingenciamento) driven by short-term fiscal priorities;[39] and b) limited political will, which sidelined climate funding amid bureaucratic inertia and competing agendas.[40]

The persistent failure to execute the Climate Fund's budget—leaving Brazil's climate agenda chronically underfunded despite legal authorization—reflects Brazil's political elites' broader inconsistencies in implementing climate policies that require trade-offs with powerful sectors such as oil and gas and agribusiness.[41] In their State of Claim, the plaintiffs in the Climate Fund case cited a report by the Senate Environmental Committee (CMA for its Portuguese acronym) that acknowledged these challenges but also noted the Fund's achievements: 'Despite these difficulties, the Climate Fund has achieved significant milestones and could benefit from enhanced strategies for both resource mobilization and allocation'.[42]

Before the *Climate Fund* case reached the Supreme Court, lawsuits in lower courts had already challenged the legality of withholding disbursements.[43] The discussion was, however, restricted to narrow legal circles, perhaps because of the technical aspects of fiscal strategies and financial mechanisms.

Under the Bolsonaro administration, the Climate Fund's underutilization intensified, prompting the issue to reach Brazil's apex court and bringing

---

36 Inesc (n 31).

37 Tigre and Setzer (n 11).

38 Patrícia Figueiredo and Vitor Sorano, 'Governo descumpre prazo e trava 86% do orçamento para enfrentar a mudança climática' *G1* (3 May 2019) <https://g1.globo.com/natureza/noticia/2019/05/03/governo-descumpre-prazo-e-trava-86percent-do-orcamento-para-enfrentar-a-mudanca-climatica.ghtml> accessed 7 October 2024.

39 Inesc (n 31).

40 ibid.

41 Eduardo Viola and Matías Franchini, 'Brazilian Climate Politics 2005–2012: Ambivalence and Paradox' (2014) 5:5 WIREs *Clim Change* 677.

42 *PSB et al v Federal Union*, *Petição Inicial* [Statement of Claim], available at Sabin Center for Climate Change Law, 'Climate Change Litigation' <https://climatecasechart.com/non-us-case/psb-et-al-v-federal-union/> accessed 21 August 2025.

43 Inesc (n 31).

climate finance into national public debate. The near-paralysis of key financial instruments—including the Climate Fund and the Amazon Fund (providing funds for reducing emissions from deforestation and forest degradation)—occurred within the broader context of Bolsonaro's systematic dismantling of environmental and climate policies. The Bolsonaro administration froze resource allocations to these funds, delayed the approval of new projects, cut the disbursement of existing resources and failed to submit mandated annual plans to congressional oversight.[44] The disbursement of Climate Fund resources sharply declined and ultimately stopped.

The Bolsonaro administration also altered the Climate Fund's governance structure, weakening mechanisms for social participation and effective management.[45] The Fund's founding law established a Steering Committee (Comitê Gestor) as the coordinating body,[46] composed of six federal government representatives and five from non-governmental sectors, with details defined by implementing regulations. These regulations allocated civil society representation to the scientific community, an NGO appointed by the Brazilian Forum on Climate Change, rural workers and traditional communities, urban labor sectors and state or municipal governments.

The Climate Fund's governance became entirely inoperative throughout 2019, when the Bolsonaro administration suspended appointments to the Steering Committee, pending regulatory reforms. These reforms ultimately modified the committee structure, eliminating broader civil society representation in favor of institutional and sectoral actors. Under the revised rules, private sector representation was reallocated to delegates from six industry confederations: the National Confederation of Industry (CNI), the National Confederation of Commerce (CNC), the National Confederation of Services (CNS), the National Confederation of Agriculture (CNA), the National Confederation of Transportation (CNT) and the Brazilian Forum on Climate Change. This restructuring significantly altered the original balance between governmental, civil society and economic stakeholders in the Fund's decision-making processes.

---

[44] Roberto Kaz, 'Fundo Amazônia está de volta após quatro anos parado' (*Observatório do Clima*, 15 February 2023) <www.oc.eco.br/fundo-amazonia-esta-de-volta-apos-quatro-anos-parado/> accessed 7 October 2024.

[45] Luiz E O C Aragão, Celso H L Silva Junior and Liana O Anderson, 'Brazil's challenge to restrain deforestation and fires in the Amazon during Covid-19 pandemic in 2020: Environmental, social implications and their governance' (2020) <https://site-antigo.socioambiental.org/sites/blog.socioambiental.org/files/nsa/arquivos/dematamento_fogo_covid19_aragao_et_al.pdf> accessed 3 January 2025.

[46] Ministerio do Planejamento e Orcamento (n 35).

The Bolsonaro administration also interfered with the Fund's priorities. While the Climate Fund Law (Article 5, §4) clearly enumerates eligible activities, Bolsonaro's implementing regulations added a broad list of priority areas aligned with his administration's political agenda. This marked a significant departure from earlier regulations, which simply mirrored the statutory language. By unilaterally redefining priorities beyond those authorized by Congress, the decree arguably distorted legislative intent and reoriented the Fund's strategic focus.[47]

Experts noted that the weakening of the Climate Fund and suspension of disbursements (mirrored in the Amazon Fund) significantly hindered the implementation of key mitigation and adaptation policies. This occurred alongside the broader dismantling of environmental agencies and reduced enforcement, which facilitated illegal environmental practices. The erosion of climate and environmental policy contributed to alarming surges in deforestation and forest fires in the Amazon, with associated emissions, undermining Brazil's reduction efforts across multiple sectors.[48]

Amid this backdrop of environmental policy inertia and regression, a group of opposition parties filed a constitutional challenge in June 2020 against the government's failure to implement the Climate Fund. They invoked key constitutional mechanisms—such as direct review of governmental omissions and the right to an ecologically balanced environment under Article 225.[49] A separate challenge targeted the parallel dismantling of the Amazon Fund.[50]

In July 2022, the Supreme Court ruled that the federal government's obstruction of the Climate Fund was unconstitutional. The decision was issued alongside a similar judgment on the Amazon Fund in 2023.[51] Following Bolsonaro's defeat in the October 2022 election, President Luís Inácio Lula da Silva (Lula) complied with the rulings, resumed the Climate Fund's operations, expanded its oversight committee and restored its original mandate.

We discuss some of the relevant aspects of the process and decision of this landmark case next.

---

47 Statement of Claim (n 42).

48 Murilo Pajolla, 'Com Bolsonaro, desmatamento na Amazônia cresce 150%; pior marca já registrada pelo Imazon' *Brasil de Fato* (Lábrea, 20 January 2023) <www.brasildefato.com.br/2023/01/20/com-bolsonaro-desmatamento-na-amazonia-cresce-150-pior-marca-ja-registrada-pelo-imazon> accessed 7 October 2024.

49 Caio Borges, 'STF realiza audiência histórica' (*climainfo*, 23 September 2020) <https://climainfo.org.br/2020/09/23/adpf-708-fundo-clima-e-politicas-ambientais/> accessed 4 October 2024.

50 Moreira (n 12).

51 ibid.

4. *PSB V BRAZIL (CLIMATE FUND)* LAWSUIT: BRAZIL'S CONSTITUTIONAL APPROACH TO CLIMATE ACCOUNTABILITY

In June 2020, a coalition of opposition political parties, including the PSB, the Socialism and Freedom Party (PSOL for its Portuguese acronym), the Workers' Party (PT for its Portuguese acronym) and the Sustainability Network Party (REDE for its Portuguese acronym), filed a constitutional challenge before Brazil's Supreme Court, arguing that the federal government's failure to operationalize the Climate Fund breached Article 225 of the Constitution, which guarantees the right to an ecologically balanced environment, and undermined both national climate policy and Brazil's international obligations under the Paris Agreement. The plaintiffs were successful in securing an order from the Supreme Court compelling the state to fulfill its constitutional duty to protect the environment by guaranteeing the regular and transparent allocation of Climate Fund resources to address climate change, as established in the Fund's legislation.[52] In this section we will discuss key aspects of this landmark lawsuit and decision to better understand the elements of Brazil's constitutionalism that we believe facilitated this outcome.

We will spotlight two aspects of the *Climate Fund* case, one substantive, the other procedural, that help explain the case's relevance to the growing global jurisprudence on climate litigation. The substantive feature is a particular type of constitutional review instrument in Brazilian constitutional law—the Arguição de Descumprimento de Preceito Fundamental (ADPF, or 'Allegation of Violation of Fundamental Principle')—which grants standing to a list of actors, including political parties, to petition the Supreme Court directly when the executive branch fails to implement laws passed by Congress with the aim to enforce fundamental rights protected under the Constitution. The ADPF permits the Supreme Court to declare governmental actions or omissions unconstitutional if they obstruct the legislative intent to give effect to constitutional guarantees. This constitutional review instrument gives a clear mandate to the Supreme Court to review cases related to positive obligations of governments in the context of violations of constitutional rights.

The second feature is a procedural innovation: the Supreme Court's authority to conduct broad public hearings in cases of major social importance. This represents a notable departure from conventional practice, as most often constitutional courts focus exclusively on legal–constitutional arguments in abstract constitutional review. By creating this exceptional forum for technical

---

[52] Jolene Lin and Jacqueline Peel, *Litigating Climate Change in the Global South* (Oxford University Press 2024).

and societal input, the Court bridges the gap between constitutional principles and ground-level realities in socially challenging cases, such as those related to climate change. The public hearings can help to galvanize attention to the broader societal impacts of legal actions or inactions.

4.1 Political Parties Challenge Climate Inaction at the Apex Court

Unlike jurisdictions that only allow constitutional questions to reach the apex court in concrete disputes, the Brazilian system allows for abstract constitutional review directly before the Supreme Court in cases concerning potential conflicts between federal or state laws (or their implementation) and the Constitution. The list of actors with standing to bring these abstract constitutional challenges includes the heads of the executive power at the federal and state levels (President or governors), the federal Attorney General, the legislative assembly of states, the Brazilian Bar Association, the Trade Union Confederation or class entities, and political parties.[53] This was a significant expansion, as in previous constitutions only the Attorney General had standing to challenge the constitutionality of laws and policies before the Supreme Court.[54]

Any political party with representation in the National Congress, interpreted as a party that holds at least one elected and currently serving member in either the Chamber of Deputies or the Federal Senate, has standing to file a constitutional challenge before the Supreme Court. This institutional arrangement highlights Brazil's distinctive approach to constitutional adjudication, which seeks to integrate democratic representation with judicial review. Some criticize this standing for allowing even minimally representative parties to challenge the constitutionality of laws passed by a legislative majority.[55] Nevertheless, the rule has consistently been interpreted according to its literal meaning.[56] This role of political parties in constitutional review reflects the political history of the country. By expanding standing to a broader range of actors, including political parties with single representation, the intention of the Constitution drafters was to democratize access to constitutional oversight and strengthen the Supreme Court's ability to fulfill its central mission: to

---

[53] Constituição Federal [CF] 88 art 103 (Braz).

[54] Oscar Vilhena Vieira, 'Supremocracia' (2008) 4:2 *Revista Direito GV* 441.

[55] A legitimidade ativa para propor ações de controle concentrado de constitucionalidade <https://www.jusbrasil.com.br/artigos/a-legitimidade-ativa-para-propor-acoes-de-controle-concentrado-de-constitucionalidade/709169328?utm_source=chatgpt.com> accessed 12 June 2025.

[56] ibid.

uphold and protect the Federal Constitution. This innovation does, however, raise risks of politicization of constitutional review in the country.[57]

Political parties have standing to file four types of constitutional review.[58] Although a detailed discussion of these instruments is beyond the scope of this chapter, a brief explanation is warranted.[59] A *Declaratory Action of Constitutionality* (ADC for its Portuguese acronym) seeks Supreme Court confirmation that a law *complies* with the Constitution; it is often used by governments to resolve legal uncertainty and pre-empt judicial challenges. The *Direct Action of Unconstitutionality* (ADI for its Portuguese acronym), the most common, seeks to invalidate laws or government acts that allegedly violate constitutional provisions. These two types of constitutional challenges are common in comparative constitutionalism. Yet Brazil's constitutional framers, acutely aware of the country's historical pattern of enacting paper norms or laws that languished unimplemented, deliberately empowered the Supreme Court to combat unconstitutional omissions by including two less common types of constitutional review instruments. The *Direct Action of Unconstitutionality by Omission* (ADO for its Portuguese acronym) and the *Allegation of Violation of Fundamental 'Precept'* (ADPF for its Portuguese acronym) enable the Supreme Court to review instances where either the executive fails to implement constitutional mandates or the legislature neglects to pass required enabling legislation to uphold fundamental principles at the core of the constitutional framework.

According to Oscar Vilhena, the Constitution did not merely represent the establishment of robust judicial protection of constitutional norms, aimed at securing the democratic transition pact enshrined in an ambitious constitutional text; it was also explicitly tasked with leading a process of social transformation. Judges would now bear responsibilities not only as *negative legislators*, in Kelsen's formulation but also as guardians of the Constitution's positive promises—including the guarantee of a broad range of fundamental rights.[60] This is the context in which the Constitution incorporated these two new instruments of direct constitutional review. By explicitly including the ADO and ADPF, the constitutional framers put to rest debates on positive

[57] Vieira (n 54). See also Sabino Fleury, 'Judicialização da política e politização da justiça: Os partidos políticos e o controle da constitucionalidade das leis no Brasil' (2010) 12:18 *Cadernos da Escola do Legislativo* 5–46.

[58] Keith S Rosenn, 'Procedural Protection of Constitutional Rights in Brazil' (2011) 59:4 *Am J Comp L* 1009.

[59] For a detailed discussion on these instruments see Luis Roberto Barroso, *O Controle de Constitucionalidade No Direito Brasileiro* (9th edn, Saraiva Jur 2022).

[60] Vieira (n 54).

versus negative obligations related to constitutional rights in Brazil. This distinguishes Brazil from other jurisdictions—such as Canada—where rights-based climate litigation may still face significant challenges due to ongoing debates over the scope of positive obligations, as discussed in Chapters 2 and 9 of this book.

Various climate lawsuits in Brazil rely on these four constitutional review instruments, challenging either laws, acts or omissions that violate Article 225's right to an ecologically balanced environment. Moreira et al analyzed 17 of these cases, filed between 2019 and 2022 (during the Bolsonaro administration's dismantling of environmental and climate laws), in which climate change was directly addressed.[61] Fifteen of these 17 cases had been filed by political parties, perhaps due to their financial and institutional capacity to overcome barriers to constitutional litigation. The *Climate Fund* case, which is an ADPF, is one of these cases.

ADPF is a versatile legal instrument for safeguarding constitutional principles and addressing gaps or abuses in governance. Not only can it address conflicts where laws or normative acts violate the Brazilian Constitution,[62] it is also applicable in cases of unconstitutional omissions or unlawful actions by public authorities that infringe upon fundamental rights. The ADPF is subsidiary in nature, meaning it can only be used when none of the other three mechanisms are adequate to address the constitutional issue at hand. This ensures that the ADPF functions as a complementary tool designed to fill gaps in the constitutional review system and provide recourse for acts or omissions that threaten fundamental constitutional principles.

The enactment of Law No. 9,882 in 1999, which established the procedural framework necessary for the application of ADPFs, led to its more frequent use since then. Nevertheless, this Law does not clearly define critical concepts such as 'act of the Public Power' or 'fundamental precept', leaving room for interpretive uncertainty. Legal scholars have since been engaged in clarifying these terms. 'Fundamental precept' generally refers to core constitutional norms and principles of special significance, including the *cláusulas pétreas* (unalterable clauses) and related provisions. Despite its increasing relevance, this constitutional review mechanism continues to face interpretative and practical challenges.

The unalterable clauses are outlined in Article 60, § 4° of the Constitution. They are provisions that safeguard core aspects of the constitutional framework not subject to constitutional amendment, including the country's federative

---

61 Moreira (n 12).

62 Gilmar Mendes and Paulo Gustavo Gonet Branco, *Curso de Direito Constitucional* (9th edn, Saraiva 2014) 1137.

structure; the separation of powers; the system of direct, secret, universal and periodic suffrage; and the rule of law. Importantly, fundamental rights and guarantees, including the right to an ecologically balanced environment, are among the constitutional clauses not subject to amendments. Even if a supermajority in Brazilian Congress would meet the stringent constitutional amendment formula, it could not abolish this fundamental right.

It is important to note that the case was initially filed as an Action of Unconstitutionality by Omission (ADO), aiming to address the failure of the government to implement necessary measures regarding the management and use of the Climate Fund. Justice Luís Roberto Barroso, who presided on the case, recognized that the government's actions included not only omissions (failure to act) but also commissions (acts taken that violated core constitutional norms). In the decision that reclassified the action as an ADPF, Justice Barroso explained

> [W]here a heterogeneous set of commissive and omissive acts harmful to the Constitution is pointed out [...] the jurisprudence [...] has tended to admit the case as an [...] ADPF. This occurs because the ADPF allows for greater heterogeneity regarding its object, as well as greater flexibility in the normative and/or concrete measures that can be granted.[63]

The central arguments in the *Climate Fund* case are that the federal government's failure to properly allocate and apply the Fund's resources, an omission, constitutes a constitutional violation. The alleged violations, however, also include actions such as the interference with the legally mandated governance structure of the Fund. Both actions and omissions would constitute infringements of the right to an ecologically balanced environment. The failure to transfer funds to BNDES (one manager of the Climate Fund), for example, and the halting of new initiatives that depended on these investments, were pointed out as clear examples of unconstitutional omissions. This, according to the plaintiffs, compromised not only the implementation of environmental policies to address climate change but also Brazil's ability to fulfill its international commitments under the Paris Agreement.

The plaintiffs underscored the essential role of the Climate Fund in Brazil's broader environmental policy. In order to argue that the paralyzation of the fund breached the right to an ecologically balanced environment, they alleged that the operation of the Climate Fund helps the state combat climate change thus a) fulfilling the duty to preserve and restore ecological processes and

63 STF, ADO 60 DF, Relator: Min Roberto Barroso, 28.06.2020, p 5, para 7 (free translation by authors from original in Portuguese). Available at: <https://litigancia.biobd.inf.puc-rio.br/visualizacao_caso/443/0/> acessed 12 June 2025.

promote the ecological management of ecosystems; b) defining territorial spaces and components to be specially protected, c) controlling the production, marketing and use of techniques, methods and substances that pose a risk to life, quality of life and the environment, and d) protecting fauna and flora.[64]

Other arguments included were the state's failure to act represented clear setbacks in environmental law regulations, violating the principle of the non-regression in matters of fundamental rights; and violating the principle of intergenerational solidarity, which provides for the responsibility to preserve the environment for future generations.[65] The federal government's lack of action was also seen as violating the principle of sustainable development, by failing to promote policies that balance economic growth, social justice and environmental protection.[66]

As remedies, the plaintiffs requested the Supreme Court issue a declaration that the federal government's actions and omissions in the management of the Climate Fund were unconstitutional. The plaintiffs also asked the Court to order the government to resume the Fund's activities, restore governance structure and adopt permanent measures to ensure the continuous allocation of the Climate Fund's resources, guaranteeing their execution in a transparent and effective manner.[67]

4.2 The Process: Expert Testimony and Social Participation in Constitutional Hearings

On 21 and 22 September 2020, the Supreme Court held a public hearing on the *Climate Fund* case. Justice Luís Roberto Barroso, the designated rapporteur for the case, requested and chaired this hearing. It is worth clarifying the role of the rapporteur in Brazil's Supreme Court proceedings, as each jurisdiction has developed its own procedural traditions. In Brazil, Supreme Court decisions are shaped by the prominent role of the *Ministro Relator* (Reporting Justice, or rapporteur). Assigned at the outset, the rapporteur leads the procedural development of the case, compiles the record and drafts the initial opinion, which typically serves as the starting point for the Court's deliberations (with other justices either following, concurring or dissenting).[68]

---

64 Statement of claim (n 42).

65 ibid.

66 ibid.

67 ibid.

68 Théo G de Sá-Kaye, 'Examining the Brazilian Supreme Federal Court's Expanded Powers in the Bolsonaro Era: A Win for Democracy or a Turn Toward Autocracy?' (2025) 56:2 *University of Miami Inter-American Law Review* 320, at 334.

While abstract constitutional review cases typically rely on legal arguments and do not involve the introduction of factual evidence, the Supreme Court has the discretion to convene public hearings that include expert testimony when complex scientific, social and legal issues are involved, and expert input is considered essential. This procedural innovation is especially relevant in cases of direct constitutional review such as the *Climate Fund* case and reflects another distinctive feature of Brazil's climate constitutionalism. By allowing for the integration of expert testimony into constitutional adjudication on climate matters, this procedural feature positions Brazil as a notable example of how higher courts can open space for expert and social participation in constitutional climate litigation.

The rapporteur convened the hearings to gain a clearer understanding of the technical dimensions involved in managing and disbursing the Climate Fund, particularly how these processes impact Brazil's capacity to combat climate change and fulfill its international climate commitments. The Supreme Court heard from 66 experts, from scientists, economists and academics, to Indigenous peoples and environmentalists, including representatives from economic sectors like agrifood and finance, from government representatives and from international experts like the UN Special Rapporteur on Human Rights and the Environment (arguing that the right to an ecologically balanced environment must include protection of climate stability as an essential component of human dignity).[69] Justice Barroso referenced relevant precedents from the Inter-American Court of Human Rights, such as Advisory Opinion No. 23/2017, which recognizes the right to a healthy environment as fundamental to human existence, and the *Lhaka Honhat v Argentina* case, which emphasizes the state's duty to prevent environmental harm and ensure food and water security.[70]

The September hearings marked a significant milestone in the national debate on the erosion of environmental and climate protections under the Bolsonaro administration. They brought climate change policy, particularly climate finance, to the forefront of Brazil's political and public discourse.[71]

---

69 Joana Setzer, 'First climate case reaches Brazil's Supreme Court' (Grantham Research Institute at LSE, 30 September 2020) <www.lse.ac.uk/granthaminstitute/news/first-climate-case-reaches-brazils-supreme-court/> accessed 23 May 2025.

70 Julia Neiva, 'A Indissociável Relação entre Direitos Humanos e a Crise Climática' in Caio Borges and Pedro Henrique Vasques (eds), *STF e as Mudanças Climáticas: contribuições para o debate sobre o Fundo Clima (ADPF 708)* (Telha 2022) 101.

71 The *Climate Fund* case (n 6). These hearings brought plurality of perspectives to the debate and reinforced the legitimacy of judicial decisions in complex and cross-sectoral issues like climate change.

Scientists offered critical insights into the impacts of climate change, warning of irreversible tipping points.[72] Central to the discussions was the Amazon Forest's vital role in climate regulation.[73] Experts warned that deforestation threatens to trigger irreversible damage to the biome, disrupting agriculture and rainfall across South America.[74] There was broad agreement that escalating deforestation and fires in the Amazon and Pantanal are deepening the environmental crisis, with serious consequences for biodiversity, local populations and regions beyond Brazil's borders.[75] Experts also emphasized Brazil's obligation to meet its international climate commitments and the global repercussions of failing to do so.[76]

The Brazilian government also presented its defence during the hearing, asserting they were fulfilling climate obligations and highlighting measures taken to reduce deforestation and carbon emissions.[77] These claims were, however, widely questioned by experts, who were able to present data contradicting the government's position and arguing that current public policies were falling short of the seriousness of the crisis.[78]

The hearings were broadcast live and also recorded, with the videos available for a broad audience on the Supreme Court's YouTube Channel.[79] Ultimately, the Supreme Court's public hearing became a democratic space for dialogue on the urgency of the climate challenge and the importance of protecting environmental law and policy against setbacks.[80] The broad participation of experts

---

72 Maurício Guetta, 'Notas sobre o Caso do Fundo Clima no Contexto da Litigância Climática perante o Supremo Tribunal Federal' in Borges and Vasques (n 70) 59.

73 ibid.

74 Julia Neiva, 'A Indissociável Relação entre Direitos Humanos e a Crise Climática' in Borges and Vasques (n 70) 96.

75 Maria Laura Canineu, 'A Crise Climático-Ambiental no Brasil sob a Perspectiva dos Direitos Humanos' in Borges and Vasques (n 70) 88.

76 João Paulo R Capobianco et al, 'Aumento do Desmatamento na Amazônia nos Anos de 2019 e 2020 como Consequência da Omissão na Implementação dos Instrumentos Legais e Recursos Disponíveis' in Borges and Vasques (n 70) 200.

77 Sergio Margulis, 'Aquecimento Global, Mudanças Climáticas e a sua Proximidade em Relação aos Tipping Points' in Borges and Vasques (n 70) 255.

78 Ricardo Abramovay, 'Fundamentalismo Sectário Explica o Pacto com a Ilegalidade' in Borges and Vasques (n 70) 157; Maria Laura Canineu, 'A Crise Climático-Ambiental no Brasil sob a Perspectiva dos Direitos Humanos' in Borges and Vasques (n 70) 94.

79 STF, 'Audiências Públicas do STF – Fundo do Clima (1/4)' <www.youtube.com/watch?v=AVXETmIp9KA>.

80 Ricardo M O Galvão, 'A Questão do Aquecimento Global e Preservação da Amazônia' in Borges and Vasques (n 70) 241.

and civil society reinforced the urgency of delivering concrete responses to the climate crisis—both nationally and globally—to align Brazil with its international commitments to combat climate change and to sustainable development. Particularly, it highlighted the crucial role of finance in addressing climate change.[81]

4.3 Decision: From Discretion to Obligation in Climate Finance

In July 2022, the Supreme Court decided in favor of the political parties. Nine of ten justices followed the rapporteur's vote, with only one dissent and one concurrence with reservations. The Court held that the Executive Branch has a constitutional duty to operationalize and disburse resources from the Climate Fund, prohibiting withholding funds for contingencies, based on a) the duty to protect an ecologically balanced environment, b) Brazil's international obligations and c) the principle of separation of powers.

The decision broke new ground by affirming the supralegal status of the Paris Agreement within Brazil's legal system, aligning international climate treaties with international human rights law and domestic constitutionally protected rights and principles. It also innovated by recognizing the principle of non-regression in environmental protection as a judicially enforceable safeguard. These features have already attracted significant scholarly attention elsewhere;[82] therefore we will just briefly discuss them here. In this section, we seek to highlight a less examined but crucial innovation: the Supreme Court's decision to declare that disbursement of climate finance from a fund created by law constitutes a constitutional obligation, rather than being an area of executive/administrative discretion. This interpretation connects the obligation to disburse climate funds to fundamental constitutional principles, reinforcing Brazil's compliance with its international climate commitments and safeguarding rights that are essential for both present and future generations. As the recognition of the key role of climate finance in meeting domestic and global climate commitments grows, the importance of this decision cannot be overstated.

The *Climate Fund* case represents a landmark in global climate litigation, marking the first time any court in the world has treated the Paris Agreement as having the same normative status as an international human rights treaty. The Supreme Court affirmed that some international environmental and climate agreements with human rights implications may hold supralegal status

---

[81] Ingo Wolfgang Sarlet and Tiago Fensterseifer, 'Direito Fundamental ao Clima Estável, Litigância Climática e ADPF n. 708/DF' in Borges and Vasques, (n 70) 338.

[82] Tigre and Setzer (n 11).

in the Brazilian legal system, something that only international human rights treaties had before. In Brazilian constitutional law, supralegal status means that international human rights treaties signed by the Executive and ratified by Congress can automatically override conflicting domestic laws, ensuring their enforceability without requiring domestication through passing of specific laws. To be clear, this supralegal effect is distinct from constitutional status. Treaties can only attain equivalence with constitutional norms if they are approved by a qualified majority of three-fifths in both houses of Congress (the Chamber of Deputies and the Senate), at which point they are treated as constitutional amendments. This distinction is crucial for understanding how Brazil integrates international obligations into its legal system while preserving the hierarchical structure of domestic law.

What the decision means is that once the Paris Agreement was signed and ratified, Brazilian laws automatically had to conform with this international treaty. Justice Barroso stated that environmental law treaties like the Paris Agreement may be considered as a subset of international human rights treaties for that matter:

> the Constitution recognizes the supralegal character of the international treaties on human rights to which Brazil is a party [...] [a]nd there is no doubt that the environmental issue fits the hypothesis. As the representative of UNEP in Brazil, during the public hearing, clearly stated: 'There are no human rights on a dead or sick planet' (p. 171). Treaties on environmental law are a species of the genus human rights treaties and enjoy, for this reason, supranational status. Thus, there is no legally valid option of simply omitting to combat climate change.[83]

By elevating the Paris Agreement to this supralegal level, the Court reinforced its binding nature within the Brazilian legal framework, emphasizing its role in shaping national climate obligations and reinforcing the idea that climate change impacts human rights. Scholars have highlighted this interpretation as a critical step in integrating international climate commitments into domestic law, ensuring stronger enforcement mechanisms for achieving climate goals.[84]

The *Climate Fund* decision also reinforced the principle that setbacks in human rights legal protections are unacceptable within the Brazilian constitutional framework, emphasizing that it is the role of constitutional courts to prevent such regressions. This principle, known as the 'prohibition of social setbacks' has played a crucial role in safeguarding rights, including environmental rights in Brazil. The principle seeks to ensure that laws created to

---

[83] Amazon Fund case (n 9) at page 9.

[84] Délton Winter De Carvalho and Rafaela Santos Martins da Rosa, 'Climate Constitutionalism as a Foundation for Climate Litigation in Latin America' (2024) 16:1 *Journal of Human Rights Practice* 71.

uphold rights guaranteed by the Constitution cannot be later suppressed or weakened by subsequent laws or administrative actions passed by legislative majorities or by powerful executives. Although the prohibition of regression is not expressly guaranteed in the Constitution, it is widely accepted by legal experts and upheld by high courts, including the Supreme Court. This principle draws its foundations from several constitutional guarantees, such as the dignity of the human person, legal security, the highest effectiveness of constitutional rules and the Social State, all of which reinforce the protection of fundamental rights.[85]

The *Climate Fund* case also broke new ground as it was the first time any court in the world has clearly affirmed a constitutional duty to operationalize a climate fund and guarantee the regular disbursements of its resources for relevant climate action. It reinforces the judiciary's role in compelling governmental accountability for lack of appropriate climate action, this time related to climate finance, adding to a number of other successful climate litigation cases that promote higher climate ambition through law.[86]

On this issue, the Rapporteur Justice Barroso rejected a preliminary motion to dismiss brought by the federal government, who argued that the case did not deal with a matter that constituted a direct offense against the constitutional text but rather engaged administrative matters of allocation of resources, which are clearly a question of administrative discretion. According to Justice Barroso, the plaintiffs were not questioning the merit of administrative decisions to allocate funds to specific projects but were rather questioning actions and mainly omissions that managed to dismantle the governance and halt the functioning of the Climate Fund, resulting in withholding fund resources in 2019 and part of 2020.

Another argument in the motion to dismiss that was rejected related to the alleged lack of constitutional grounds: the government had argued that the plaintiffs were challenging administrative acts and decisions, which are regulated by infra-constitutional law, and that the ADPF was not a suitable instrument. Justice Barroso ruled that the ground for action was the constitutionally protected right to an ecologically balanced environment, and any infra-constitutional law or governmental act—or omission—arguably infringing such a right could be challenged via this instrument.

Justice Barroso emphasized that the case engaged a constitutional law question due to its close relation with environmental protection. He referred to Article 225 of the Brazilian Constitution as imposing a duty on the government

---

85 Ingo Wolfgang Sarlet, *A Eficácia dos Direitos Fundamentais: Uma Teoria Geral dos Direitos Fundamentais na Perspectiva Constitucional* (11th edn, Livraria do Advogado 2012) 407.

86 For example, *Urgenda v Netherlands* [2015] HAZA C/09/00456689.

to protect, preserve and restore the environment for the present and future generations, and that duty extends to preserving a stable climate. Justice Barroso underscored that environmental protection is therefore not a matter of political discretion; rather, it constitutes a binding obligation on executive authorities, including the President, which in Brazil is the Chief of the Executive Power. On this broader interpretation of Article 225, Tiago and Ingo note:

> The recognition of a new climate dimension inherent to [. . .] art. 225 of the CF/1988 gives rise to the characterization of specific duties of protection and promotion, including those of an organizational and procedural nature, with regard to combating, containing and reducing the causes and consequences of climate change, implying, in the case of non-compliance by action and/or omission (general and partial), the possibility of judicial control (in addition to the indispensable and permanent social control) and, in this context, operating as a material parameter for the application of the principle of the prohibition of climate retrogression.[87]

This decision underscores the constitutional significance of environmental protection but invites reflection on the scope of this obligation, particularly when weighed against other constitutional objectives competing for limited resources, such as the rights to health and education. In this case, the Supreme Court's reasoning relied on the pre-existing legal framework of the Climate Fund, where funds that had been allocated by law were available but the government decided not to disburse them. Thus, the decision did not impose new obligations with financial impacts on the legislature or executive, but reinforced compliance with established laws, minimizing discretionary balancing. By enforcing these frameworks, the Court delineated a baseline for action, aligning with the prohibition of climate retrogression and ensuring adherence to constitutional and international commitments.

The Court underscored that the principle of separation of powers limits the Executive's discretion vis-à-vis clear legislative mandates established in laws, particularly when constitutional rights are at stake. Scholars have noted that this principle, as applied in this case, reinforces the judiciary's role in ensuring the allocation of resources aligns with constitutional and international climate obligations. By prohibiting withholding Climate Fund resources for contingencies, the STF established clear limits to Executive inaction in environmental governance.

Justice Barroso accepted the plaintiffs' request to order the federal administration to resume the operation of the Climate Fund in subsequent years. He also ruled that the federal administration is prohibited from refusing to

---

[87] Ingo Wolfgang Sarlet and Tiago Fensterseifer, 'Direito Fundamental ao Clima Estável, Litigância Climática e ADPF n. 708/DF' in Borges and Vasques (n 70) 332.

disburse the Fund's allocated resources to climate projects, emphasizing that such legal obligations—requiring the application of resources for specific activities—are the result of complex legislative processes, and cannot be unilaterally disregarded by the Executive. Justice Barroso noted that failure to comply with these obligations established by the Legislature constitutes a violation of the principle of separation of powers, enshrined in the Brazilian Constitution. In other words, Justice Luís Roberto Barroso reversed a traditional government defense based on the principle of separation of powers, which argues that courts should not interfere in polycentric matters better left to political powers (legislative and executive). Instead, he invoked the same principle to assert that courts have a duty to uphold valid legislative objectives grounded in constitutional norms, when decisions by the Executive attempt to frustrate or undermine them.

Furthermore, Justice Barroso emphasized that the obligation to channel resources from the Climate Fund to climate projects is not only a constitutional duty but also supported by Article 9, § 2, of the Fiscal Responsibility Law (LRF for its Portugues acronym). He highlighted that this legislation prohibits the limitation of expenses related to constitutional obligations, such as the Climate Fund. In this context, the LRF was used to reinforce that revenues earmarked for specific purposes—especially those aimed at environmental protection and combating climate change—cannot be withheld for contingencies. Therefore, Justice Barroso stressed that by withholding Climate Fund resources for contingencies to achieve fiscal goals, even before the Bolsonaro administration, the Executive violated both constitutional and legal norms, justifying the judiciary's intervention to correct the omission.[88]

One of the ten Justices in the Supreme Court, Justice Fachin, followed the Rapporteur's vote with reservations. In his vote, he highlighted the importance of addressing what he termed a 'state of unconstitutional affairs', emphasizing that the situation involved a grave, systematic and continuous violation of fundamental rights, affecting a large and indeterminate number of people. He also accepted all of the pleas made by the plaintiffs, however suggested that the Court should also require the federal administration to a) publish quarterly statistical reports to be prepared by the Brazilian Institute of Geography and Statistics, a branch of the Science, Technology and Innovation Ministry, giving evidence of the percentages of expenditure of the Climate Fund's resources on five distinct sectors: energy, industry, farming, land use and forestry, and waste; and b) create within a reasonable timeframe the National Inventory on Greenhouse Gas Emissions and Removal, with detailed numbers by State and Municipality, giving it high publicity.

---

[88] LC 101/00.

Such measures, while potentially beneficial in promoting transparency and accountability, may be viewed as exceeding the judiciary's role by delving into detailed administrative decisions. This approach could be criticized for judicial overreach, as it imposes specific actions and priorities that are typically within the executive's discretion.[89] Nevertheless, it highlights a growing trend in environmental litigation where courts take an active stance to ensure compliance with constitutional and legal obligations, particularly in contexts where executive inaction has been alleged.[90]

Only one justice, Nunes Marques, voted to reject the plaintiffs' pleas. Justice Nunes Marques, who had been appointed to the Supreme Court by President Jair Bolsonaro and is generally aligned with more conservative agendas, argued that the Climate Fund is not the only tool available to address climate change and that the executive retains discretion to determine which measures it will use to achieve climate objectives. Furthermore, he emphasized that the allocation and disbursement of resources from the Climate Fund should align with the government's broader priorities and administrative criteria of convenience and opportunity. He concluded that the federal government's actions indicated no unconstitutional action or omission, asserting that rejecting the pleas would respect the principle of separation of powers.

5. A NEW ERA FOR CLIMATE LITIGATION IN BRAZIL?

The *Climate Fund* case shows that the Brazilian constitutional framework and judicial tradition is receptive to bold claims related to government accountability for climate action or inaction that infringe the constitutional right to an ecologically balanced environment, which now encompasses a right to a stable climate. The system of constitutional review of government actions and omissions, created by the framers of the Constitution in a very different context post-dictatorship, and a history of innovative tools like public hearings, is now facilitating government accountability before the Supreme Court in the context of the climate challenge.

The decision holds remarkable potential to open new doors for other successful climate lawsuits in Brazil. By recognizing the Paris Agreement as a supranational treaty and affirming the constitutional right to an ecologically

---

89 Teresa Kramarz, David Cosolo and Alejandro Rossi. 'Judicialization of environmental policy and the crisis of democratic accountability' *Review of Policy Research* 34.1 (2017): 31–49.

90 For example *Mathur v. Ontario*, discussed in Chapter 2, and the Neubauer case against Germany. Neubauer, et al. v. Germany, online <https://climatecasechart.com/non-us-case/neubauer-et-al-v-germany/> accessed 21 August 2021.

balanced environment—including a stable climate—the Supreme Court has set a groundbreaking precedent. This decision strengthens the role of the judiciary in advancing climate action and highlights the importance of international commitments within the Brazilian legal system. Beyond its immediate impact, this ruling could inspire broader legal strategies, encouraging lower courts to adopt climate-conscious approaches. However, Brazil's judiciary is vast and fragmented, shaped by the country's federal structure. Decisions from the STF often fail to filter consistently through first-instance and appellate courts. While the STF sets the tone for legal interpretation, the day-to-day functioning of the judiciary is marked by uneven application of its rulings. This systemic complexity creates barriers to translating landmark decisions into widespread judicial practice.

At the same time, the judiciary cannot drive climate progress in isolation. Brazil's separation and balance of powers demands collaboration between the executive, legislative and judicial branches. Yet this cooperation is often absent. While the STF may push for bold advancements, the executive and legislative branches frequently fail to align with the same urgency or vision. This disconnect undermines Brazil's ability to implement effective climate policies, even when judicial milestones are achieved.

The challenges are evident even within the Supreme Court itself. Justice Edson Fachin, for example, proposed an annual review of how the government allocates and uses resources from the Climate Fund—a measure that could have introduced stronger accountability mechanisms. However, this idea did not gain full support among the other justices. Such internal disagreements reflect the broader limits of judicial influence in addressing systemic issues and underscore the need for more cohesive action across all branches of government.

The *Climate Fund* case represents a transformative moment in both Brazilian constitutional law and global climate litigation. By affirming the federal government's constitutional duty to operationalize the Climate Fund and disburse its resources effectively, the Supreme Court bridged the gap between national environmental governance and Brazil's international commitments under instruments such as the Paris Agreement. This landmark ruling underscores that climate protection is not a mere policy prerogative, but a constitutional mandate grounded in the right to an ecologically balanced environment, as enshrined in Article 225 of the Brazilian Constitution.

Crucially, the Supreme Court elevated the Paris Agreement to supralegal status, recognizing its binding force within Brazil's legal framework, akin to international human rights treaties. This interpretation not only aligns Brazilian law with global climate goals but also sets a powerful precedent for integrating international obligations into domestic governance. The judgment demonstrates the judiciary's capacity to address governmental inaction, compelling the state to fulfill its constitutional and international responsibilities while protecting the rights of current and future generations. The Supreme Court's articulation of the principle of non-regression in environmental law further strengthens the jurisprudence against setbacks in fundamental rights. This principle establishes a baseline for climate governance, confirming that any reversal in environmental protections or climate commitments is unconstitutional.

By tying climate finance directly to constitutional and human rights, the decision broadens the scope of climate litigation, opening new avenues for judicial intervention in cases of policy stagnation or regression. Moreover, the decision reaffirms the judiciary's critical role in holding the Executive accountable for resource mismanagement, highlighting that budgetary discretion cannot be exercised at the expense of fundamental rights. The Supreme Court's directive for transparent and effective resource allocation from the Climate Fund redefines climate finance as an indispensable tool for achieving Brazil's emissions targets and supporting vulnerable communities, emphasizing its dual role in mitigation and adaptation efforts.

At a global level, the *Climate Fund* case sets a significant precedent by framing climate finance as a constitutional obligation rather than a policy instrument subject to political discretion. This innovative approach could inspire courts worldwide to adopt similar frameworks, reinforcing the intersection of constitutional rights, environmental protections and international climate commitments.

Finally, this decision reflects a growing recognition of the interdependence between environmental preservation, climate balance, human dignity and sustainable development. The ruling not only rectifies past administrative failings but also lays a robust foundation for Brazil's future climate governance, ensuring that the country is equipped to confront the escalating challenges of the climate crisis with legal and institutional coherence.

8. *Held v Montana*: climate science takes the stand

Jarryd Page and Helene Krasnoff

1. INTRODUCTION

In March 2020, with global attention focused not on the climate crisis but on another pressing issue of international concern—the spread of COVID-19—16 Montana youth filed a constitutional challenge in Montana state court against the state, the Governor, and several state agencies. That filing would lead to one of the most well-known climate decisions in the world, *Held v Montana*.[1]

The youth, led by fifth-generation Montanan Rikki Held, alleged that the State Energy Policy Act and a provision of the Montana Environmental Policy Act (MEPA), as well as the way the defendants were implementing those laws, violated their inalienable rights under the Montana Constitution, including the express right to a 'clean and healthful environment'. The case centered on a MEPA provision that prohibited state officials from considering greenhouse gas (GHG) emissions or climate change impacts when conducting environmental reviews (the MEPA Limitation), including for reviews associated with permits related to fossil fuels and fossil fuel infrastructure.

After extensive discovery, motions, requests for interlocutory relief, and multiple attempts by the state legislature to derail the case, the Montana District Court in the capital city of Helena held a seven-day trial from June 12–20, 2023. It was the first youth constitutional climate case to go to trial in the United States. The plaintiffs presented testimony from 24 witnesses, including a delegate from Montana's 1972 constitutional convention that adopted the rights provision, the plaintiffs themselves, and multiple climate and health experts. Defendants offered only testimony from two state agency officials and an economist.

On August 14, 2023, the trial court issued a landmark 103-page decision with 289 findings of fact holding unconstitutional both the MEPA Limitation and a

[1] 2024 MT 312, 419 Mont 403 (2024).

related provision that prevented courts from providing equitable remedies for violations of MEPA.[2] The decision declared that the Montana Constitution's enumerated right to a clean and healthful environment 'includes climate as part of the environmental life-support system'.[3] The decision was remarkable for its extensive discussion of climate science and recognition of how climate change is impacting Montana and the youth plaintiffs specifically.

On December 18, 2024, in a 6–1 majority decision, the Montana Supreme Court affirmed the lower court.[4] As a result, not only will agencies in Montana be able to consider climate change in environmental reviews, but the state Supreme Court made clear that the state's constitutional promise of a clean and healthful climate is indeed a fundamental protection enjoyed by all Montanans and includes the right to a stable climate system.

2. THE BACKSTORY

Held v Montana did not come out of a vacuum. Rather, it exists within a context of similar youth-led constitutional climate and environmental rights cases across the country and world, many brought by the U.S.-based organization Our Children's Trust.[5] While its factual allegations are similar to many of these other cases, *Held*'s unique legal circumstances set it apart. Montana has one of the nation's strongest environmental rights amendments, added to the state constitution in 1972 shortly after the first Earth Day in 1970. However, Montana law also had a provision that expressly forbade state officials from considering climate change (the MEPA Limitation). Perhaps due to this stark contradiction, the youth plaintiffs' case succeeded where similar cases had failed to date.[6]

---

[2] *Held v Montana* No CDV-2020-307, 2023 WL 5229257 (Mont Dist Ct Aug 14 2023).

[3] ibid [7].

[4] *Held* 2024 MT 312 (n 2).

[5] For a discussion of youth climate cases in the United States and around the world, see Randall S Abate, 'Youth and Indigenous Voices in Climate Justice: Leveraging Best Practices from U.S. and Canadian Legislation' (2022) 45 *Pub Land & Res L Rev* 78.

[6] See e.g., *Sagoonick v State of Alaska II* 503 P3d 777 (Alaska 2022); *Aji P v Washington* 198 Wash 2d 1025 (Wash 2021); *Reynolds v State of Florida* 316 So 3d 813 (Fla Dist Ct App 2021). *Cherniak v Brown* 475 P3d 68 (Or 2021); *Foster v Wash Dept of Ecology* No 75374-6-1 (Wash Ct App, Sept 5, 2017); *Texas Comm'n on Env't Quality v Bonser-Lain* 438 SW3d 887 (Tex App 2014); *Funk v Wolf* 158 A3d 642 (Pa 2013); *Farb v Kansas* No 12-C-1133 (Kan Dist Ct, Jun 4, 2013); *Aronow v Minnesota* No A12-0585, 2012 WL 4476642 (Minn Ct App 2012).

3. SETTING THE STAGE: THE FILING

In March 2020, Rikki Held and 15 other youth plaintiffs filed for declaratory and injunctive relief against the State of Montana, the Governor, and several state agencies.[7] Their complaint alleged that the State Energy Policy Act, the MEPA Limitation, and the defendants' actions taken pursuant to those laws perpetuate a fossil fuel-based energy system in Montana that causes dangerous levels of GHG emissions and violates the plaintiffs' rights guaranteed by several provisions of the Montana Constitution. They also argued that the state's laws and actions violated the Public Trust Doctrine.[8]

While no doubt a climate case, this complaint was firmly rooted in the Montana Constitution. In addition to the broad right to seek safety, health, and happiness;[9] the right to individual dignity and equal protection;[10] and preambular language stating that it is designed for 'this and future generations', plaintiffs relied on two specific provisions of the Montana Constitution related to the environment:

- All persons are born free and have certain inalienable rights. They include the right to a clean and healthful environment …[11]
- The state and each person shall maintain and improve a clean and healthful environment in Montana for present and future generations.[12]

Provisions such as these are usually referred to as 'Environmental Rights Amendments' (ERAs) or 'Green Amendments'. While the United States Constitution has no similar provisions, seven states do,[13] and more than a

---

7 Compl 1, *Held v Montana* 2020 WL 13801049 (Mont Dist Ct Mar 13, 2020) (No CDV-2020-307).

8 Mont Const art IX, § 3.

9 Mont Const art II, §§ 3, 15, 17, art IX, § 1.

10 Mont Const art II, §§ 4, 15.

11 Mont Const art II, § 3.

12 Mont Const art IX, § 1(1).

13 The number of states with ERAs varies depending on how they are counted. See Romany Webb, 'Environmental Rights in State Constitutions' (*Climate Law Blog*, 31 August 2021) <https://blogs.law.columbia.edu/climatechange/2021/08/31/environmental-rights-in-state-constitutions/#:~:text=However%2C%20the%20constitutions%20of%20six,Supreme%20Court%2C%20Robinson%20Township%20v> accessed 5 December 2024 (writing, before New York's ERA, that the 'constitutions of six states do have provisions with explicit environmental rights'); but, there are 'now 3 states with Green Amendments'. 'September 2024 Newsletter' (*Green Amendments for the Generations*, 21 September 2024) <https://forthegenerations.org/blog/2024/09/21/september-2024-newsletter/> accessed 5 December 2024.

dozen states are in various stages of considering their adoption.[14] The ERAs of Pennsylvania, Montana, and Hawai'i are often referenced as the most protective, although the exact language varies from state to state and differs as to things like who can enforce the right, against whom, and for which environmental media (air, water, climate, etc.).[15] New York is the most recent state to add an ERA to its state constitution in 2021,[16] and litigants have already invoked it in litigation.[17]

In addition to the provisions quoted, the Montana Constitution's Declaration of Rights includes a specific provision directed at youth. This provision ensures that '[t]he rights of persons under 18 years of age shall include, but not be limited to, all the fundamental rights of this Article unless specifically precluded by laws which enhance the protection of such persons'.[18]

3.1 The Plaintiffs and Their Counsel

At filing, the 16 *Held* plaintiffs ranged in age from two to 18 years old and resided in different parts of Montana. Sariel S., a member of the Confederated Salish and Kootenai Tribes, lived on the Flathead Indian Reservation; several lived in more urban settings, like Bozeman and Missoula; others were scattered throughout the state.[19]

More than twenty pages of the youths' complaint details specific harms the plaintiffs faced from a changing climate, both because of their age and because of their and their families' circumstances. For example, Taleah H., 16, as with many others, attested to losing many types of recreational opportunities because of being forced indoors more days by wildfire smoke and poor air quality. Ruby, 11, described exacerbated asthma conditions and impacts to her cultural practices as a member of the Crow Tribe of Montana. These

---

[14] Amendments have been proposed in 18 states. Green Amendments for the Generations (n 14).

[15] See Barry E Hill and Jarryd C Page, 'Fundamental Rights', *Climate Science and Law Curriculum for Judges* (January 2023) <https://cjp.eli.org/curriculum/fundamental-rights> accessed 5 December 2024.

[16] See NY Const art I, § 19; see also 'New York's Environmental Rights Repository' (*Pace University Elisabeth Haub School of Law*) <https://nygreen.pace.edu> accessed 25 October 2024.

[17] Michael B Gerrard and Edward McTiernan, 'New York's Green Amendment: The First Decisions' (2023) *NYLJ* <https://scholarship.law.columbia.edu/faculty_scholarship/3906> accessed 28 February 2025.

[18] Mont Const art II, §15.

[19] Compl 1 (n 8) [14]–[81].

included fewer opportunities to pick wild chokecherries, huckleberries, and Oregon grapes, and fewer lodgepole pines to build tipis due to increased wildfire damage.

Other harms were directly economic. Claire, a ski instructor at Big Sky Resort, alleged that a disrupted climate meant less snow and fewer clients, and thus less income. Rikki Held described similar challenges with a decreased snowpack, but perhaps less directly. For the Held family's ranching business, no spring runoff means less water for cattle and reduced profits at market.

The *Held* plaintiffs were represented pro bono by Our Children's Trust, a non-profit public interest law firm based in Oregon, as well as lawyers from the Western Environment Law Center in Helena, Montana, and McGarvey Law, in Kalispell, Montana. Our Children's Trust has brought more than two dozen similar rights-based actions in the United States and globally starting in 2011, when it filed complaints in nine states.[20] Before *Held* was filed, 12 such cases had been dismissed on procedural or legal grounds, including standing and the political question doctrine.[21]

The firm's most well-known case, *Juliana v United States*, was arguably both a precursor and a blueprint for *Held*. A precursor, because Our Children's Trust filed the *Juliana* litigation on behalf of Kelsey Juliana and 20 other youth plaintiffs in federal court in Oregon in 2015. A blueprint, because while *Juliana* was not based on an express ERA, it alleged similar theories about plaintiffs' federal rights to the constitutional protections of a safe and stable climate. *Juliana* was dismissed by the U.S. Court of Appeals for the Ninth Circuit in 2020 on standing grounds, specifically that the court lacked authority to fashion relief that would redress the plaintiffs' injuries.[22]

---

20 Links to legal actions in the United States and globally can be found online at Our Children's Trust. Our Children's Trust <https://www.ourchildrenstrust.org/> accessed 28 September 2024.

21 See *Sagoonick* (n 6); *Layla H v Commonwealth* 81 Va App 116 (Va Ct App 2024); *Reynolds* (n 6); *Natalie R. v State of Utah* 576 P3d 550 (Utah 2025).

22 *Juliana v United States* 947 F3d 1159 (9th Cir 2020). Following the Ninth Circuit's dismissal, the District Court for the District of Oregon allowed the plaintiffs to amend their complaint to include exclusively declaratory relief, but the Ninth Circuit again ordered dismissal. In *re United States v United States District Court for the District of Oregon* No 24-684, 2024 WL 5102489 (9th Cir May 1, 2024). The plaintiffs sought relief from the U.S. Supreme Court, which was denied in November 2024. In *re Kelsey Cascadia Rose Juliana*, No 24-298. Plaintiffs subsequently filed a writ of certiorari in December 2024, which the U.S. Supreme Court denied in March 2025. *Juliana v United States* No 24-645.

3.2 The Defendants

Held was brought against the state government and its officials whose actions were alleged to have contributed to climate change and harmed the plaintiffs. That list included the state, as sovereign trustee over public resources; the Governor, in his executive supervisory capacity; the Department of Environmental Quality, as the primary state environmental regulator; the Department of Natural Resources and Conservation, as manager of state trust lands; the Department of Transportation, responsible for transportation infrastructure and fuel taxes; and the Public Service Commission, the state's regulator of utilities, railroads, and pipelines.

4. THE PATH TO TRIAL

In many ways the case followed a predictable pattern, including evidence-gathering, motions practice, and hearings. In others, the path was anything but predictable, given several attempts by the defendants to have the Montana Supreme Court intervene before trial and the state legislature's repeated efforts to derail or dispose of the case through legislative changes that would alter or eliminate the statutory provisions at issue.

The complaint requested declaratory relief, including declarations that the State Energy Policy Act and the MEPA Limitation are unconstitutional and violated plaintiffs' fundamental rights, and that the Montana Constitution's fundamental right to a clean and healthful environment 'includes a stable climate system that sustains human lives and liberties'.[23] Plaintiffs further requested injunctive relief that included requiring an accounting of Montana's GHG emissions, a remedial plan to reduce emissions 'consistent with the best available science', and retention of court jurisdiction over such a plan.[24]

The case was filed in Lewis and Clark County, where the state capital is located, and assigned to Judge Kathy Seeley, one of four trial judges in the First Judicial District.[25] The court is one of general jurisdiction, hearing a wide variety of cases, including civil and criminal; it does not specialize in any one type of dispute.[26]

---

23 Compl 1 (n 8) [4].

24 ibid [7], plaintiffs requested the appointment of a special master 'with appropriate expertise' if one would be necessary to review the remedial plan for efficacy.

25 Montana Judicial Branch, '1st Judicial District (Broadwater and Lewis & Clark Counties)' <https://courts.mt.gov/CourtLocator/1stJudicialDistrict> accessed 25 October 2024.

26 Montana Judicial Branch, 'District Courts' <https://courts.mt.gov/courts/dcourt/> accessed 25 October 2024.

The state's motion to dismiss, filed in April 2020, presented the trial Court with its first invitation to dispose of the case. The state alleged that the plaintiffs lacked 'case-or-controversy' standing, that they had failed to exhaust their administrative remedies, and that their request for a remedial plan was non-justiciable because it posed a political question.[27] Although plaintiffs opposed that motion one month later,[28] the district court did not rule until August 2021, when it allowed the case to proceed but disallowed most of the plaintiffs' requests for injunctive relief (e.g., the GHG accounting and remedial plan).[29]

In its order, the district court discussed *Juliana* at length, noting that it was dismissing the request for most of the injunctive relief for similar reasons (i.e., 'Such expansive relief presents a political question and exceeds the court's powers').[30] But the Court also explained how the standing analysis in *Held* was different, because of both the legal standard for standing in Montana state courts and the specific Montana statutes at issue. This case, therefore, would be allowed to proceed where others had not.

Ten months later, on June 10, 2022, the state made its first (of several) requests asking the Montana Supreme Court to step in to take supervisory control of the case and to dismiss the plaintiffs' remaining request for injunctive relief.[31] On June 14, the state high court declined to do so, allowing the case to keep moving forward in the district court.[32]

The parties engaged in discovery through January 2023, including dozens of interrogatories, 22 expert reports, 36 depositions, and more than 50,000 pages of documents. Both parties filed motions to exclude certain evidence, some of which were granted and some of which were denied for each, and the state also sought an independent medical examination of eight of the youth plaintiffs, which the Court denied. Following the close of discovery, the state moved for summary judgment in early February, claiming that all of the plaintiffs' claims failed as a matter of law.

---

27 Defs' Mot to Dismiss Under M R Civ P 12(b)(1), 12(b)(6) and 12(h)(3), *Held v Montana*, 2023 WL 13801048 (Mont Dist Ct Apr 24 2020) (No CDV-2020-307).

28 Pls' Resp to Defs' Mot to Dismiss, *Held v Montana* 2023 WL 1997864 (Mont Dist Ct May 29, 2020) (No CDV-2020-307).

29 Ord on Mot to Dismiss, *Held v Montana* 2021 WL 11506414 (Mont Dist Ct Aug 4, 2021) (No CDV-2020-307).

30 ibid 17.

31 Petition for Writ of Supervisory Control, *State of Montana v Mont First Judicial Dist Court* No 22-0315 (Mont Jun 10 2022).

32 *State of Montana v Mont First Judicial Dist Court* 409 Mont 557, 512 P.3d 1178 (2022).

4.1 Legislative Interventions

As the case was heading to trial, state legislators took action directed at derailing the lawsuit. On March 16, 2023, the Governor signed HB 170, which repealed the State Energy Policy Act.[33] Based on those changes, the state claimed the case was now moot and asked the Court to dismiss the claims related to that now-repealed act.[34]

The legislature also passed HB 971, which clarified that the MEPA Limitation prohibited the state from considering GHGs in MEPA decisions. The state moved to dismiss all of the plaintiffs' claims related to the MEPA Limitation based on that change. Shortly after, the Governor signed a law that eliminated equitable remedies for MEPA challengers raising climate change issues,[35] yet another legislative effort clearly intended to circumvent plaintiffs' claims.

The district court ruled on a number of pending motions on May 23, 2023. It granted defendants' motion to partially dismiss the State Energy Policy Act claims, finding that those claims should not go forward given its repeal, citing issues of redressability and prudential standing. However, it denied the state's request for summary judgment and to dismiss the MEPA claims, making clear the case would go to trial the following month.[36]

Unhappy with that result, less than two weeks before trial the state again sought an emergency writ asking the Montana Supreme Court to take control of the case, repeating its arguments from the motion to dismiss and summary judgment motion.[37] Days later, the Supreme Court again declined to intervene.[38] The historic trial would proceed.

---

33 The law repealed the energy policy goals and the energy policy development process.

34 Defs' Mot to Partially Dismiss for Mootness, *Held v Montana* 2023 WL 4136052 (Mont Dist Ct Apr 3, 2023) (No CDV-2020-307).

35 This was enacted by SB 557. Mont Code Ann Sec 75-1-201(6)(a)(ii).

36 Ord on Defs' Mot to Partially Dismiss for Mootness and For Summ J, *Held v Montana* 2023 WL 1997864 (Mont Dist Ct May 23, 2023) (No CDV-2020-307).

37 Emergency Petition for Writ of Supervisory Control and Stay of June 12, 2023 Trial, *State of Montana v First Judicial Dist Court* No 23-0311 (Mont Jun 5 2023).

38 *State v Mont First Judicial Dist Court* 531 P.3d 546 (2023).

5. THE TRIAL

The court scheduled a two-week trial, allotting one week each to the plaintiffs and defendants, but Judge Kathy Seeley ultimately presided over a seven-day trial between June 12 and 20, 2023.[39]

The first week, plaintiffs called 24 witnesses and had 168 exhibits admitted. In contrast, defendants presented evidence for only one day of their allotted week, calling three witnesses and having four documents admitted. In addition to the witness testimony and exhibits, counsel for the plaintiffs wove in detailed and sophisticated visual demonstrations throughout, for example, using maps showing where reduced snowpack has affected tribal and cultural practices or the locations and dates of wildfires.

5.1 Plaintiffs' Case

Plaintiffs' witnesses fell primarily into two categories: the plaintiffs themselves and experts in various climate science disciplines. Their first witness, Mae Nan Ellingson, a delegate from Montana's 1972 constitutional convention that adopted the environmental rights amendment, testified that among the 'whole convention from every part of the state was a love of their unique—a very unique heritage through the landscape of Montana'.[40]

5.1.1 Plaintiffs

Most of the plaintiffs testified. Rikki Held was first, talking about her family's ranch and motel, the main source of family income, and how she has experienced climate impacts like 'wildfires, drought, flooding, more extreme weather events, such as windstorms and hail, yeah, changes in wildlife behavior'.[41] She recalled one wildfire from 2012 that 'burned like 70 miles of power lines so we lost electricity for a month'.[42]

On the third day, Taleah H. testified about having headaches from wildfire smoke and how seeing the impacts was 'scary, and it's sad, and it's disappointing'.[43] Wildfire smoke meant missed work training horses and fewer days out hiking or at the lake, and fewer cold days meant fewer opportunities to

---

[39] Our Children's Trust provides public access to video recording of the trial, <https://heldvmontana.ourchildrenstrust.org/full-trial-video/> accessed 28 September 2024.

[40] Transcript of Trial, *Held v Montana* No CDV-2020-307, 2023 WL 1997864 (Mont Dist Ct 2023) 41 (on file with authors).

[41] ibid [64].

[42] ibid [65].

[43] ibid [719].

ice skate.[44] In one remarkable moment, she noted how, one day after a rare rainstorm, she looked up to a blue sky and 'realized that I'd forgotten what the sky normally looked like because I was so used to such a smoky environment'.[45]

5.1.2 Plaintiffs' experts

The bulk of the testimony, however, was devoted to scientific experts on climate change, its causes, impacts in Montana and on Montanans, including the plaintiffs, the state's GHG emissions contributions, and the feasibility of Montana's energy transition away from fossil fuels.[46] Several scientists were local, including professors from the University of Montana and Montana State University. These experts repeatedly relied on consensus climate science reports, including the Intergovernmental Panel on Climate Change's (IPCC) assessment reports, the most recent Montana Climate Assessment's (2017) main report, and a 2021 Climate Change and Human Health in Montana report.[47]

Dr. Steven Running, professor emeritus at the University of Montana, lead chapter author for the IPCC's 4th assessment report, and world-leading climate scientist opened the expert testimony with basic climate science facts, including the greenhouse effect, earth's energy balance, and how we know that human activity is responsible for the current climate change. He also explained the IPCC, its scope and process, and how the U.S. national climate assessments differ. When asked about the scientific consensus on what is causing climate change, his answer was clear and concise:

> At this point, there's really no controversy in the formal literature at all ... [M]ost developed nations have some level of a formal science body for their nation['s] science. And I think every single one of them in the last 10 years have put out official

---

[44] ibid [724].

[45] ibid [720], [721]–[24].

[46] For an overview of the scientific evidence presented at trial, see Jarryd Page, 'Climate Science on the Docket: How Held v Montana is Bridging Science and Law', (*Vibrant Environment*, 11 August 2023) <www.eli.org/vibrant-environment-blog/climate-science-docket-how-held-v-montana-bridging-science-and-law> accessed 5 December 2024.

[47] See e.g., H Lee and J Romero, eds, 'Climate Change 2023: Synthesis Report. Contribution of Working Groups I, II and III to the Sixth Assessment Report of the Intergovernmental Panel on Climate Change' (IPCC 2023) 35–115 <www.ipcc.ch/assessment-report/ar6/>; Cathy Whitlock and others, '2017 Montana Climate Assessment' (Montana Institute on Ecosystems 2017) <https://scholarworks.montana.edu/handle/1/13584> accessed 5 December 2024; Alexandra Adams and others, 'Climate change and human health in Montana: A special report of the Montana Climate Assessment' (Montana Institute on Ecosystems 2021) <https://doi.org/10.15788/c2h22021> accessed 5 December 2024.

> statements recognizing the reality of climate change and the causality of fossil fuel burning.[48]

Dr. Running also testified about what actions are necessary to reduce the impacts of climate change, including returning to the concentration of carbon dioxide that scientists have found necessary to stabilize the planet's climate—350 parts per million:[49]

> As long as we keep emitting fossil fuels it'll keep getting worse …I think Montana and really everywhere else needs to as rapidly as possible quit burning fossil fuels. It's really quite straightforward.[50]

Along the way, defendants objected to Running's reliance on the IPCC reports on hearsay grounds, their counsel claiming it did not fall within the exception for government data because, 'I don't think it's our government'.[51] Judge Seeley overruled the objection and admitted the document, a decision consistent with courts from around the world that have relied on IPCC reports as reliable and authoritative.[52]

Running's testimony was supplemented by Dr. Cathy Whitlock, professor at Montana State University and lead author of the 2017 Montana Climate Assessment. Dr. Whitlock testified about 'recent trends and future projections in temperature, precipitation, snow accumulation and snowmelt, and stream runoff in Montana', and explained how they affect ecosystems, communities, and livelihoods of people that depend on them.[53] Asked about impacts on the plaintiffs, Dr. Whitlock said:

> [I]t's going to affect them in all kinds of ways. If you think about Rikki, for example, and her ranch, warmer temperatures, longer summers, drier summers are going to affect their ability to ranch and their livelihoods. If you think about Eva, who we heard from, and Grace, they're worried about hot temperatures in the summertime and how they're going to deal with that. So it's going to affect them. And then you have plaintiffs that depend on snowpack for training to be skiers and for

---

48 Transcript of Trial (n 41) 117–18.

49 This references a proposal that 'human changes to atmospheric CO2 concentrations should not exceed 350 parts per million by volume'. See e.g., Johan Rockström and others, 'A Safe Operating Space for Humanity' (2009) 461 *Nature* 472, 472–75.

50 Transcript of Trial (n 41) 106.

51 ibid 140.

52 See e.g., Sandra Nichols Thiam and others, 'Weathering the Storm of Global Climate Litigation: Enabling Judges to Make Sense of Science' (2023) 54 *Geo J Int'l L* 563, 563–92.

53 Transcript of Trial (n 41) 238–39.

> their livelihoods as ski instructors, and that'll be impacted as we get less snow and more rain.[54]

Dr. Whitlock also testified that the state climate assessment was presented to and distributed among government officials, and that she personally had presented to the Governor, his staff, legislators, and the State Interim Water Policy Committee.[55] She submitted that the state has 'been aware of climate change for several decades', including its consequences 'on the economy, the environment, and well-being in Montana'.[56]

A pediatrician (Dr. Lori Bryon) and a psychologist (Dr. Lise Van Susteren) testified about the impacts of climate change on the youth plaintiffs. Dr. Byron, a contributing author of the 2021 report on Climate Change and Human Health in Montana, explained climate change's impacts on the children's physical health. She testified that 'youth are being harmed by climate disproportionately when compared to adults, both due to their physiologic features and their psychological development … that also means, though, that the kids with chronic medical problems or disabilities are even more at risk'.[57] These impacts manifest in various ways, including through heat-related illness, and the exacerbation of underlying respiratory health conditions, like asthma, from which some of the plaintiffs suffer.[58]

Many studies have also documented the psychological toll of climate change. On this point, plaintiffs offered testimony from Dr. Van Susteren, a general and forensics clinical psychiatrist, about the psychological harms of the MEPA Limitation.[59]

Additional witnesses testified to Montana's emissions profile. Peter Erickson, an expert in GHG accounting—the field of study related to quantifying the emissions of a specific source like the state of Montana—showed Montana's role in fuel consumption, extraction, and infrastructure. This testimony was key, because one of the few defenses the state mounted was based on the idea that Montana's emissions are insignificant.

In his testimony, Erickson used a 'full lifecycle' approach, which includes emissions not only from the burning of fossil fuels, but from its extraction, transportation, processing, and consumption for energy.[60] Using this approach, Erickson, describing the state as 'a linchpin of fossil fuel economy in western

54 ibid 284.
55 ibid 243–44.
56 ibid 256–57.
57 ibid 473.
58 ibid 476.
59 Transcript of Trial (n 41) 1163–64.
60 ibid 914.

North America',[61] found that Montana has the sixth largest per capita GHG emissions of all U.S. states.[62] Finally, Erickson concluded that Montana has only 'scratched the surface of the fossil fuels that remain and are still underground within the state'.[63]

Stanford Professor Mark Jacobson then testified about the feasibility of Montana swiftly transitioning from fossil fuels to clean and renewable energy, pointing to a 2018 Montana Department of Environmental Quality (MDEQ) report that noted Montana has significant potential for solar energy.[64] He went on to discuss wind energy potential and the co-benefits of renewables like cleaner air. Importantly, Jacobson argued that the barriers to renewable energy are political, not technical or economic barriers—a point that surfaced in Judge Seeley's opinion.

5.2 Defendants' Case

In stark contrast to the plaintiffs' trial strategy, the state chose to call only three witnesses. Moreover, their strategy was apparently under revision during the trial, as at the end of the first week they notified the Court that their scientific expert, Dr. Judith Curry, would not be testifying. This was notable, since Dr. Curry was proffered as the expert who would dispute some of the climate science offered by plaintiffs' experts. In particular, the opinions she had planned to offer included that:

> [t]he climate-related concerns observed by the Plaintiffs are well within the range of historical natural weather and climate variability, with worse occurrences of weather and climate extremes observed during the early 20th century;
>
> [p]laintiffs' concerns about climate change in the 21st century are greatly exaggerated, and not consistent with the most recent assessment reports and research publications; and
>
> [e]missions from fossil fuels generated in Montana provide a miniscule [sic] contribution to global greenhouse gas emissions and do not influence directly Montana's weather and climate.[65]

---

61 ibid 931.

62 ibid 938.

63 For example, the state extracts 34 million tons of coal per year, with 707 million tons in reserve. ibid 945–46.

64 Transcript of Trial (n 41) 1086–87.

65 Report of Judith Curry, PhD, 1 (Sabin Center for Climate Change Law, 27 October 2022) <https://climatecasechart.com/wp-content/uploads/case-documents/2022/20221027_docket-CDV-2020-307_expert-report-1.pdf> accessed 30 March 2025.

While the defendants did not supply a reason why Dr. Curry was not ultimately called to testify, she subsequently published a blog post that suggested that the decision can perhaps be explained by the state's strategy of pressing largely legal (rather than factual) arguments.[66] Dr. Curry's blog also expresses frustration with how the attorneys for the state handled the case.[67]

The three defense witnesses who did take the stand were two Montana Department of Environmental Quality (DEQ) employees (Chris Dorrington, then-director, and Sonja Nowakowski, the bureau chief for fossil fuel permitting) and Dr. Terry Anderson, an economist who testified about GHG emissions worldwide and in Montana. Dorrington, designated as a hybrid witness to provide both factual testimony and serve as an expert in environmental and public policy matters, had reviewed hundreds of MEPA documents. He focused on the statute's process and how the agency understood its responsibilities. In one insightful exchange on cross examination, Dorrington conceded that the state constitution, including the environmental provisions, governed DEQ's actions but maintained that the MEPA Limitation constrained the agency's environmental reviews.[68]

Nowakowski, also a hybrid witness, offered more detailed information on how the air quality program implements its statutory responsibilities. She maintained that reviews conducted under MEPA served a procedural function and that absent additional regulatory authority for DEQ, the outcome of the review 'would not change the act of issuing, conditioning, modifying, or denying a permit'.[69] In her view, 'plaintiffs would need to challenge these permitting decisions' on an individual basis.[70] Nowakowski also countered some of the plaintiffs' clean energy experts, testifying that the state is making progress—distributing $2 million for EV infrastructure, with tens of millions in applications for federal funding to do more, distributing $18 million for solar generation, and other actions.[71]

Finally, Dr. Terry Anderson, economist and a senior fellow at Stanford's Hoover Institution, who had previously taught at Montana State University, offered extremely limited testimony, simply providing numbers that

---

66 Curry stated that 'the Montana legal team presented this case as one about procedural law'. Judith Curry, 'Held v Montana Climate Lawsuit' (*Climate Etc*, 21 June 2023) <https://judithcurry.com/2023/06/21/held-v-montana-climate-lawsuit/> accessed 5 December 2024.

67 ibid. Curry expressed frustration with turnover among the state lawyers, calling them 'totally unprepared', and explaining that she was 'deeply critical of how the cross-examination of the Plaintiffs' climate witnesses was going'.

68 Transcript of Trial (n 41) 1289–92.

69 ibid 1392.

70 ibid.

71 ibid 1362–63.

represented: (1) the 2020 and 2022 worldwide GHG emissions; (2) the 2020 and 2022 GHG consumption emissions for the state of Montana; and (3) the percentage that those Montana figures represented of the worldwide total.[72] This again highlights that the State relied much more on legal than factual arguments; Dr. Anderson's testimony in support of the State (on direct and redirect combined) totaled a mere ten minutes.

6. THE RULING

On August 14, 2023, less than two months after trial, but 1,249 days after the complaint was filed, the Court ruled in favor of the plaintiffs, finding that the MEPA Limitation violated the rights guaranteed to them by Montana's Constitution. Headlines described the opinion as 'historic',[73] 'landmark',[74] and a 'watershed moment'.[75] With nearly 1,000 references to various aspects of climate science throughout the 103-page opinion, the case represents a breakthrough in the judicial process enabling development of a factual record about the causes of climate change and its impacts on people and communities.[76] Since most of the facts presented by the plaintiffs were uncontested, the Court's findings of fact largely mirrored the plaintiffs' expert evidence, starting with general climate science and climate change projections, moving into the impacts on Montana and the plaintiffs and how Montana's actions contribute to climate change.

---

[72] Dr. Anderson also answered two questions about the percent of land in Montana that is tribal and federal. His cross-examination focused largely on a series of errors contained in his original expert report.

[73] Jennifer Hijazi, 'Montana Youth Win Historic Case on Harm from Climate Change' (*Bloomberg Law*, 14 August 2023) <https://news.bloomberglaw.com/environment-and-energy/montana-youth-climate-plaintiffs-get-historic-win-in-state-case> accessed 5 December 2024.

[74] David Gelles and Mike Baker, 'Judge Rules in Favor of Montana Youths in a Landmark Climate Case' *The New York Times* (New York, updated 16 August 2023) <www.nytimes.com/2023/08/14/us/montana-youth-climate-ruling.html> accessed 5 December 2024.

[75] Isabel Hicks, '"Watershed Moment": Judge Rules in Favor of Youth in Landmark Held v Montana Climate Change Case' *Bozeman Daily Chronicle* (Bozeman, 14 August 2023) <www.bozemandailychronicle.com/news/environment/watershed-moment-judge-rules-in-favor-of-youth-in-landmark-held-v-montana-climate-change/article_e3ad55bc-3acd-11ee-a924-c3f6301a199f.html> accessed 5 December 2024.

[76] Jarryd Page, 'Unpacking the Headline: Climate Science and Held v State of Montana' (*Climate Judiciary Project*, 13 September 2023) <https://cjp.eli.org/news/230913-unpacking-headline-climate-science-and-held-v-state-montana> accessed 5 December 2024.

6.1 The Facts of Climate Change

6.1.1 Climate change is real, and it is caused by humans

Drawing on Dr. Running's and Dr. Whitlock's testimony, the opinion sets out the fundamentals of climate science, stating that '[t]here is overwhelming scientific consensus that Earth is warming as a direct result of human GHG emissions, primarily from the burning of fossil fuels'.[77] The '[s]cience is unequivocal that dangerous impacts to the climate are occurring due to human activities, primarily from the extraction and burning of fossil fuels'.[78] The Court found that 'CO2 levels have fluctuated throughout history, but the rate of increase in atmospheric CO2 is 100 times faster than in natural CO2 fluctuations and cycles, and it is happening in a very short timeframe that is unprecedented in the geologic record'.[79]

6.1.2 Montana is experiencing climate impacts now, and they are projected to worsen

Making things local, the decision found that '[t]he science is clear that there are catastrophic harms to the natural environment of Montana and Plaintiffs and future generations of the State due to anthropogenic climate change'.[80] These impacts vary and include '[i]ncreasing temperatures, changing precipitation patterns, increasing droughts and aridification, increasing extreme weather events, increasing severity and intensity of wildfires, and increasing glacial melt and loss'.[81] In other words, '[c]limate change impacts result in hardship to every sector of Montana's economy, including recreation, agriculture, and tourism'.[82]

The opinion specifically called attention to the multiple ways that Montanans recreate, highlighting how access to hunting and fishing resources are impacted:

---

77 *Held* No CDV-2020-307 (n 3) [67].

78 ibid [70]. The 'unequivocal' language comes from the IPCC's Sixth Assessment Report, which found that it is 'unequivocal that human influence has warmed the atmosphere, ocean and land', Valérie Masson-Delmotte and others, eds, 'Climate Change 2021: The Physical Science Basis. Contribution of Working Group I to the Sixth Assessment Report of the Intergovernmental Panel on Climate Change' (CUP 2021) <www.ipcc.ch/report/ar6/wg1/>.

79 *Held* No CDV-2020-307 (n 3) [77].

80 ibid [193].

81 ibid [140], citing testimony from Dr. Jack Stanford, Dr. Cathy Whitlock, and Dr. Daniel Fagre.

82 ibid [141], citing testimony from Dr. Steven Running and Dr. Cathy Whitlock.

> Access to boating and fishing on certain rivers and lakes in Montana has been limited, and in some instance [sic] completely foreclosed, because of river flows or high-water temperatures. These changes limit the ability of some Plaintiffs to fish and access the State's rivers and lakes for sport or recreation.[83]
>
> Wildfires resulting from climate change have caused nitrogen levels in Montana's lakes to increase. This has caused nutrient imbalances that threaten the plant and animal life in the lakes. If GHG emissions continue to rise, impacts to the climate will further harm Montana's wildlife and fisheries, and the ability of Plaintiffs to hunt and fish.[84]

The court found that the degradation to Montana's environment and the resulting harm to the plaintiffs will worsen if the State continues ignoring GHG emissions and climate change. Put simply, 'extreme weather events and other climactic [sic] events such as droughts and heatwaves will occur more frequently and in greater magnitude, and Plaintiffs will be unable to live clean and healthy lives in Montana'.[85]

6.1.3 Children, including the youth plaintiffs, are vulnerable to climate impacts

The Court made findings about the impacts of climate change on young Montanans and connected those findings to specific impacts described by plaintiffs in their testimony.[86] Citing multiple experts, the Court stated that '[c]hildren are uniquely vulnerable to the consequences of climate change, which harms their physical and psychological health and safety, interferes with family and cultural foundations and integrity, and causes economic deprivations'.[87] Moreover, '[b]ecause of their unique vulnerabilities, their stages of development as youth, and their average longevity on the planet in the future, Plaintiffs face lifelong hardships resulting from climate change'.[88]

---

83 ibid [174], citing testimony from Dr. Steven Running and Dr. Jack Stanford.

84 ibid [175]–[176].

85 ibid [89], citing testimony from Dr. Steven Running and Dr. Lise van Susteren.

86 'Children who have pre-existing respiratory conditions, including asthma, are especially vulnerable to climate impacts, including increasing air pollution and rising temperatures. Wildfire smoke has harmed the health of Plaintiffs Olivia, Jeffrey, and Nate, all who have pre-existing health conditions, and other Plaintiffs, including Badge and Eva', *Held* No CDV-2020-307 (n 3) [128].

87 The experts cited are Dr. Lori Byron, Dr. Lisa van Susteren, and Michael Durglo Jr, ibid [104].

88 ibid [133].

While no event attribution studies were referenced,[89] the decision connected the dots between climate impacts and the specific injuries faced by the plaintiffs. For example, Rikki Held testified to seeing cattle struggling at her ranch following things like wildfires and dry seasons. Testimony from experts Cathy Whitlock, Daniel Fagre, Jack Stanford, and others had linked global climate change to increased wildfires, drought, and reduced snowpack, including in Montana.[90] The Court put these pieces together:

> Climate change has impacted the snowpack on the [Held] ranch in recent years, with snow typically not lasting through the winter. Reduced winter snowpack means less natural water available for cattle. As a result, the cattle must rely on water tanks, which are far apart and expensive to install. With less water, there is also less grass available for the cattle to eat. With less water and grasses, cattle travel further for water and food, and lose weight. This means the cattle are not as valuable and the ranch profits and income declined.[91]

6.1.4 Montana contributes to global climate change

Since these impacts went unrefuted at trial, the question was whether defendants' actions contribute to them. The Court answered affirmatively, finding that emissions in Montana are increasing as a result of defendants issuing permits without evaluating the impacts of climate change and GHG emissions when doing so. The Court went on to explain that those emissions 'exacerbate anthropogenic climate change and cause further harms to Montana's environment and its citizens, especially its youth'.[92]

On the question of whether Montana's emissions are sufficient to make a difference in the global carbon equation—an issue that the State did dispute, the Court found Mr. Erickson was 'well-qualified' and 'his testimony informative and credible', but it found that testimony from the defendants' expert, Dr. Anderson, 'was not well-supported [and] contained errors', and, therefore 'was not given weight by the Court'.[93] Relying on Dr. Erickson, the Court found that '[i]t is possible to calculate the amount of CO2 and GHG emissions that results from fossil fuel extraction, processing and transportation, and consumption

---

89 For an overview of attribution science see Michael Burger, Jessica Wentz, and Radley Horton, 'The Law and Science of Climate Change Attribution' (2020) 45 *Colum J Envt L* 57. See Sabin Center for Climate Change Law, 'Climate Attribution Database' <https://climateattribution.org/> accessed 20 October 2024 (compiling event, impact, and source attribution studies). See also Curry (n 66).

90 *Held* No CDV-2020-307 (n 3) [140].

91 ibid [195] (i)–(j).

92 ibid [268].

93 ibid [210]–[211].

activities that are authorized by Defendants'.[94] These calculations showed that 'Montana is a major emitter of GHG emissions in the world in absolute terms, in per person terms, and historically'.[95]

Just how major? In addition to being the sixth highest-emitting U.S. state, '[t]he 166 million tons of CO_2 due to Montana's fossil fuel-based economy is equivalent to the emissions from' the Netherlands (pop. 18 million), Argentina (pop. 47 million), or Pakistan (pop. 248 million).[96] The district court summed it up—'[w]hat happens in Montana has a real impact on fossil fuel energy systems, CO2 emissions, and global warming'.[97]

The Court also made findings about the feasibility of Montana transitioning to low-carbon and no-carbon energy sources. Namely, that 'it is technically and economically feasible for Montana to replace 80% of existing fossil fuel energy by 2030 and 100% by no later than 2050, but as early as 2035'.[98] This is 'across all sectors, including electricity, transportation, heating/cooling, and industry'.[99] Any barriers to accomplishing this 'are not technical or economic, but social and political', and '[t]ransitioning to [wind, water, and sunlight] will keep Montana's lights on while saving money, lives, and cleaning up the air and the environment, and ultimately using less of Montana's land resources'.[100]

6.2 Conclusions of Law and Constitutional Climate Rights

With the bulk of the opinion devoted to fact-finding, the Court's legal conclusions comprise less than 20 of the 100-plus pages. Concise and to the point, the conclusions of law found the plaintiffs established standing by proving injury, causation, and redressability, and held unconstitutional both the MEPA Limitation and a related provision that limited MEPA challenges and enjoined enforcement of those provisions.

6.2.1 Plaintiffs established standing

The district court found that the plaintiffs demonstrated each element of standing—an injury in fact, traceable to the defendants' actions, that could be redressed by a favorable court decision. While the State vigorously disputed the plaintiffs' standing throughout the case, the Court found:

---

94 ibid [214].
95 This cites Peter Erickson, ibid [222].
96 ibid [219].
97 ibid [237].
98 ibid [272].
99 ibid [271].
100 ibid [281]–[282].

> Youth Plaintiffs have experienced past and ongoing injuries resulting from the State's failure to consider GHGs and climate change, including injuries to their physical and mental health, homes and property, recreational, spiritual, and aesthetic interests, tribal and cultural traditions, economic security, and happiness.[101]
>
> There is a fairly traceable connection between the State's disregard of GHG emissions and climate change, pursuant to the MEPA Limitation, GHG emissions over which the State has control, climate change impacts, and Plaintiffs' proven injuries.[102]

On redressability, the decisive factor in *Juliana*'s dismissal in federal court, Judge Seeley explained that while mere psychological satisfaction of prevailing at trial is not enough to establish redressability, the MEPA Limitation changes the analysis:

> Defendants can alleviate the harmful environmental effects of Montana's fossil fuel activities through the lawful exercise of their authority if they are allowed to consider GHG emissions and climate change during MEPA review, which would provide the clear information needed to conform their decision-making to the best science and their constitutional duties and constraints, and give them the necessary information to deny permits for fossil fuel activities when inconsistent with protecting Plaintiffs' constitutional rights.[103]

6.2.2 The Montana Constitution guarantees the right to a stable climate

In the district court's estimation, all signs pointed to the State's environmental rights amendment extending to cover climate: 'Based on the plain language of the implicated constitutional provisions, the intent of the Framers, and Montana Supreme Court precedent, climate is included in the "clean and healthful environment" and "environmental life support system."'[104] The Court found, moreover, that this right is 'forward-looking and preventative', and 'requires enhancement of Montana's environment'.[105]

6.2.3 The MEPA Limitation Violates Plaintiffs' Constitutional Rights

With a justiciable controversy, and establishment of a constitutional guarantee to a stable climate, the remaining question was whether the defendants' actions violated that guarantee.

Because the right at issue was a fundamental one, under Montana law, the Court applies the most rigorous level of judicial review, strict scrutiny. In cases

---

101 ibid [4].
102 ibid [13].
103 ibid [18].
104 ibid [49].
105 ibid [43].

that involve allegations that a particular law violates a fundamental right, the burden is on the government to meet a high legal bar and demonstrate the law is 'narrowly tailored' to serve a 'compelling interest'. Here, it may have been impossible for the government to prevail since, notably, '[t]he State did not put forward any evidence of a compelling governmental interest for the MEPA Limitation'.[106] But the Court made clear that even if 'the State had established a compelling interest for the statute, the MEPA Limitation is not narrowly tailored to serve any interest'.[107]

Because the MEPA Limitation foreclosed an understanding and examination of 'how additional GHG emissions will contribute to climate change or be consistent with the Montana Constitution', the Court declared it 'facially unconstitutional'.[108] The Court's ultimate conclusion was that '[b]y enacting and enforcing the MEPA Limitation, the State is failing to meet their affirmative duty to protect Plaintiffs' right to a clean and healthful environment, and to protect Montana's natural resources from unreasonable depletion'.[109]

7. THE MONTANA SUPREME COURT GETS ITS SAY

Without an intermediate appellate court in Montana, the State appealed directly to the state Supreme Court.[110] The State also sought a stay of the trial court's decision to keep the MEPA Limitation in place pending resolution of their appeal. The trial court denied that request,[111] and the Montana Supreme Court likewise refused to grant a stay in early 2024.[112] Thus, since August 2023, state agencies are no longer barred from considering climate impacts and greenhouse gas emissions when conducting environmental reviews.

---

106 ibid [63].

107 ibid [66].

108 ibid [59].

109 ibid [55]. The Court did not rule on the plaintiffs' other alleged constitutional violations, including the right to seek safety, health, and happiness, the right to individual dignity and equal protection, or on the public trust claim, finding they were all 'predicated on degradation of Montana's clean and healthful environment', ibid, 92.

110 Seven justices sit on Montana's Supreme Court, six associates and one chief. Mont Code Ann 3-2-101.

111 Order Denying Defs' Mot for Clarification and for Stay of Judgment Pending Appeal, *Held v Montana* No CDV-2020-307 (Mont Dist Ct Jan 16, 2024).

112 Order Denying Mot for Relief from the Dist Court's Order Denying State's Motion to Stay, *Held v Montana* 2024 MT 312 (Mont 2024) (No DA 23-0575).

7.1 The State's Arguments on Appeal

On appeal, the defendants continued to press a number of legal arguments revolving around standing, namely causation and redressability.[113] The State repeated its claims, which the district court had rejected, that 'Montana cannot influence [the] global climate to any appreciable or measurable degree', and that 'GHG emissions in Montana do not come close to having a meaningful influence on global climate'.[114] As to redressability, the state agencies and Governor charged that '[s]triking one subsection of MEPA—a procedural statute that does not give agencies any authority to condition, modify, or deny permits—cannot stop, alter, or meaningfully slow global climate change'.[115] In short, they argued that MEPA does not cause the plaintiffs' injuries and getting rid of the limitation will not fix them.

7.2 Friends of the Court

Although none were filed with the district court, 20 amicus briefs were submitted at the Montana Supreme Court, on behalf of federal, state, and local interests, including governments, industry and industry associations, academics, and others. Amici filed on wide-ranging topics, including the Montana economy, public and pediatric health, recreation, and tribal interests. Notably, six former Montana Supreme Court justices whose tenures on the bench ranged from 1990 to 2017 filed a joint amicus brief, in support of neither party, asking the Court to 'strike down the MEPA Limitations which purport to limit equitable remedies for violations of constitutional rights'.[116]

Officers from the state Senate and House of Representatives, and a coalition of states led by North Dakota, filed in support of Montana. Many others, including the Montana and U.S. Chambers of Commerce[117] and a coalition of

---

[113] Def/Appellants State of Mont's Reply Brief, 2024 WL 1223277, 2-11, *Held v Montana* 2024 MT 312 (No DA 23-0575). The State also argued that the constitutional amendments were not self-executing, ibid 11–13.

[114] Defs/Appellants State of Mont's Reply Brief, 2024 WL 2215302, 1-2, *Held v Montana* 2024 MT 312 (No DA 23-0575).

[115] Appellant State Agencies' and Governor's Reply Brief, 2024 WL 2215320, 2, *Held v Montana* 2024 MT 312 (No DA 23-0575).

[116] Former Justices' Amicus Curiae Brief, 2024 WL 1601713, *Held v Montana* 2024 MT 312 (No DA 23-0575).

[117] The local Helena, Billings, and Kalispell chambers of commerce joined this filing.

coal and oil trade associations,[118] advanced arguments similar to the state on causation, redressability, and the contours of MDEQ's authority.

Others filed in support of the plaintiffs, including environmental law professors, the national American Civil Liberties Union and its local chapter, child rights advocates, an interfaith group, a coalition of Montana's outdoor athletes led by Conrad Anker, outdoor recreation companies, and several public health experts and doctors, whose brief was cited by the majority. Numerous tribes and tribal coalitions submitted briefs too, advancing arguments similar to the plaintiffs', focused on the importance of constitutional protections for a clean and healthy environment and state agencies' responsibility to uphold those protections.[119]

7.3 Oral Argument and Decision

On July 10, 2024, the Montana Supreme Court held oral argument in Helena.[120] The oral arguments focused largely on the arguments in the briefing, but at one point, counsel for Montana declared, unprompted, that:

> the state never did contest, and I hope this is clear, that there are anthropogenic human factors that contribute to this problem. And we were concerned that that point sort of got lost at trial. But that doesn't mean that we feel that this global problem can be influenced in any way by a state district court judge in Montana.[121]

This moment emphasized the consensus around the basic scientific concepts of anthropogenic climate change, and the State's decision in this case not to contest that consensus. Indeed, the Montana Supreme Court noted the same in its opinion, explaining that '[t]he State acknowledges that it does not dispute or challenge the District Court's findings of fact on the science and impacts of climate change'.[122]

---

118 These organizations include Frontier Institute, Buckeye Institute, Montana Association of Oil, Gas, & Coal Counties, Montana Coal Council, Montana Mining Association, Montana Taxpayers Association, United Property Owners of Montana, and Westmoreland Mining LLC.

119 *Held* 2024 MT 312 (n 2).

120 A recording is available on YouTube. Montana Courts, 'DA 23-0575 HELD, et al., Plaintiffs and Appellees, v State, et al., Respondents and Appellants' (12 July 2024) <www.youtube.com/watch?v=_yfdS-DmXA8> accessed 5 December 2024.

121 ibid 29 <www.youtube.com/watch?v=_yfdS-DmXA8>.

122 *Held* 2024 MT 312 (n 1).

The high court handed down its 6–1 opinion affirming the district court decision on December 18, 2024.[123] The majority affirmed the judgment for the plaintiffs on each issue: that (1) the plaintiffs had standing, (2) the 'clean and healthful environment' constitutional provision includes the right to a stable climate system, and (3) the MEPA Limitation violated that right.[124] Focused on these legal questions, the Court did not foray into the factual record in detail, but did acknowledge that '[t]he District Court made extensive, undisputed findings of fact that GHG emissions are drastically altering and degrading Montana's climate, rivers, lakes, groundwater, atmospheric waters, forests, glaciers, fish, wildfire, air quality, and ecosystem.'[125]

The lone dissent argued that the plaintiffs were deficient on each element of standing, with injuries that amounted to generalized grievances borne by all Montanans. But the majority disagreed, explaining that the injury in this case is a law that infringes on a constitutional right, not climate change writ large.[126] The plaintiffs 'showed that the State's policies, including the MEPA Limitation … impacts their right by prohibiting an analysis of GHG emissions, which blindfolded the State, its agencies, the public, and permittees' from carrying out their responsibilities to 'realize the right to a clean and healthful environment.'[127] Stated simply, 'Plaintiffs have standing for the declaratory and injunctive relief they seek because they allege that the MEPA Limitation violates their right to a clean and healthful environment and declaring it unconstitutional will alleviate the harm that statute causes to their constitutional right.'[128]

The majority also emphatically rejected Montana's primary contention that the state's emissions were too insignificant to matter globally. They disagreed 'that the delegates … would grant the State a free pass to pollute the Montana environment just because the rest of the world insisted on doing so'.[129] The Court explained that if other countries' failure to reduce emissions were a

---

123 A brief concurrence agreed with the majority's conclusions but raised concerns about the degree of particularized injury alleged, ibid [49]–[51].

124 A fourth issue related to the district court's denial of the state's request for a psychiatric evaluation of the plaintiffs, with the majority concluding that the trial court had not abused its discretion, ibid [47]–[48].

125 ibid [19].

126 ibid [27].

127 ibid [28]–[29]. The majority rejected the dissent's contention that plaintiffs' grievances were more appropriate in response to a specific permitting decision, recognizing that constitutional challenges to statutes were not only proper, but common, ibid [36]–[37].

128 ibid [36].

129 ibid [20].

sufficient reason to deny relief, it 'would effectively immunize the State from any litigation' on whether its laws violated the environmental rights provision of the constitution.[130] However, the Court was careful to limit its ruling to the statutes at issue here: 'We decide only that the Constitution does not permit the Legislature to prohibit environmental reviews from evaluating GHG emissions'.[131]

8. *HELD*'S LEGACY

Held exemplifies the broader rights turn in climate litigation across the world, but its long-term impact may be difficult to quantify. The legal precedent it establishes applies only within a single U.S. state, meaning other state courts may look to it but are not bound to follow either its result or its reasoning. Consequently, *Held* may not immediately reverberate in courts throughout the United States; no decision outside of Montana has yet to rely on its reasoning.

In addition, the circumstances of the case were unique from the outset. While other states have similar constitutional provisions, no other state has a similar combination of a strong constitutional right and a statute that outright bars state agencies from even considering climate change during the permitting process. Observers have argued that these circumstances could limit *Held*'s applicability in other jurisdictions.[132] If so, the case may turn out to be an outlier.[133]

In Montana, the case has had at least two immediate consequences. First, the Montana Supreme Court's decision has reaffirmed and expounded upon the importance of the state constitution's guarantee of a right to a clean and healthful climate. The precise boundaries of this right are not yet well understood, and future cases are likely to explore the scope and application of this responsibility for other agencies and actors. Litigants will surely begin to test them, filing broad challenges aimed at the entire legal framework governing certain activities and more narrow ones targeted at an agency's decision-making applied to a discrete project. The process and outcomes in these cases will offer the first signs about how sharp the right's teeth will ultimately become.

The Montana Supreme Court's subsequent ruling in *Montana Environmental Information Center v MDEQ* (MEIC), which involved an environmental

---

130 ibid [33]–[34].

131 ibid [46].

132 Erin Ferguson, '*Held v State of Montana*: A Constitutional Rights Turn in Climate Change Litigation?' (2024) 36 *J of Env't L* 453.

133 Don C Smith, '*Held v Montana*: The beginning of a climate change lawsuit trend in US state level courts or a one-shot wonder?' (2023) 41 *J Energy & Nat Res L* 369, 369–78.

review of a natural gas-fired power plant, offers an early data point. Decided just days after *Held*, the *MEIC* decision clarified that '[w]e did not hold in *Held*, and do not hold here, that DEQ is required to analyze GHG emissions for every potential state action'.[134] However, 'in a case like this one, which undisputedly involves a significant amount of CO2 emissions (nearly 770,000 tons annually) … MEPA requires DEQ to analyze the direct, secondary, and cumulative impacts'.[135]

A second immediate consequence is that state agencies, no longer barred by statute, can again consider climate change in environmental reviews, and according to the *MEIC* case may need to in certain instances. MDEQ's first post-*Held* reviews provide some indication of its implementation.[136] For example, in evaluating a landfill expansion, MDEQ included a 'Greenhouse Gas Assessment' section, but the only direct and indirect emissions considered were from the increased fuel from diesel trucks to and from the site; emissions associated with construction, or the emissions from the landfill itself, were not included.[137] A disposal facility for automobile shredding byproduct got similar treatment, but MDEQ briefly acknowledged there that secondary impacts of increased emissions might mean 'drier summers, reduced soil moisture, and increased chance of wildfires'.[138] Additionally, a draft environmental assessment issued in connection with a permit to expand a coal mine analyzed the emissions associated with the mining of the coal itself but not the subsequent transportation or the ultimate downstream combustion.[139]

These examples show that notwithstanding *Held*, MDEQ is not yet considering the full scope of GHG emissions or climate change impacts in environmental reviews. However, shortly after the trial court's *Held* decision, MDEQ

---

134 *Mont Envt Info Ctr v Mont Dept of Envt Qual* 2025 MT 3 (2025) [30].

135 ibid.

136 Montana Department of Environmental Quality, 'MEPA Documents' <https://deq.mt.gov/public/mepa> accessed 31 December 2024.

137 Montana Department of Environmental Quality, 'Final Environmental Assessment, Class II Solid Waste Management System, Lake County Landfill Expansion' (5 September 2024) <https://deq.mt.gov/files/Land/SolidWaste/Documents/EA'S/Lake_County/FINALEA_LakeCountyExpansion_6_2024.pdf> 28–29.

138 Montana Department of Environmental Quality, 'Final Environmental Assessment, Proposed Class II Solid Waste Management System Auto Shred Residue (ASR) Repository' (11 June 2024) <https://deq.mt.gov/files/Land/SolidWaste/Documents/DRAFTEA_PSR_5_2023_EXTENSION.pdf> 38–39.

139 Montana Department of Environmental Quality, 'Draft Environmental Assessment, Surface Mining Permit C1993017' (4 April 2024) <https://deq.mt.gov/files/Land/CoalUranium/Pending%20Applications/BullMtnAM6DraftEA.pdf> 43–45.

initiated a broader review of MEPA implementation. MDEQ assembled a climate analysis subtask group to help with that review, which issued a final report in June 2024.[140] The group recommended that the agency develop a short-term framework for climate analysis, study long-term implementation, and draft a bill that would study various approaches to climate analysis.[141] Public comments on the draft report also overwhelmingly supported reviews that evaluate climate impacts.[142] The report does not indicate a timeline for implementation of these recommendations, and as of this writing it is unclear whether any actions pursuant to the report have been taken.

Held's legacy may be harder to trace outside Montana, but Hawai'i may offer some insight. There, a suit based on the state's constitutional environmental rights, *Navahine F. v Hawai'i Department of Transportation*, brought by Our Children's Trust on behalf of youth against the Hawai'i Department of Transportation and other state defendants, was headed to trial in the summer of 2024. While plaintiffs did not rely on *Held* in their briefs, at the eleventh hour, the Court approved a settlement that provided a path for setting interim targets for the state to achieve an emissions-free transportation sector by 2045.[143] Although the decision in *Held* did not directly cause the *Navahine* settlement, the parties were no doubt aware of both the ruling and the Montana Supreme Court arguments held just days prior. More recently, Our Children's Trust did cite *Held* in a case against the Florida Public Service Commission, arguing that the uncontroverted fact-finding in *Held* showed that every ton of emissions matters in the global climate system.[144]

While observers have argued that *Held* may have sway outside the United States,[145] it has not yet been cited in prominent international climate decisions. This includes a pair of Canadian cases, *La Rose v Her Majesty the Queen*

---

140 Montana Environmental Policy Act Workgroup, 'Recommendations and Strategies to Improve MEPA Implementation' (June 2024) <https://deq.mt.gov/files/about/mepa/MEPAWorkgroup_DraftFinalReport_061224.pdf>.

141 ibid 10–15.

142 ibid 4.

143 *Navahine F v Hawai'i Department of Transportation* No 1CCV-24-000269 (Haw Cir Ct Jun 24, 2024); see also Dara Albrecht, 'Landmark Climate Settlement Highlights Relevance of Climate Science for Judges' (*Vibrant Environment Blog*, 19 September 2024) <www.eli.org/vibrant-environment-blog/landmark-climate-settlement-highlights-relevance-climate-science-judges>.

144 Amended Compl for Declaratory Relief, *Reynolds v Florida Public Service Commission* 2024-019966-CA-01 (Fla Cir St Dec 12, 2024) [64].

145 Chhaya Bhardwaj and Daniel Stein, '*Held v State of Montana*: Environmental constitutionalism in the US and its comparison with India' (2024) 15 *Jindal Global L Rev* 213.

(Federal Court of Appeals decision in December 2023) and *Mathur v Ontario* (Ontario's appellate court decision in October 2024) (see Chapter 2),[146] the European Court of Human Right's decision in *KlimaSeniorinnen* (April 2024),[147] and the South Korea Constitutional Court's decision in *Do-Hyun Kim v South Korea* (August 2024), though those cases were decided before the Montana Supreme Court opinion.[148]

While *Held*'s long-term, national, and global impact remain to be seen, the Montana court's clear findings about the facts of climate change and its impact on the youth plaintiffs cannot be ignored. These findings will no doubt influence future climate litigation in the United States and around the world. In these ways, *Held* is likely to become a template not only for youth climate cases and cases involving fundamental or constitutional rights but also for any dispute that involves the consideration of climate science. In that respect, *Held* may represent the leading edge of a deep judicial engagement with climate change and its impacts in the United States.

---

[146] *La Rose v His Majesty the King in Right of Canada* 2023 FCA 241; *Mathur v Ontario* 2024 ONCA 762.

[147] *Verein KlimaSeniorinnen Schweiz and Others v Switzerland* App no 53600/20 (ECHR, 9 April 2024).

[148] *Do-Hyun Kim et al v South Korea* No 2020HunMa389 (Constitutional Court of Korea, 29 August 2024).

9. *Misdzi Yikh v Canada*: indigenizing rights-based climate litigation

Benjamin Ralston

1. INTRODUCTION

Climate litigation is a global phenomenon. The geographic scope of this publication provides a rough approximation of just how broadly it has proliferated across the world. This is particularly true for cases exemplifying the '(human) rights turn' in climate litigation since 2015.[1] In the wake of the Paris Agreement, a growing number of litigants are seeking legal accountability from governments and corporate actors for their contributions to climate change based on its impacts on human rights, especially the rights to life and a healthy environment.[2] As the case of *Misdzi Yikh v Canada* shows, Canadian courts are no longer strangers to this phenomenon.[3] However, Canadian engagement with climate litigation remains 'nascent'.[4]

---

1 Jacqueline Peel and Hari M Osofsky, 'A Rights Turn in Climate Litigation?' (2018) 7 Transnat'l Envtl L 37.

2 See e.g. Meinhard Doelle, Sara Seck and Rose Klug, 'The Use of Elements of the Paris Agreement and IPCC Reports in Climate Litigation' in Alexander Zahar (ed), *Research Handbook on the Law of the Paris Agreement* (Edward Elgar 2024); Brian J Preston, 'The Influence of the Paris Agreement on Climate Litigation: Legal Obligations and Norms (Part I)' (2021) 33 J Envtl L 1; Charles Beauregard and others, 'Climate Justice and Rights-Based Litigation in a Post-Paris World' (2021) 21 Climate Pol'y 652.

3 Hassan M Ahmad and Heera E Sen, 'Climate Change Litigation in Canada' in Robert C Brears and Jade Lindley (eds), *The Palgrave Handbook of Environmental Policy and Law* (Palgrave Macmillan (forthcoming)); Lisa Benjamin and Sara L Seck, 'Mapping Human Rights-Based Climate Litigation in Canada' (2022) 13 J HR & Envtl 178; Camille Cameron and Riley Weyman, 'Recent Youth-Led and Rights-Based Climate Change Litigation in Canada: Reconciling Justiciability, Charter Claims and Procedural Choices' (2022) 34 J Envtl L 195.

4 Benjamin and Seck (n 3).

Indigenous peoples have long been at the forefront of climate litigation, highlighting the human and social dimensions of climate change through legal proceedings. For instance, one of the earliest examples of human rights-based climate litigation was a petition that the Inuit Circumpolar Conference filed with the Inter-American Commission of Human Rights in 2005. This petition aimed to hold the United States accountable for impacts of its historic greenhouse gas emissions on Inuit Nunaat (the Inuit's Arctic homeland).[5] Similarly, one of the earliest examples of tort-based climate litigation was initiated by an Iñupiat village in Alaska in 2008: *Kivalina v ExxonMobil Corp.*[6] More recently, in 2022, the UN Human Rights Committee issued a ground-breaking decision in *Daniel Billy v Australia*, finding Australia in breach of the rights of a group of Torres Strait Islanders through its inadequate response to climate change impacts.[7] These are just three cases among others in which Indigenous peoples have used litigation to highlight their heightened vulnerability to climate change's impacts and their unique capacities for crafting and contributing to effective solutions.[8] Yet these cases remain few and far between amidst a more exponential growth of rights-based climate cases internationally.[9]

Plaintiffs have been attempting to frame climate change and greenhouse gas emissions as legal issues amenable to judicial determination for at least three decades.[10] However, it is only recently that climate litigation has evolved into a clearly defined field of law with a sizeable body of legal precedents, albeit spread across myriad jurisdictions.[11] These cases share a number of common causes and concerns, including the jurisprudential and evidentiary

---

[5] Hari M Osofsky, 'Inuit Petition as a Bridge? Beyond Dialectics of Climate Change and Indigenous Peoples' Rights' (2007) 31 Am Indian L Rev 675; Sébastien Jodoin, Shannon Snow and Arielle Corobow, 'Realizing the Right to Be Cold? Framing Processes and Outcomes Associated with the Inuit Petition on Human Rights and Global Warming' (2020) 54 Law & Soc'y Rev 168.

[6] *Native Village of Kivalina v ExxonMobil Corp* 663 F Supp 2d 863 (ND Cal 2009), affd 696 F3d 849 (9th Cir 2012).

[7] Human Rights Committee, *Views Adopted by the Committee under Article 5(3) of the Optional Protocol, Concerning Communication No. 3624/2019*, UN Doc CCPR/C/135/D/3624/2019 (22 September 2022) (*Daniel Billy* Decision).

[8] Maria Antonia Tigre, 'Climate Litigation and Indigenous Groups: The Rise of Indigenous Voices in Climate Litigation' (2022) 9 E-Publica 214.

[9] ibid 247–248.

[10] See e.g. *City of Los Angeles v National Highway Traffic Safety Administration* 912 F2d 478 (DC Cir 1990); *MacMillan Bloedel Ltd v Russow* 1994 CanLII 2633; *Environmental Resource Centre v Canada (Minister of Environment)* 2001 FCT 1423.

[11] United Nations Environment Programme, *Global Climate Litigation Report: 2023 Status Review* (United Nations Environment Programme 2023) XII.

barriers they face, many of which transcend jurisdictional boundaries.[12] In many ways these cases are entangled with one another, including through citation practices and shared strategies, framings, and narratives.[13]

This chapter focuses on one unique instance of rights-based climate litigation brought by two Indigenous collectives. This case reflects a number of opportunities and challenges associated with transnational climate litigation strategies in the Canadian context. In parallel with several other climate cases across the world—many of which take inspiration from the *Juliana* litigation in the United States—its plaintiffs seek remedies against a government for inadequate mitigation of greenhouse gases and the impacts of climate change on their constitutional rights. At the same time, this familiar framing of rights-based climate litigation has been reworked to fit the circumstances of the Indigenous rights-holding plaintiffs in unprecedented ways.[14]

The plaintiffs are two Wet'suwet'en kin groups and their traditional leaders. They are challenging government-sanctioned emissions in a representative capacity based on climate change's impacts as particularized through the lens of the Wet'suwet'en legal order.[15] They assert Canada is breaching their rights to life, liberty, and security of the person under the *Canadian Charter of Rights and Freedoms*. They also argue that these individual rights are dependent on a communal interest in sustaining healthy and liveable territories. A close analysis of their strategy illustrates a mix of risks and opportunities that may be of comparative interest to litigants and scholars working on climate litigation elsewhere across the globe.

This chapter will provide a detailed analysis of the *Misdzi Yikh* litigation, including its parties, cause of action, and procedural history. It will focus on

---

[12] Gianni Ghinelli, 'Standing, Justiciability, and Burden of Proof in Climate Litigation: Challenges and Proposals' in Eva Storskrubb (ed), *YSEC Yearbook of Socio-Economic Constitutions 2022: Funding of Justice* (Springer Cham 2023); Beauregard (n 2).

[13] Natasha Affolder and Godwin E K Dzah, 'The Transnational Exchange of Law through Climate Change Litigation' in Francesco Sindico and others (eds), *Research Handbook on Climate Change Litigation* (Edward Elgar 2024); Phillip Paiement, 'Urgent Agenda: How Climate Litigation Builds Transnational Narratives' (2020) 11 Transnat'l L Theory 121.

[14] The term 'Indigenous rights' is used in this chapter to encompass the distinct rights that Indigenous peoples hold under both international law and Canada's domestic law.

[15] The term 'legal order' is used in this chapter to holistically encompass Wet'suwet'en norms embedded in non-state social, political, economic, and spiritual institutions. See Val Napoleon, 'Thinking About Indigenous Legal Orders' in René Provost and Colleen Sheppard (eds), *Dialogues on Human Rights and Legal Pluralism* (Springer 2013).

challenges the plaintiffs face in bringing transnational climate litigation strategies into dialogue with Canadian law more generally, as well as specifically in relation to Indigenous rights and law. In this light, *Misdzi Yikh* can be seen as an attempt to overcome access to justice challenges that have hindered the uptake of rights-based climate litigation in Canada. At the same time, this lawsuit is poised to 'indigenize' rights-based climate litigation by bringing its more established practices, frames, and narratives into dialogue with an Indigenous people's own legal order and worldview.[16] Each of these subtopics will be carefully unpacked by reference to the case's substantive and procedural characteristics.

2. THE PLAINTIFFS: WET'SUWET'EN COLLECTIVES AND THEIR REPRESENTATIVES UNDER WET'SUWET'EN LAW

One unique feature of the *Misdzi Yikh* litigation is how it is being brought as a representative proceeding premised on Indigenous law. In February of 2020, an initial Statement of Claim[17] was filed in Canada's Federal Court to commence this lawsuit against the federal government.[18] It was filed by two Wet'suwet'en hereditary chiefs, Dini Ze' Lho'imggin and Dini Ze' Smogilhgim.[19] The lawsuit is constituted as a representative action, meaning it is brought by representative plaintiffs who speak on behalf of groups.

Representative proceedings are a historic precursor to class action litigation in Canada, but they remain distinct.[20] Due to the collective nature of most Indigenous rights claims, representative proceedings are the most common procedural vehicle for litigating these rights.[21] Coincidentally, they are becoming a common vehicle for rights-based climate litigation, including youth-led lawsuits.[22] Representative proceedings can avoid procedural complexity and

---

16 David Rosenberg, 'Indigenization of Civil Litigation: Barriers and Opportunities' (2024) 6 Lakehead LJ 26, 29.

17 This refers to the type of pleading used to initiate an action.

18 Statement of Claim, *Misdzi Yikh et al v Her Majesty the Queen in Right of Canada*, Court File T-211-20, filed 10 February 2020, 1.

19 ibid 1.

20 Michael A Eizenga and Emrys Davis, 'A History of Class Actions: Modern Lessons from Deep Roots' (2011) 7 Can Class Action Rev 5.

21 See discussion in *Soldier v Canada* 2009 MBCA 12; *Horseman v Canada* 2015 FC 1149 [24]–[40].

22 Camille Cameron and Riley Weyman, 'Class Actions and Climate Loss and Damage Litigation' in Meinhard Doelle and Sara L Seck (eds), *Research Handbook on Climate Change Law and Loss & Damage* (Edward Elgar 2021).

substantive obstacles to environmental class actions.[23] Most notably, they avoid the costly and complicated requirements for class action certification. Likewise, they avoid potential mismatches between class actions and the collective nature of many Indigenous rights.[24] For example, as class actions are typically used to litigate multiple individual claims in aggregate, class members can opt out of these lawsuits and sue individually if they wish. In contrast, collective Indigenous rights can only be litigated by authorized representatives of the rights-holding group.[25] The outcomes of these cases bind all rights-holders as a group,[26] making individual opt-out rights inappropriate.[27]

The Wet'suwet'en govern themselves as a federation of collectives. As Wet'suwet'en hereditary chiefs are not recognized as representatives for Wet'suwet'en rights-holders under any Canadian statute, the plaintiffs need to carefully define the groups they claim to speak for and demonstrate clear authority to speak on their behalf.[28] Lho'imggin and Smogilhgim have initiated this lawsuit on their own behalf and on behalf of all members of the collectives they represent as hereditary head chiefs (*dini ze'*) within the Wet'suwet'en legal order.[29] More specifically, Lho'imggin seeks to represent Misdzi Yikh (Owl House) and Smogilhgim seeks to represent Sa Yikh (Sun House). Reinforcing the representative nature of this case, the pleadings generally refer to Misdzi Yikh and Sa Yikh as its plaintiffs rather than the *dini ze'* who speak for them.[30]

In order to grasp the nature of the claims made in *Misdzi Yikh*, some knowledge of Wet'suwet'en law will be instructive.[31] The Wet'suwet'en form a

---

23 ibid; Rudiger Tscherning, 'Climate Litigation and the Class Action Mechanism—The Potential Role of the UN Sustainable Development Goals' (2020) 33 J Envtl L & Prac 133, 141–142.

24 Allan Lutfy and Emily McCarthy, 'Rule-Making in a Mixed Jurisdiction: The Federal Court (Canada)' (2010) 49 Sup Ct L Rev 313, 323–324.

25 *Behn v Moulton Contracting Ltd* 2013 SCC 26 [30].

26 See e.g. *Goodswimmer v Canada (A-G)* 2017 ABCA 365.

27 *Enge v Canada* 2017 FC 932 [93]–[99].

28 *Watson v Canada* 2020 FC 129 [412], discussing requirements under Rule 114 of the *Federal Court Rules* for representative proceedings brought by 'unrecognized rights-bearing collective[s]'.

29 Amended Statement of Claim, *Misdzi Yikh et al v Her Majesty the Queen in Right of Canada*, Court File T-211-20, filed 26 August 2024 [9].

30 ibid [2], [9].

31 Existing scholarship provides expressions for 'law' in the Wet'suwet'en language. In Antonia Mills, *Eagle Down Is Our Law* (UBC Press 1994), two are referenced: *yinkadinii' ha baa ten* ('the ways of the people on the surface of the earth') and *deni biits wa aden* ('the way the feast works'). Some publications use another term, *'anuc niwh'it'en* ('the proper conduct of lawful relations'): e.g. Tyler

self-identified Indigenous people who have occupied and governed their territories (*yintah*) since time beyond memory.[32] Taken together, the Wet'suwet'en *yintah* are in an area defined by the Bulkley River drainage, the headwaters of the Fraser River drainage, and the Babine Lake drainage in British Columbia, Canada.[33]

Wet'suwet'en society is organized into five matrilineal descent groups called *dïdikh* (clans).[34] Each *dïdikh* is further divided up into other matrilineal descent groups known as *yikh* (houses).[35] Wet'suwet'en society is further structured through hereditary titles that establish leadership for each *yikh*.[36] Each *yikh* has a head chief (*dini ze'*) who governs fishing sites and large tracts of territory (*yintah*) associated with their *dïdikh* and *yikh*, in addition to subchiefs with other responsibilities under Wet'suwet'en law.[37] Finally, Wet'suwet'en law is enacted and upheld through the institution of *balhats* (feasts), which provide a forum for resolving disputes, demonstrating ownership over contested territory, and clarifying succession to hereditary titles, among other functions.[38] While many Indigenous rights-holding collectives across Canada are represented by municipal-style elected councils imposed on them by a federal statute called the *Indian Act*, the Wet'suwet'en have long insisted on engaging with the Canadian government through their hereditary chiefs for recognition of their Indigenous rights.[39]

According to the Statement of Claim, each *dini ze'* 'speaks for his House and is responsible for the welfare of his House members and for the protection

McCreary, *Indigenous Legalities, Pipeline Viscosities: Colonial Extractivism and Wet'suwet'en Resistance* (University of Alberta Press 2024), 256. Rather than arbitrarily choosing one of these options, 'Wet'suwet'en law' is used throughout this chapter.

32 Karen Whonnock, 'Every One Counts: Examining Aboriginal Youth Incarceration by Gender, Statistics and Witsuwit'en Laws' (PhD dissertation, Royal Roads University 2018), 50.

33 Mills (n 31) 37.

34 Mills (n 31) 102. Modern orthographic conventions for Wet'suwet'en words have been adopted for this chapter: McCreary (n 31), XII, 255–56.

35 Mills (n 31) 106–107.

36 Mills (n 31) 112–114.

37 Mills (n 31) 112–113.

38 Mills (n 31) 38; McCreary (n 31) 9–10.

39 See e.g. *Delgamuukw v British Columbia* [1997] 3 SCR 1010, a leading Aboriginal title lawsuit brought by Wet'suwet'en and Gitxsan hereditary chiefs. See also Canada, *Gitksan and Wet'suwet'en Self-Government Project* (Ottawa: Department of Indian and Northern Affairs, 1989), collating documents from self-government negotiations with Wet'suwet'en and Gitxsan hereditary chiefs.

of his House's possessions, including its territories'.[40] Likewise, each 'House has a unique set of possessions under Wet'suwet'en law, which it manages for the benefit of the House as a whole'.[41] These include exclusive land and riverine fishing territories, as well as oral histories, crests, and hereditary titles that play integral roles within the Wet'suwet'en legal order.[42] The Statement of Claim lists several duties of the *dini ze'*, including protection of the welfare and health of their *yikh* members; protection of the possessions of their *yikh*, including its territories; ensuring their *yikh* meets its legal obligations; and speaking for the *yikh*.[43] The pleading details numerous other aspects of Wet'suwet'en law that speak to the identification of members of each *yikh*, as well as their leaders and territories, presumably in support of the representative nature of the proceeding.

Relying on an Indigenous legal order for rights-based climate litigation distinguishes this case from almost every other climate case filed to date.[44] Yet Indigenous peoples across the globe tend to be identified by common traits, including having distinct legal orders, deep connections to their ancestral homelands, and place-based, ecological worldviews.[45] In this light, *Misdzi Yikh* could set an important precedent for Indigenous-led climate cases the world over.

Notably, this case is by no means the first instance of Wet'suwet'en *dini ze'* arguing claims based in Wet'suwet'en law before Canadian courts. One of Canada's best-known cases on Indigenous rights, *Delgamuukw v British Columbia*, was brought by Wet'suwet'en and Gitxsan hereditary chiefs on behalf of their respective houses and territories.[46] There is also a consistent line of jurisprudence in which Wet'suwet'en law has been recognized by Canadian courts, including the central role played by *bahlats* and the *dini ze'* who govern and speak on behalf of their *yikh*.[47] If nothing else, existing jurisprudence

40 Amended Statement of Claim (n 29) [9].

41 ibid [14].

42 ibid [14].

43 ibid [16].

44 The author is aware of only one other example: *Smith v Fonterra Co-Operative Group Limited* [2024] NZSC (7 February 2024), discussed in Chapter 4.

45 See e.g. James Youngblood Henderson and others, *Aboriginal Tenure in the Constitution of Canada* (Thomson Reuters 2022), §7.11; Gregory Cajete, *Native Science: Natural Laws of Interdependence* (Clear Light 1999).

46 *Delgamuukw* (n 39).

47 See e.g. *Forsythe v Collingwood Sales Ltd* [1987] BCJ No 40 (BC Co Ct), affd [1988] BCJ No 683 (CA) (dealing with a successful claim of damages for the cost of holding a feast under Wet'suwet'en law).

supports the standing of *dini ze'* to speak on behalf of their *yikh*.[48] Likewise, there are many cases in which Canadian courts assert that Indigenous law should be treated as part of Canadian law, at least in principle.[49] Yet Canadian courts have been far more reluctant to recognize substantive applications of Indigenous law.[50] It therefore awaits to be seen how this innovative framing fares as a basis for the particularization of climate harms.

3. THE OTHER PARTY: CANADA, DEFENDING GOVERNANCE-RELATED CONTRIBUTIONS TO THE CLIMATE CRISIS

The only other party to this lawsuit is the defendant: the Government of Canada. The matter has not seen applications from potential intervenors at the time of writing. This is not surprising since the pleadings have yet to close.[51] So far, the lawsuit has been scrutinized by two levels of court based on the plaintiffs' pleadings and whether they make out viable claims for trial. At this stage, prior to hearing any evidence, Canadian courts rarely authorize intervenors to participate.[52] The choice of defendant is, however, worthy of a deeper discussion due to Canada's unique model of federalism, which structures and confounds Canadian environmental governance.

Canada is one of the most decentralized federations in the world.[53] Its division of legislative powers between one federal government and ten provincial governments remains almost entirely unchanged from the country's first constitution in 1867.[54] Given that 'the environment as we know it today was

---

[48] See e.g. *Delgamuukw* (n 39); *Canadian Forest Products Inc v Sam* 2011 BCSC 676, revd on other grounds, 2013 BCCA 58.

[49] See e.g. *Mitchell v MNR* 2001 SCC 33 [10]; *Dickson v Vuntut Gwitchin First Nation* 2024 SCC 10 [322].

[50] See e.g. *Coastal GasLink Pipeline Ltd v Huson*, 2019 BCSC 2264, rejecting a defence based on the defendants' alleged compliance with Wet'suwet'en law.

[51] Federal Court of Canada, Court Files, Court Number T-211-20 (*Dini Ze' Lho'imggin and others v. Her Majesty the Queen*) <https://www.fct-cf.gc.ca/en/court-files-and-decisions/court-files#cont> accessed 1 December 2024.

[52] *Tanudjaja v Attorney General (Canada)* 2013 ONSC 1878 [15].

[53] Philippe Chassé, Olivier Jacques and Colin Scott, 'Between Decentralization and Asymmetry: Explaining Preferences Toward the Division of Power in Canada' (2024) Publius: J Federalism 1, 5.

[54] Canada also has three territories, but they do not have constitutionally entrenched legislative powers. While proposals to clarify jurisdiction over environmental protection and add environmental rights to the Canadian Constitution were considered in the 70s and 80s, the only explicit amendment of relevance was s 92A, fortifying provincial governments' authority over non-renewable resources.

not a major legislative concern' in the late 19th century, legislative authority over environmental protection was not explicitly assigned to either level of government.[55] Instead, Canada's federal and provincial governments draw their legislative and regulatory powers over the environment from a patchwork of more specific sources, including federal powers over criminal law, fisheries, and navigation and shipping; provincial ownership of most natural resources; and provincial authority over property and civil rights and local or private matters, among others. Considerable uncertainty over the constitutionality of novel forms of environmental regulation in Canada flows from this decentralization.[56]

In principle, Canadian courts seek to interpret the Canadian model of federalism in a way that 'afford[s] both levels of government ample means to protect the environment while maintaining the general structure of the Constitution'.[57] However, whether they achieve this balance in practice is at least up for debate.[58] Moreover, Canada's constitutional law on the books does not always reflect how different levels of government put the law into action. More specifically, the federal government has tended to limit the exercise of its authority over environmental regulation in deference to provincial governments taking the lead role.[59] As a result, focusing on the federal government as the sole defendant in *Misdzi Yikh* is by no means an uncontroversial choice.

In keeping with several well-known international examples of rights-based climate litigation, the plaintiffs take aim at the actions and omissions of the government with respect to its regulation of greenhouse gas emissions. More specifically, they argue Canada has the authority to be far more ambitious in its efforts to mitigate greenhouse gas emissions, including through its direct jurisdiction over emissions from road vehicles, fossil fuel infrastructure, and

---

55 *British Columbia (Environmental Management Act) v Canadian National Railway Company* 2022 BCSC 135 [255].

56 See e.g. *Responsible Plastic Use Coalition v Canada (Environment and Climate Change)* 2023 FC 1511 (finding a ban on select single-use plastics exceeded federal constitutional powers).

57 *R v Hydro-Québec* [1997] 3 SCR 213 [116].

58 See e.g. the contrasting views of the majority and dissent in *Reference re Impact Assessment Act* 2023 SCC 23 regarding federal jurisdiction over environmental assessments.

59 Inger Weibust, 'The Great Green North? Canada's Bad Environmental Record and How the Feds Can Fix It' in Gordon DiGiacomo and Marantonett Flumian (eds), *The Case for Centralized Federalism* (University of Ottawa Press 2010); Kathryn Harrison, *Passing the Buck: Federalism and Canadian Environmental Policy* (UBC Press 1996).

developments in Canada's northern territories.[60] The plaintiffs also point to Canada's influence over emissions from other sources via subsidization or taxation of fossil fuel use and production, its ability to purchase fossil fuel infrastructure, and its jurisdiction over inter-provincial and export-related transportation infrastructure.[61]

Since the filing of the first pleading in *Misdzi Yikh*, the Supreme Court of Canada has released a pair of decisions that clarify an additional role for the federal government in setting minimum national standards for carbon pricing. However, in doing so they emphasized that the federal government must carefully tailor its climate policies to respect concurrent provincial authority over a wide range of sources of emissions.[62] In other words, Canadian federalism treats the regulation of greenhouse gas emissions as a matter over which federal and provincial governments each have constitutional roles but with a high potential for conflicts. This may complicate the plaintiffs' ability to attribute emissions to the federal government's actions.

While it is impossible to know for certain why this lawsuit takes sole aim at the Government of Canada based on the pleadings, they do provide clues. Most notably, the theory of the case is premised on liability for Canada's failure to effectively implement its international commitments to mitigate its greenhouse gas emissions, 'including those made at the 1988 International Conference on the Changing Atmosphere, the 1992 United Nations Framework Convention on Climate Change, the 1998 Kyoto Protocol, the 2009 Copenhagen Accord, and the 2010 Cancun Agreement'.[63] Consistent with rights-based climate litigation strategies employed the world over, it places special emphasis on Canada's adoption of the Paris Agreement, citing both its Nationally Determined Contributions and the agreement's overarching goal of holding global temperature increases to between 1.5°C and 2°C.[64] Of particular note, Canada's Parliament has ratified the Paris Agreement, passed a declaration in support of its intention to comply with the objective of this agreement, and affirmed it in national legislation.[65]

The plaintiffs argue Canada's Nationally Determined Contributions are inconsistent with the temperature goal underpinning the Paris Agreement, and Canada is not even on track to achieve its own targets.[66] At least two

60 Amended Statement of Claim (n 29) [38]–[39], [41].

61 ibid [40].

62 *Reference re Greenhouse Gas Pollution Pricing Act* 2021 SCC 11 [196]–[206]; *Reference re Impact Assessment Act* (n 57) [184]–[189].

63 Amended Statement of Claim (n 29) [42].

64 ibid [44]–[52].

65 ibid [47]–[49].

66 ibid [50]–[57].

other contemporaneous climate lawsuits in Canada challenge the adequacy of provincial mitigation targets instead. They face an additional hurdle of having to establish a normative standard for how Canada's carbon budget should be carved up among sub-state levels of government.[67] In contrast, Canada's Nationally Determined Contributions provide a clear standard that is readily attributable to the federal level of government.[68]

A second reason for focusing on the federal government may relate to its permissive approach to oil and gas infrastructure affecting the plaintiffs' *yintah* (territories).[69] They claim Canada has the power to better manage fossil fuel infrastructure developments that substantially contribute to the country's emissions. The plaintiffs raise particular concerns over infrastructure for liquefied natural gas (LNG) exports, including an LNG export terminal at Kitimat, British Columbia and the Coastal GasLink Pipeline that feeds it.[70] As discussed later in this chapter, the latter pipeline project faced an unsuccessful legal challenge from Wet'suwet'en *dini ze'* in another proceeding.[71] Wet'suwet'en objections to this pipeline have also been raised in proceedings stemming from injunctions and prosecutions over direct action against the project.[72] The *Misdzi Yikh* case links these localized concerns with national and international dimensions of Canadian climate policy. While large natural resource projects tend to require a combination of federal and provincial authorizations under Canada's model of environmental federalism, the Canadian government abruptly cancelled a federal environmental assessment

67 See especially *Mathur v His Majesty the King in Right of Ontario* 2023 ONSC 2316 [109], finding a determination of Ontario's 'fair' share of the remaining carbon budget is not justiciable. However, the Ontario Court of Appeal has reopened this question, opining that a provincial share can potentially be judicially determined. For another example of provincial-level climate litigation in Canada that may face similar challenges, see *Dykstra et al v Saskatchewan Power Corporation*, Originating Application, Court File KB No 848 of 2023 (filed 31 Mar 23) <https://www.sustainablesask.ca/uploads/1/2/3/8/123825013/issued_-_originating_application.pdf> accessed 1 December 2024. Among other things, the applicants seek an order compelling the Saskatchewan government to set emissions reduction targets that are consistent with Saskatchewan's share of emissions reductions needed to limit global warming below the upper limit of the temperature range in the Paris Agreement (2°C).

68 *La Rose v Canada* 2023 FCA 241 [38].

69 Amended Statement of Claim (n 29) [59]–[70].

70 ibid [62]–[66].

71 *Wet'suwet'en Treaty Office Society v British Columbia (Environmental Assessment Office)* 2021 BCSC 717.

72 *Huson* (n 50); *Coastal GasLink Pipeline Ltd v Huson* 2024 BCSC 509.

of the Coastal GasLink Pipeline in 2013.[73] The Wet'suwet'en plaintiffs may plan to rely on this as a key example of the federal government's permissive attitude towards fossil fuel infrastructure.

For all these reasons, Canada's distinct model of federalism complicates the task faced by plaintiffs who wish to hold Canadian governments accountable for inadequate responses to the climate crisis. So far, the national focus of cases like *Misdzi Yikh* has seen benefits and pitfalls in comparison to its provincially-focused contemporaries. Most notably, legal obligations under international law are more readily attributed to the federal level of government that represents Canada in international relations.[74] However, the federal government's less certain authority over key sources of emissions could remain a challenge. It is still too early to tell where Canadian climate cases will end up following final determinations on their merits.

4. THE MERITS: AS TESTED IN THE ABSTRACT, BUT NOT YET BASED ON AN EVIDENTIARY RECORD

The *Misdzi Yikh* case has yet to reach trial for a hearing on its merits. Nevertheless, two levels of court have already weighed in on the Wet'suwet'en plaintiffs' proposed causes of action through preliminary challenges. Much of the legal basis for their claims will look familiar to those who follow the transnational literature on rights-based climate litigation, at least at first blush. The plaintiffs originally pleaded that the Government of Canada's contributions to climate change breached three distinct constitutional obligations. They asserted the defendant's actions infringed two constitutional rights protected under the *Charter*: the right to life, liberty, and security of the person enshrined in s 7; and the right to equality, as protected under s 15.[75] Likewise, they pleaded a novel argument that Canada had breached an alleged constitutional duty to make laws for peace, order, and good government under s

---

73 Impact Assessment Agency of Canada, Canadian Impact Assessment Registry, Coastal GasLink Pipeline Project <https://iaac-aeic.gc.ca/050/evaluations/proj/80020> accessed 1 December 2024.

74 See however *Mathur v Ontario* 2024 ONCA 762 [70]. The Court of Appeal expressed the view that a provincial share of necessary emissions reductions can be defined by international standards and scientific evidence, albeit without elaborating on how this will be done.

75 Statement of Claim (n 18) [86]–[94]. For discussion of Canadian climate cases invoking the same causes of action, see Ahmad and Sen (n 3). See also *Dykstra* (n 67).

91 of the *Constitution Act, 1867*.[76] The latter two causes of action have since been struck, so the case will proceed on the sole basis of alleged infringements of s 7 rights.[77] Before closely analyzing this remaining cause of action, a more detailed discussion of the case's procedural history will provide helpful context.

4.1 Procedural History: Knocked Down, but Not Yet Knocked Out

Canada's first response to *Misdzi Yikh* was to apply to strike the lawsuit on the basis that it had no reasonable prospect of success. In 2020, the Federal Court accepted this position. It struck the pleading without leave to amend, finding the issues it raised were not justiciable.[78]

First, the Federal Court gave short shrift to the plaintiffs' argument that Canada had breached an alleged duty to make laws for the peace, order, and good government of the country by enacting laws that allow for approval and long-term operation of high greenhouse gas emitting projects.[79] The Court pointed out that this allegation relied on a constitutional provision that has only ever been interpreted as empowering the creation of laws, as opposed to requiring the enactment of laws 'for particular ends'.[80] The Court also took issue more generally with what it viewed as being 'ask[ed] to tell the legislature to enact particular laws'.[81]

Second, the Federal Court concluded the plaintiffs' claims under the *Charter* were not justiciable without finding it necessary to distinguish between the rights they relied upon.[82] The Court held that their claims were too 'broad and diffuse' to be adjudicated as they impugn an assortment of environmental assessment legislation, natural resource project approvals, and domestic climate change policy.[83] In its view, the plaintiffs failed to identify specific laws or actions that can be assessed for constitutional compliance. The Court also expressed a concern that the plaintiffs were demanding adjudication of an essentially 'political' issue in that climate action 'spans across various government, [and] involves issues of economics and foreign policy, trade, and a host of other issues'.[84] Likewise, the Court took issue with some of the rem-

---

76 Statement of Claim (n 18) [82]–[85].
77 Amended Statement of Claim (n 29) [82]–[89].
78 *Misdzi Yikh v Canada* 2020 FC 1059.
79 ibid [26]–[47].
80 ibid [27].
81 ibid [47].
82 ibid [48]–[58].
83 ibid [54].
84 ibid [56].

edies sought by the plaintiffs, including a demand for judicial supervision of Canada's enactment of laws that would be more responsive to the plaintiffs' claimed rights against climate change.[85] For these reasons, among others, the action was struck without leave to amend.

As a parallel youth-led climate lawsuit, *La Rose v Canada*, met the same fate before the Federal Court on similar grounds,[86] plaintiff-side appeals from both decisions were heard jointly by the Federal Court of Appeal in 2023. These appeals were partially successful and the plaintiffs have since been given an opportunity to amend and clarify their pleadings, though only to the extent of refocusing them on the multi-faceted *Charter* right to life, liberty, and security of the person.[87]

In contrast to the lower courts, the Federal Court of Appeal firmly rejected the proposition that the plaintiffs' claims are not justiciable 'simply because the question of climate change is complex' or because legislation enacted in response 'reflects a political choice on how to address the problem'.[88] From the appellate court's perspective, legislating invariably requires political choices that are informed by a variety of public policy considerations.[89] The same may be said of at least 'some' exercises of discretion too.[90] Once policy trade-offs have been considered and a legislative or executive response has crystallized, the resulting law or action can be judicially scrutinized for consistency with constitutional rights.[91]

The Court of Appeal also found objective legal standards against which *Charter* rights can be assessed in each case. The plaintiffs pleaded Orders in Council and legislation that they alleged to be contributing to the climate harms they face.[92] They also pointed to an objective standard in Canada's Nationally Determined Contributions under the Paris Agreement, which are ratified by Parliament and thus legally defined.[93] The Court was satisfied that these elements provide an adequate basis to assess their *Charter* claims since Canada is alleged to have failed to meet its Nationally Determined Contributions, among other shortcomings.

This was not the first appellate decision on climate litigation in Canada. In 2021, the Quebec Court of Appeal struck a youth-led climate class action over

---

85 ibid [64].
86 *La Rose v Canada* 2020 FC 1008.
87 *La Rose* FCA (n 68).
88 ibid [32].
89 ibid [33].
90 ibid [33].
91 ibid [33].
92 ibid [45].
93 ibid [38].

broadly expressed justiciability concerns.[94] In stark contrast to that ruling, the Federal Court of Appeal's reasons in this case give strong support to the justiciability of climate litigation, at least in theory. The Court clearly rejected the premise that government responses to climate change are distinguishable from other policy-laden state actions and decisions that have faced *Charter* scrutiny in the past, such as limits on private medical treatment or the availability of supervised injection sites.[95] Yet the Court did echo the lower courts' concerns with the range of remedies sought by the plaintiffs in *Midszi Yikh* and its companion case, *La Rose*, as well as the diffuse and broad scope of their claims.[96] It simply disagreed that these were incurable flaws.

Notably, the Federal Court of Appeal upheld the decision to strike the plaintiffs' equality-based arguments without leave to amend while allowing them to revise and clarify their approach to claiming their rights to life, liberty, and security of the person have been breached. Assessing state liability for climate harms under anti-discrimination law would have aligned these cases with a well-established strategy for rights-based climate litigation: highlighting disproportionate impacts of climate change on children and youth as proxies for future generations.[97] However, the Court of Appeal declined to allow these arguments to proceed, dedicating only a few lines of conclusory reasons to this aspect of its decision.

The plaintiffs initially pleaded that Canada's contributions to climate change deprived them of their right to equal protection and benefit of the law based on the disproportionate impacts of climate change on child and youth members from each *yikh*, in addition to future generations.[98] They claimed climate change will deprive younger and future members of rights 'to good health, to knowledge of their territories, fisheries, social relations and laws,

---

[94] *Environnement Jeunesse v Procureur général du Canada* 2021 QCCA 1871, leave to appeal to SCC ref. 40042 (28 July 2022).

[95] *La Rose* FCA (n 68) [30].

[96] ibid [46]–[52].

[97] See especially *Juliana v United States* 217 F Supp 3d 1224 (D Or 2016). Among other causes of action, the plaintiffs argued children and unborn generations are denied equal protection of the law because they have no voting rights and little or no political power to bring to bear on the issue of climate change. While this aspect of the case was not addressed on its merits, intergenerational equity has since been integrated into other leading rights-based climate lawsuits. See eg Lydia Slobodian, 'Defending the Future: Intergenerational Equity in Climate Litigation' (2020) 32 Georgetown Envtl LR 569; James A Goldston, 'Climate Litigation through an Equality Lens' in César Rodríguez-Garavito (ed), *Litigating the Climate Emergency* (CUP 2023) at 142–143.

[98] Statement of Claim (n 18) [90]–[94].

to fully participate in their society's institutions and decision-making, and to develop their full potential as heirs to their millennia-old culture and society'.[99] The Court of Appeal held that these arguments overshot the limits of equality jurisprudence under the *Charter* by asking the courts to adjudicate future rather than present inequalities.[100] The Court also characterized the plaintiffs' equality claims as asking for an assessment of how legislation will affect children and youths when they are older rather than how it affects them today.[101] Furthermore, it held that the plaintiffs were asking the judiciary to make 'policy choices around resource allocation', stepping into the domain of the legislative and executive branches.[102] Confusingly, this line of reasoning seems to mirror the justiciability arguments that the Court of Appeal firmly rejected elsewhere in its reasons.[103]

While the Federal Court of Appeal allowed the plaintiffs in *Misdzi Yikh* and *La Rose* to amend their pleadings and proceed with claims based on alleged breaches of their rights to life, liberty, and security of the person, it reiterated the lower courts' concerns with the open-ended way in which these had been pleaded. While the plaintiffs identified laws and actions by the federal government that permit or encourage emissions, they failed to zero in on specific provisions.[104] Instead, the Court found their pleadings 'effectively put the entirety of Canada's response to climate change up for scrutiny'.[105] In allowing the appeal, the Court pointed to the plausibility of seeking declaratory relief based on more specific contributions to emissions.[106] However, it is unclear whether the plaintiffs' amended pleadings adequately address these concerns on their face, so this may remain a significant point of contention.

4.2 Cause of Action: A Right to Life, Liberty, and Security of the (Indigenous) Person

The plaintiffs in *Misdzi Yikh* have framed their remaining cause of action in a way that deserves careful analysis. They plead that Canada's alleged failure to adequately mitigate its greenhouse gas emissions deprives the plaintiffs, their

---

99 ibid [92].

100 *La Rose* FCA (n 68) [83].

101 ibid [86].

102 ibid [83].

103 Nigel Bankes and others, 'What Does *La Rose* Tell Us About Climate Change Litigation in Canada?' (February 2, 2024), online: <https://ablawg.ca/wp-content/uploads/2024/02/Blog_NB_JK_JWK_MO_LaRose.pdf> at 8.

104 *La Rose* FCA (n 68) [128].

105 ibid [128].

106 ibid [48]–[52], [126]–[134].

members, and their future members of the right to life, liberty, and security of the person on three bases. First, with respect to the right to life, climate change increases their risk of premature death from air pollution, extreme weather events, and vector-borne disease, among other consequences.[107] Second, with respect to the right to liberty, it increases the risk to their individual and collective autonomy, including by restricting their freedom to choose where they move and live on their territories and in their communities, which is exacerbated by the 'kinship-based and place-dependent' nature of these groups.[108] Third, with respect to the right to security of the person, it increases the risk of injury, disease, and mental health conditions, including through its impacts on food security, air pollution, extreme weather events, vector-borne disease, and psychological and social trauma for 'an already vulnerable society and community'.[109]

As these claims broadly mirror arguments explored in the academic literature on the right to life, both in Canada and internationally, it is worth exploring their potential merits with reference to that literature, as well as the relevant case law the plaintiffs will need to engage with.

It is well-understood that the right to life is no ordinary right; it is a pre-condition to the enjoyment of all other rights.[110] Arguing that climate change poses a risk to this right brings *Misdzi Yikh* into close alignment with existing transnational strategies. In fact, the right to life has become the most common basis for rights-based climate litigation across the globe.[111] The equivalent right under the *Charter* is engaged whenever a law or state action imposes death or an increased risk of death on the rights-holder, directly or indirectly.[112] Scholars have suggested that this right is in tension with Canada's lax approach to environmental regulation.[113] Unlike at least 100 other countries across the world, Canada lacks an explicit right to a healthy environment in

---

107 Amended Statement of Claim (n 29) [85(a)].

108 ibid [85(b)].

109 ibid [85(c)].

110 Christian Tomuschat, 'The Right to Life—Legal and Political Foundations' in Christian Tomuschat and others (eds), *The Right to Life* (Martinus Nijhoff, 2010) 3; Amitai Etzioni, 'Life: The Most Basic Right' (2010) 9 J HR 100.

111 Samvel Varvastian, 'The Advent of International Human Rights Law in Climate Change Litigation' (2021) 38 Wisconsin Int'l LJ 369 at 383.

112 *Carter v Canada (A-G)* 2015 SCC 5 [62].

113 See eg Noel Lyon, "Canadian Law Meets the Seventh Generation" (1993) 19 Queen's LJ 350, 370, fn 38 ('... the *Charter*'s guarantee of the right to life and security of the person, in s. 7, makes no sense if courts of law are going to watch helplessly as governments continue to licence environmental destruction in its many forms.').

its national constitution.[114] However, courts in at least a dozen other countries have found a right to a healthy environment to be implicit in the more general right to life.[115] Cases like *Misdzi Yikh* give Canadian courts an opportunity to follow this lead, but it is far from clear that they will seize this opportunity.

Section 7 also protects a right to liberty that protects 'an irreducible sphere of personal autonomy wherein individuals may make inherently private choices free from state interference', at least to the extent that these choices go 'to the core of what it means to enjoy individual dignity and independence'.[116] This provides further alignment with international rights-based climate cases, many of which assert breaches of the right to respect for private life as well, especially those argued pursuant to the *European Convention on Human Rights* (ECHR).[117]

The ECHR includes a right to 'respect for private and family life' that limits public authorities' ability to interfere with one's private and family life, home, and correspondence.[118] While the Canadian *Charter* lacks explicit mention of a right to private life, s 7 has been interpreted as protecting a person's right to choose where to establish their home as an example of private choices that go to the core of one's dignity and independence.[119] Canadian courts have yet to extend this interpretation to displacement due to state-sanctioned pollution. However, it is not difficult to make an analogy with cases from the European Court of Human Rights that have found breaches of the right to private life on this basis.[120] Doing so may help link Canadian climate cases to the European Court of Human Rights' framework set out in *KlimaSeniorinnen*, where it found a state obligation to adopt and effectively apply greenhouse gas

---

114 At least 100 countries have constitutional protection for the right to a healthy environment: David Boyd, 'Catalyst for Change: Evaluating Forty Years of Experience in Implementing the Right to a Healthy Environment' in John H Knox and Ramin Pejan (eds), *The Human Right to a Healthy Environment* (CUP 2018) at 18. The Canadian province of Quebec also enshrines an explicit version of this right in s 46.1 of its *Charter of Human Rights and Freedoms*, CQLR, c C-12.

115 Boyd (n 114) 18.

116 *R v Malmo-Levine* 2003 SCC 74 [85].

117 See especially *Verein KlimaSeniorinnen Schweiz and Others v Switzerland*, No 53600/200 [2024] ECHR 304, discussed in Chapter 6.

118 *Convention for the Protection of Human Rights and Fundamental Freedoms* (European Convention on Human Rights, as amended) art 8.

119 *Godbout v Longueuil (City)* [1997] 3 SCR 844 [66].

120 See e.g. David WL Wu, 'Embedding Environmental Rights in Section 7 of the Canadian Charter: Resolving the Tension Between the Need for Precaution and the Need for Harm' (2014) 33 NJCL 191, 209–211.

mitigation regulations and measures to protect the right to private life of its citizens.[121]

Section 7 can also anchor the plaintiffs' arguments to the *International Covenant on Civil and Political Rights* (ICCPR), a binding treaty to which Canada is a signatory.[122] The ICCPR protects a right to be free from arbitrary or unlawful interference with one's privacy, family, home, or correspondence.[123] This has been interpreted in pollution-related cases as extending to the protection of distinct ways of life that include 'special attachment to and dependency on the land'.[124] Of particular note, the United Nations Human Rights Committee found a breach of this right in Australia's failure to implement adequate mitigation and adaptation measures to combat the effects of climate change on an Indigenous people in the *Daniel Billy* decision.[125] Given the presumption that Canadian law conforms to the ICCPR, s 7's interpretation in cases like *Misdzi Yikh* could feasibly be guided by these precedents as well.[126]

Of further relevance, s 7 encompasses a right to security of the person that is engaged by state interference with one's physical or psychological integrity.[127] This could be effectively linked to physical and psychosocial harms from climate change. In fact, climate litigants have initially succeeded in doing so in another leading Canadian case.[128]

For various reasons, scholars have singled out s 7 as 'the most obvious home for environmental rights' in the *Canadian Charter of Rights and Freedoms*.[129] Several academics have posited that a healthy environment is implicitly guaranteed by this provision as a necessary condition for life, as well as that displacement or impacts on one's psychological or physical health from state-sponsored environmental harm could make out viable claims under this provision.[130] It is thus little surprise that the Federal Court of Appeal concluded that

---

121 *KlimaSeniorinnen* (n 117).

122 *International Covenant on Civil and Political Rights*, 19 December 1966, 999 UNTS 171, Can TS 1976, 6 ILM 368 (entered into force 23 March 1976, accession by Canada 19 May 1976).

123 ibid.

124 Varvastian (n 111) 391.

125 *Daniel Billy* (n 7).

126 Wu (n 120) 211.

127 *Carter* (n 112) [64].

128 *Mathur* ONSC (n 67) [23(f)], appeal allowed on other grounds *Mathur* ONCA (n 74), discussed in Chapter 2.

129 Lynda M Collins, 'Safeguarding the Longue Durée: Environmental Rights in the Canadian Constitution' (2015) 71 Sup Ct L Rev 519, 529–530.

130 See e.g. Lynda M. Collins, 'Constitutional Eco-Literacy in Canada: Environmental Rights and Obligations in the Canadian Constitution' (2022) 26 Rev Const Stud at 227, 237–238; Laura Worstman, ''Greening' the Charter:

this section of the *Charter* provides the most viable cause of action for *Misdzi Yikh* to proceed.[131] Indeed, both levels of court in another leading example of Canadian climate litigation, *Mathur v Ontario*, expressed the view that climate change's contributions to increased risks of death and security of the person are 'indisputable' based on the expert evidence they considered.[132]

4.3 Collective Dimensions of a Right to Life, Liberty, and Security of the (Indigenous) People

While the *Misdzi Yikh* plaintiffs' s 7 arguments have much in common with the arguments being raised in youth-led climate litigation in Canada, they also invoke a more complex interplay between collective and individual dimensions to the right to life, liberty, and security of the person. They speak to some individualized harms to the members of each *yikh* in general terms, such as elevated risks of death and injury associated with dangerous levels of climate change. Yet they also address threats to the survival of the *yikh* plaintiffs as territorially-bounded, kin-based collectives. By framing breaches of s 7 around continuing access to their *yintah* (territories) and the kin-based governance system rooted in those *yintah*, the plaintiffs' arguments propose more than just an aggregation of individual claims. They assert a more communal dimension to the rights to life, liberty, and security of the person. In their own words, the identity, culture, legal order, and sustenance of the *yikh* plaintiffs and their members are so tightly bound up with their *yintah* that '[t]hey cannot be who they are at some other place'.[133]

Interestingly, the plaintiffs cite the United Nations Declaration on the Rights of Indigenous Peoples (UNDRIP)[134] as part of the legal basis for their claims.[135] UNDRIP's art 7 confirms that 'Indigenous *individuals* have the rights to life, physical and mental integrity, liberty and security of the

Section 7 and the Right to A Healthy Environment' (2019) 28 Dal J Legal Stud 245, 249–256; Wu (n 120); David R Boyd, *The Right to a Healthy Environment: Revitalizing Canada's Constitution* (UBC Press 2012), 212–215.

131 *La Rose* FCA (n 68) [89]–[119]. See also Dustin W Klaudt, 'Can Canada's "Living Tree" Constitution and Lessons from Foreign Climate Litigation Seed Climate Justice and Remedy Climate Change?' (2018) 31 J Envtl L & Prac 185, 233.

132 *Mathur* ONCA (n 74) [44], citing with approval *Mathur* ONSC (n 67) [120].

133 Amended Statement of Claim (n 29) [5].

134 United Nations Declaration on the Rights of Indigenous Peoples, UN GA Resolution 61/295 of 13 September 2007.

135 Amended Statement of Claim (n 29) [89].

person'.[136] It also specifies that 'Indigenous *peoples* have the *collective* right to live in freedom, peace and security as distinct peoples'.[137] Arguably, s 7 of the *Charter* could be interpreted as protecting a similar mix of individual and collective aspects of the right to life, liberty, and security of the person. Other provisions in UNDRIP, such as art 29's recognition of Indigenous peoples' 'right to the conservation and protection of the environment and the productive capacity of their lands or territories and resources' could further buttress the plaintiffs' proposed approach.[138]

While UNDRIP does not bind Canada as an international treaty, the Government of Canada and the Government of British Columbia have enacted statutes to affirm UNDRIP's application within the Canada's positive law.[139] Likewise, the Supreme Court of Canada has cited articles in UNDRIP to support its interpretation of how the *Charter* applies to Indigenous peoples.[140] This reflects the presumption that the *Charter* provides equal protection to provisions in UNDRIP.[141]

One of UNDRIP's roles in international law is to clarify how human rights elaborated in treaties apply to the more specific cultural, historical, social, and economic circumstances of Indigenous peoples.[142] A clear example of this can be seen in the *Daniel Billy* decision, where the UN Human Rights Committee interpreted the right to enjoy one's own culture under the ICCPR as 'enshrin[ing] the inalienable right of [I]ndigenous peoples to enjoy the territories and natural resources that they have traditionally used for their subsistence and cultural identity'.[143] While the Committee acknowledged that the ICCPR articulates the right to culture in individual rather than collective terms, it held that an individual's right to culture is dependent on the ability of the group to which they belong to maintain its culture.[144] The plaintiffs in *Misdzi Yikh*

---

136 *UNDRIP* (n 134) art 7(1) [emphasis added].

137 ibid [emphasis added].

138 ibid art 29.

139 See *Reference re An Act respecting First Nations, Inuit and Métis children, youth and families*, 2024 SCC 5 [4], interpreting *United Nations Declaration on the Rights of Indigenous Peoples Act*, SC 2021, c 4, s 4(a); *Declaration on the Rights of Indigenous Peoples Act*, SBC 2019, c 44, s 2(a).

140 *Dickson* (n 49) [117], [171] (Kasirer and Jamal JJ) [283], [289], [293], [317]–[319] (Martin and O'Bonsawin JJ).

141 ibid [317] (Martin and O'Bonsawin JJ).

142 Sylvanus Gbendazhi Barnabas, 'The Legal Status of the United Nations Declaration on the Rights of Indigenous Peoples (2007) in Contemporary International Human Rights Law' (2017) 6 Intl HRL Rev 242, 254.

143 *Daniel Billy* (n 7) [8.13].

144 ibid [8.13].

allege a similar relationship between individual and collective impacts of climate change. They argue that the liberty interests of *yikh* members in moving around and living on their *yintah* can only be preserved if their *yikh* and *yintah* survive as collective entities and entitlements.

As for the role of Wet'suwet'en law in establishing violations of s 7, this is not fully spelled out in the pleadings. The plaintiffs make it clear that they view global warming both 'as a threat and as a responsibility'.[145] They highlight duties of the *dini ze'* to protect their *yikh* and *yintah*, and to ensure their *yikh* meet all their legal obligations.[146] More specifically, the plaintiffs state that each *yikh* is responsible for any harm that comes to others because of the actions of its members or third parties on its *yintah*.[147] While these points provide support for the representative framing of the case, they also speak to the jurisdiction of the *dini ze'* and *yikh* in Wet'suwet'en law, which may transcend procedural questions. Literature on Wet'suwet'en law supports the interpretation that *dini ze'* and *yikh* have positive obligations to protect their *yintah*, as do all Wet'suwet'en.[148] Arguably, the ways in which climate change threatens the ability of the *dini ze'* and *yikh* to fulfill their obligations under Wet'suwet'en law may particularize ways in which it poses an existential threat to their collective way of life, as well as the liberty of their members to meaningfully take part in and uphold that distinct way of life. But how these obligations will be linked to the alleged breaches of the plaintiffs' rights will only be apparent after a trial.

In sum, the *Misdzi Yikh* plaintiffs have blended a relatively familiar constitutional argument in support of their case with a number of features specific to Wet'suwet'en law and society. While success in this lawsuit is by no means guaranteed, it could be transformative for how the *Charter* rights of Indigenous individuals and collectives are understood, in addition to its potential to impact climate policy in Canada. The merits of the plaintiffs' approach can be seen from the fact that this case survived a preliminary application to strike the pleadings. However, it will be up to a trial court to navigate in greater detail whether and how collective and individual aspects of the right to life, liberty, and security of the person can be brought to bear on Canada's contributions to climate change, as well as the more specific role that Wet'suwet'en law can play in defining and addressing this challenging question.

---

145 Amended Statement of Claim (n 29) [2].

146 ibid [16].

147 ibid [15].

148 See discussion of the Wet'suwet'en law of *wiggus* in McCreary (n 31) 182; See also Whonnock (n 32), canvassing the meanings of *yintah* and *wiggus*.

5. CLIMATE CHANGE'S IMPACTS ON ABORIGINAL RIGHTS: A LITIGATION ROAD NOT TAKEN

One striking omission from *Misdzi Yikh* is the lack of any mention of Aboriginal rights. This is notable since the Canadian constitution explicitly provides for protection of Aboriginal and treaty rights in s 35(1) of the *Constitution Act, 1982*. Some scholars have argued that this provision provides a solid foundation for Indigenous peoples to pursue climate litigation in response to the unique impacts they face from climate change.[149] It encompasses Aboriginal title, which confers exclusive rights to decide how tracts of land will be used: to enjoy, occupy, and possess them; to enjoy their economic benefits; and to proactively use and manage them.[150] It also protects a range of other territorially-defined rights, including rights to hunt, trap, and harvest.[151] It is not difficult to imagine a hypothetical case in which the contributions of government-sanctioned emissions to climate change are challenged based on how they erode these rights through worsening environmental conditions. The *dini ze'* also have clear precedents to support their ability to assert s 35 rights on behalf of their *yikhs*.[152] So why is s 35 conspicuously absent from their pleadings?[153]

There could be any number of plausible explanations for this litigation road not taken. For example, the Wet'suwet'en *dini ze'* appear to be in active negotiations towards an agreement for recognition of their Aboriginal rights and title.[154] In the past, Canadian governments have refused to negotiate

149 See e.g. Tyler Paquette, 'The Inhabitants of an Imagined Body: The Crown's Duty to Consult and Accommodate Indigenous Communities in the Arctic Adversely Affected by Climate Change' (2020) 15 JSDLP 139; Andrew Stobo Sniderman and Adam Shedletzky, 'Aboriginal Peoples and Legal Challenges to Canadian Climate Change Policy' (2014) 4 West J Legal Stud 1.

150 *Tsilhqot'in Nation v British Columbia* 2014 SCC 44 [73].

151 ibid [8].

152 See e.g. *Delgamuukw* (n 39), an Aboriginal title case brought by Wet'suwet'en and Gitxsan hereditary chiefs; *Sam* (n 48), where Wet'suwet'en hereditary chiefs sought injunctive relief based on Aboriginal rights.

153 At least two publications note this omission: Colin Feasby, David de Vlieger and Matthew Huys, 'Climate Change and the Right to a Healthy Environment in the Canadian Constitution' (2020) 58 Alta L Rev 213 226; Kierra Parker, 'Navigating the Duty of Active Protection of Maori Interests in a Climate Crisis' (2020) 24 NJ J Envtl L 209, 238.

154 See e.g. Canada, 'Memorandum of Understanding Between Canada, British Columbia and Wet'suwet'en as agreed on February 29, 2020' <https://www.rcaanc-cirnac.gc.ca/eng/1589478905863/1589478945624> accessed 1 December 2024; Government of Canada, 'Negotiations in Progress' <https://www.rcaanc-cirnac.gc.ca/eng/1100100030285/1529354158736> accessed 1 December 2024.

Indigenous land claims that are actively before the courts.[155] Therefore, it may be the case that the *dini ze'* declined to plead s 35 rights so as to avoid prejudicing these negotiations.

Another explanation may come down to timeliness. Section 35 litigation tends to proceed at a snail's pace compared to almost any other form of Canadian lawsuit, arguably making it an ill-fitting response to a climate crisis that leaves far too little time left to act. Climate litigation has more generally been criticized on this basis. Arguably the sluggishness and incremental nature of any judicial development of the law can be a poor fit for the urgency of the climate crisis.[156] Yet Aboriginal and treaty rights cases before Canadian courts have been far more time-consuming than most other forms of litigation. In contrast to *Charter* rights, which tend to have an abstract legal foundation, s 35 rights typically require detailed proof of site-specific activities, extensive expert and oral history evidence on Indigenous cultures and laws, and historic evidence.

To illustrate this, Canada's leading decision on Aboriginal title, *Tsilhqot'in Nation*, was issued by the Supreme Court of Canada in 2014, marking the final stage of a lawsuit first filed in 1983.[157] It required a 339-day trial spread over five years in addition to numerous preliminary rulings and appeals.[158] More recent examples of section 35 litigation have progressed more quickly, but still require well over 100 days of trial.[159] In stark contrast to s 35 litigation, the first *Charter*-based climate case to proceed to trial in Canada required just three days of court time in spite of the novel and complex evidentiary and legal questions it raised.[160]

Canadian jurisprudence on Aboriginal and treaty rights does contemplate procedural rights that can be raised in more timely judicial review proceedings,

---

155 For discussion of the Canadian government's 'litigate-or-negotiate' policy, see: BC Treaty Commission, *Treaty Commission Annual Report 2004* (BC Treaty Commission 2004) <https://bctreaty.ca/wp-content/uploads/2016/09/Annual_Report_04.pdf> accessed 1 December 2024; Tony Penikett, *Reconciliation: First Nations Treaty Making in British Columbia* (Douglas & McIntyre Ltd 2006) 90, 222, 272.

156 Guy Dwyer, 'Climate Litigation: A Red Herring among Climate Mitigation Tools' in Benoit Mayer and Alexander Zahar (eds), *Debating Climate Law* (CUP 2021) 129.

157 *Tsilhqot'in* (n 150) [5].

158 ibid [7].

159 See e.g. *Yahey v British Columbia* 2021 BCSC 1287.

160 *Mathur v His Majesty the King in Right of Ontario* 2023 ONSC 2316, discussed in Chapter 2.

which rarely require more than a few days of court time.[161] Of particular note, Canadian governments owe Indigenous peoples a constitutional duty to consult and accommodate their rights, whether those rights are proven or merely asserted, prior to taking actions or making decisions that may adversely affect them.[162] In theory, this duty could be triggered by government decisions or actions that contribute to climate change's adverse impacts on Aboriginal and treaty rights.[163] However, legal challenges on this basis focus almost exclusively on reviewing the processes through which governments make their decisions for evidence of meaningful dialogue with Indigenous peoples rather than second-guessing their substantive outcomes.[164] This emphasis on proceduralism may be a heavy price to pay for the avoidance of the more onerous evidentiary requirements for proving rights under s 35 as courts are reluctant to impose outcomes on governments based on breaches of the duty to consult and accommodate.[165] This form of litigation can delay government actions and decisions that potentially impact rights, but it only rarely puts a final stop to them.[166]

This is not to say that s 35 has no future as a basis for climate litigation in Canada. In fact, the Aboriginal and treaty rights it protects have several strategic advantages as potential causes of action for climate cases. Most notably, there is a long line of jurisprudence in which these rights have been successfully relied on as a way to challenge unsustainable government actions and decisions, especially harvesting rights that are dependent on environmental

161 See e.g. *Gitxaala Nation v Canada* 2016 FCA 187. This decision reflects consolidated proceedings for 15 applications for judicial review and four statutory appeals challenging a pipeline project's approvals. Still, the proceedings only required six days of court time.

162 For a detailed overview, see Isabelle Brideau, *The Duty to Consult Indigenous Peoples* (Library of Parliament 2019).

163 Paquette (n 149).

164 See e.g. *Ktunaxa Nation v British Columbia (Forests, Lands and Natural Resource Operations)* 2017 SCC 54 [79] ('[...] Section 35 guarantees a process, not a particular result. [...] There is no guarantee that, in the end, the specific accommodation sought will be warranted or possible. The ultimate obligation is that the Crown act honourably'.). See also Stephen Young, 'The Deification of Process in Canada's Duty to Consult: Tsleil-Waututh Nation v Canada (Attorney General)' (2019) 52 UBC L Rev 1065.

165 Brideau (n 162) 6.

166 See e.g. Lynda Collins, 'The Unwritten Constitutional Principle of Ecological Sustainability: A Solution to the Pipelines Puzzle' (2019) 70 UNBLJ 30, 55 (discussing outcomes from cases in which fossil fuel infrastructure has been challenged based on the duty to consult and accommodate).

conditions.[167] The enduring, intergenerational character of these rights may even impose inherent limits on how rights-holders can exercise them so as to sustain them for future generations.[168] Canadian courts have also shown a greater willingness to contemplate positive government obligations to take action and legislate in order to protect these rights.[169] Furthermore, the courts have allowed lawsuits to proceed where Aboriginal and treaty rights are invoked in support of private law claims against resource developers in addition to public law claims against governments.[170] These distinct advantages over *Charter* rights may encourage further creativity from Indigenous litigants and their legal counsel in the future. However, they will need to be carefully weighed against the countervailing challenges faced by s 35 claims.

Whatever the *Misdzi Yikh* plaintiffs' reasons might be for avoiding s 35, the Aboriginal rights of the Wet'suwet'en have been sidestepped through a focus on their *yikh* and *yintah* as aspects of the Wet'suwet'en legal order that contextualize their rights under the *Charter*. This may have advantages in that it could avoid access to justice barriers that tend to bog down litigation of rights under s 35—particularly the onerous evidentiary requirements. Yet it could come at the cost of having to convince courts to be more creative in their engagement with the legal arguments presented to them.[171]

6. WHEN CLIMATE CASES FORM ONE ELEMENT OF A BROADER LEGAL STRATEGY

The *Misdzi Yikh* case also hints at how climate litigation research could benefit from a broader, more inclusive approach to studying this phenomenon. This

---

167 Lynda Collins and Meghan Murtha, 'Indigenous Environmental Rights in Canada: The Right to Conservation Implicit in Treaty and Aboriginal Rights to Hunt, Fish and Trap' (2010) 47 Alta L Rev 959; Collins, 'Constitutional Eco-Literacy' (n 130) 231.

168 *Tsilhqot'in* (n 157) [86]. See discussion in Collins, 'Constitutional Eco-Literacy' (n 130) 232; Klaudt (n 13) 233 fn 246.

169 Sniderman and Shedletzky (n 149) 11; Erika Chamberlain, 'The Crown's Fiduciary Duties to Aboriginal Peoples as an Aspect of Climate Justice' (2012) 30 Windsor YB Access Just 289, 312–313. See e.g. *Thomas v Rio Tinto Alcan Inc*, 2024 BCCA 62 [417]–[418] (confirming governments have a positive obligation to protect established s 35 rights).

170 *Ominayak v Penn West Petroleum Ltd* 2015 ABQB 342; *Saik'uz First Nation and Stellat'en First Nation v Rio Tinto Alcan Inc* 2015 BCCA 154, leave to appeal to SCC refd, 36480 (15 October 2015); *Uashaunnuat c Compagnie minière IOC inc* 2014 QCCS 4403, affd 2015 QCCA 2, leave to appeal to SCC refd, 36332 (15 October 2015).

171 See e.g. *Huson* (n 50) [139].

case is by no means the only one in which Wet'suwet'en *dini ze'* challenge fossil fuel infrastructure in their *yintah*. In particular, the *dini ze'* have already challenged an approval for the Coastal GasLink Pipeline project referenced in their pleadings.[172] Some *dini ze'* have also opposed that pipeline through direct action and related proceedings, inspiring solidarity actions across the country.[173] Yet standard definitions of climate litigation would not encompass these other proceedings as they make no explicit mention of climate change mitigation, adaptation, or science as material issues.[174] At best they could be characterized as 'peripheral' instances of climate litigation—cases that aim to have a practical impact on climate policy without directly relying on climate science or law.[175]

The fact that the *dini ze'* plaintiffs take issue with the Coastal GasLink Pipeline in their pleadings in *Misdzi Yikh* suggests that this action could form just one element of a broader legal strategy against this project and related infrastructure. Interestingly, one interview-based study found that the plaintiffs in *Misdzi Yikh* are hoping to leverage this case in support of the argument that project-specific emissions must be considered before approving high-emitting fossil fuel infrastructure projects like the Coastal GasLink Pipeline in the future.[176] While it is not entirely clear how this strategy would play out, it is not hard to imagine litigants relying on a finding that the protection of *Charter* rights requires reduced emissions when challenging future projects.

A more detailed analysis of the links between *Misdzi Yikh* and other legal challenges to fossil fuel infrastructure affecting Wet'suwet'en *yintah* would go well beyond the scope of this publication. Nevertheless, it is worth noting that a more inclusive approach to climate litigation research could help map out how some litigants employ a broader mix of direct and indirect legal strategies to influence climate policy through the courts.

---

172 *Wet'suwet'en Treaty Office* (n 71).

173 See e.g. discussion in *Canadian Pacific Railway Limited v Doe* 2020 BCSC 388; *Canadian National Railway Company v John Doe, Jane Doe and Persons Unknown* 2020 ONSC 8225.

174 Kate McKenzie and others, 'Climate change litigation: One definition to rule them all…?' in Francesco Sindico and others (eds), *Research Handbook on Climate Change Litigation* (Edward Elgar 2024).

175 For discussion of why these 'peripheral claims' should matter to climate litigation research, see Maria Antonia Tigre, Natalia Urzola and Alexandra Goodman, 'Climate Litigation in Latin America: Is the Region Leading a Quiet Revolution?' (2023) 14:1 JHRE 67.

176 Kevin P Beck, 'Fighting Climate Change with the Charter: An Inquiry into the Effects of Litigating the Right to a Healthy Environment' (LLM thesis, York University 2024), 72–73.

7. INDIGENIZING CLIMATE LITIGATION IN FORM OR SUBSTANCE?

Returning to the title of this chapter, it may be controversial to suggest that *Misdzi Yikh* is poised to 'indigenize' climate litigation. Indigenization implies more than just inclusion of Indigenous content or interests. It has been described as the unconditional welcoming of Indigenous peoples' distinct worldviews and philosophies into spaces that have been exclusively reflective of Western worldviews and philosophies in the past.[177] Specific to law, it can be understood as acceptance of Indigenous legal traditions as separate but equal to the common and civil law traditions.[178] The Statement of Claim filed by the Wet'suwet'en plaintiffs in *Misdzi Yikh* dedicates significant space to articulating aspects of the Wet'suwet'en legal order. However, Wet'suwet'en law's role in the case could be limited to establishing the standing of the *dini ze'* to represent their *yikh*. It could also have relevance to more substantive questions, helping flesh out the alleged breaches of the plaintiffs' constitutional rights through evidence at trial. Only after a full hearing of the plaintiffs' evidence will it become clear whether *Misdzi Yikh* can meaningfully indigenize climate litigation in the sense of affording equal standing to Wet'suwet'en law, as well as the distinct worldviews and philosophies underpinning it, in the way this matter is heard and adjudicated.

Several legal scholars have advocated for deeper engagement with Indigenous law in climate litigation. Doing so could provide a forum for Indigenous law to contribute to the reform of Western legal approaches to climate change. It could also help guard against the erasure of Indigenous ways of knowing and seeing the world in climate litigation through sole reliance on the dominant frames of Western systems of law.[179] Indigenous-led climate litigation could even resonate more deeply with Indigenous peoples when it is framed in a manner that incorporates and responds to their own legal traditions, increasing the likelihood that these cases spur political and legal mobilization on the ground.[180] The case of *Misdzi Yikh* may at the very least furnish an opportunity to test these theories once the plaintiffs' arguments are fully developed through trial and any subsequent appeals.

---

177 Rauna Johanna Kuokkanen, *Reshaping the University: Responsibility, Indigenous Epistemes, and the Logic of the Gift* (UBC Press 2007) 2–3.

178 Larry Chartrand, "Indigenizing the Legal Academy from a Decolonizing Perspective" (2015) Ottawa Faculty of Law Working Paper 2015–22, <https://papers.ssrn.com/sol3/papers.cfm?abstract_id=2631163> accessed 15 October 2024.

179 Affolder and Dzah (n 13) at 227–228.

180 Jodoin, Snow and Corobow (n 5) 189–192.

Based on the existing literature on Wet'suwet'en law, it is easy to see how deeper substantive engagement with this distinct legal order could be transformative. For instance, the Wet'suwet'en understand the world and forces within it to be sentient beings, and they see themselves as part of the environment, resulting in an emphasis on reciprocity with the environment.[181] Likewise, the concept of *yintah* encompasses more than just 'territories' or 'land'.[182] It encompasses a broader sacred relationship and responsibility that all Wet'suwet'en share with respect to their *yintah*.[183] It has been described as their sacred relationship with their ancestors, lands, air, water, plants, medicines, fish, and wildlife, providing the source for their distinct and collective identity, laws, and relationships.[184] Whether the Federal Court is willing and able to meaningfully engage with concepts like these embedded in the Wet'suwet'en legal order on their own terms will likely determine its fate as either an approach to climate litigation worthy of emulation or a source of hard-earned lessons for other Indigenous litigants.

8. CONCLUSION

Misdzi Yikh v Canada poses several challenges for the Federal Court of Canada through its merger of transnational climate litigation strategies with Wet'suwet'en law. While it is not possible to predict where this and other Canadian climate cases will ultimately lead, *Misdzi Yikh* has potential to find success where climate cases have failed given its unique features, including the way it localizes and contextualizes climate change's impacts. At the same time, the plaintiffs' decision to persist in challenging a broad array of government actions and laws in their pleadings may risk ultimately derailing their claims.

---

[181] Verity Thomson, 'Decolonization, Feminism and Climate Change: A Commentary on Misdzi Yikh v Canada' (2022) 18 JSDLP 158, 168, citing Leslie M Johnson Gottesfeld, 'Conservation, Territory, and Traditional Beliefs: An Analysis of Gitksan and Wet'suwet'en Subsistence, Northwest British Columbia, Canada' (1994) 22 Hum Ecol 443.

[182] Whonnock (n 32) 225.

[183] ibid 51.

[184] ibid 51.

APPENDIX

A significant development has occurred in the *Misdzi Yikh* matter since this chapter was finalized and accepted for publication. After the plaintiffs amended their pleadings in 2024, the federal government brought a motion to the Federal Court to strike their claims a second time and was once again successful in doing so (*Lho'imggin v Canada*). In this latest decision, the Court described the central focus of the plaintiffs' amended pleadings as the allegation that Canada has no existing or planned legislative or policy initiatives that will enable it to achieve the emission reductions required to meet either its Nationally Determined Contribution or the commitment under the Paris Agreement to keep mean global warming well below 2°C above pre-industrial levels.[185] The plaintiffs remain set on challenging Canada's facilitation of the development and operation of high greenhouse gas emitting projects as breaches of their rights to life, liberty, and security. The Federal Court distinguished this approach from the parallel youth-led case of *La Rose v Canada* that will proceed to trial in late 2026. While the latter case now challenges the constitutionality of targets and plans established under one specific federal statute (the Canadian Net-Zero Emissions Accountability Act), the *Misdzi Yikh* plaintiffs continue to attack the constitutionality of the federal government's overall legislative response to its own voluntary commitments. The Federal Court struck the amended pleadings on the basis that the plaintiffs failed to adequately particularize their claims. They seek to challenge the cumulative effect of numerous federal laws and actions, potentially putting almost two thousand provisions at issue. As a result of this broad scope, they failed to plead the elements of their cause of action in relation to each alleged source of infringement. Yet the Court has left the plaintiffs with leave to amend their pleadings once more. Whether the *Misdzi Yikh* plaintiffs are willing and able to do so in a manner that gets them to a hearing on the merits remains to be seen.

---

185 *Lho'imggin v Canada* 2025 FC 1586.

10. Promoting climate litigation via industrial law and policy? Chinese climate litigation in context

Mingzhe Zhu

1. INTRODUCTION

In 2016, the Friends of Nature Institute (FON), an environmental non-governmental organization (NGO), brought environmental public interest litigation (PIL) against two state-owned provincial electric power companies in Gansu and Ningxia, respectively, for their high rates of renewable energy 'curtailment'—a term describing grid companies' reduction in renewable power transmission.[1] These cases single out as 'core' climate cases because mitigation features as the central issue in the pleadings.[2] In comparison, the Supreme People's Court of China (SPC) uses the term 'carbon-related cases' instead of climate cases and reports that over 1 million such cases have been decided by Chinese courts since the Paris Agreement.[3] This broader definition includes

---

1 The Mining Area Court in Gansu, the Friends of Nature Environmental Institute of Chaoyang District, Beijing v State Grid Gansu Electric Power Corporation (17 April 2023) [(2019) Gan 95 Minchu No.7], 甘肃矿区人民法院公告 [(2019)甘95民初7号] (cited hereafter as 'Gansu case'); The Yinchuan Intermediate People's Court in Ningxia, the Friends of Nature Environmental Institute of Chaoyang District, Beijing v State Grid Ningxia Electric Power Corporation (20 April 2023) [(2018) Ning 01 Minchu No.109], 宁夏回族自治区银川市中级人民法院民事判决书 [(2018)宁01民初109号] (cited hereafter as 'Ningxia case').

2 David L Markell and JB Ruhl, 'An Empirical Assessment of Climate Change in the Courts: A New Jurisprudence or Business as Usual?' (2012) 64 *Fla L Rev* 15; Jacqueline Peel and Hari M Osofsky, *Climate Change Litigation: Regulatory Pathways to Cleaner Energy* (CUP 2015) 8.

3 Supreme People's Court, Providing Judicial Service to Carbon Peaking and Carbon Neutrality, (最高人民法院：《为积极稳妥推进碳达峰碳中和提供有力司法服务》), 〈www.court.gov.cn/zixunxiangqing-389371.html〉.

even contractual disputes involving companies specializing in low-carbon technology.[4]

The judicial system in China can be viewed as an archetypal example of what Mirjan Damaška identifies as the hierarchical officialdom operating within an activist state.[5] In the vast majority of climate litigation cases and 'carbon-related cases', judges act as state functionaries entrusted with the task of advancing climate goals. They use ambitious climate policy documents to motivate statutory or contractual interpretation even when litigants did not put forward climate considerations.[6] For instance, courts can annul cryptocurrency mining contracts because they consider these contracts contrary to the public interest or public order, relying on non-binding policy documents warning of the risks of cryptocurrency-related business to financial market stability and energy consumption.[7] However, references to human rights and ethics-based justifications are absent in the climate cases argued before Chinese courts. Although citizens' environmental rights were contemplated when the Constitution was amended in 2018, it was the state's duty of protection that became enshrined in the Constitution as amended.[8]

China is the largest greenhouse gas (GHG) emitter and a global leader in renewable energy, already surpassing its 2030 target of installing 1,200 gigawatts (GW) of wind and solar capacity by 2024.[9] This rapid increase in renewable energy capacity is largely due to the construction of massive wind and solar farms in China's underdeveloped western interior provinces, in contrast to the more industrialized coastal areas.[10] However, it is one thing to generate renewable electricity but quite another to fully use it to replace thermal power.

---

4 SPC, *White Paper on China's Environment and Resources Adjudication (2019)* (People's Court Press 2020) 16–18 (最高人民法院：《中国环境资源审判（2019）》，人民法院出版社2020 年版，第16–18 页).

5 Donald Clarke, 'Order and Law in China' [2022] *U of Illinois L Rev* 541; Mirjan R Damaška, *The Faces of Justice and State Authority: A Comparative Approach to the Legal Process* (Yale University Press 1986).

6 Mingzhe Zhu, 'The Rule of Climate Policy: How Do Chinese Judges Contribute to Climate Governance without Climate Law?' (2021) 11 TEL 119.

7 Mingzhe Zhu, 'The "Bitcoin Judgements" in China: Promoting Climate Awareness by Judicial Reasoning?' (2023) 32 *Rev of Eur, Comparative & Intl Environmental L* 158.

8 Zhu, 'The Rule of Climate Policy' (n 6) 125.

9 Jillian Ambrose, 'China to Head Green Energy Boom with 60% of New Projects in Next Six Years' *The Guardian* (London, 9 October 2024) <www.theguardian.com/environment/2024/oct/09/china-to-head-green-energy-boom-with-60-of-new-projects-in-next-six-years> accessed 17 October 2024.

10 Tyler Harlan, 'Low-Carbon Frontier: Renewable Energy and the New Resource Boom in Western China' [2023] 225 *China Q* 591.

Until recently, grid companies in China have been reluctant to fully integrate renewable power into its power grids.

The technical reason for this reluctance is straightforward: a conventional power grid must balance mismatches between generation and consumption within a reasonable timeframe to maintain frequency stability and prevent power outages. Solar and wind energy, which vary significantly in space and time, can cause abnormal frequency and voltage, overload transmission lines, and exacerbate supply-demand mismatches.[11]

As this chapter will discuss, grid companies' reluctance is also driven by economic and policy factors. Interprovincial electricity transmission in China has been uncommon due to the lack of infrastructure, resulting in a mismatch between supply and demand in China's western renewable energy hubs where industrial energy consumption remains low. As a result, before 2015, renewable energy generated in the western regions was neither fully consumed locally nor transmitted to the energy-hungry eastern regions, leading to high curtailment rates.

This chapter first offers an in-depth description of two cases brought against state-owned provincial electric power companies in Gansu and Ningxia, respectively. It then identifies the possible innovations in China's legal and policy frameworks as well as the access to justice challenges. Finally, it assesses the potential domestic and transnational implications of these cases.

2. DETAILED CASE ANALYSIS

2.1 Case Summaries

In the Gansu case and the Ningxia case, the plaintiff, Friends of Nature (FON), argued that the defendants'—State Grid Gansu Electric Power Corporation (Gansu State Grid) and State Grid Ningxia Electric Power Corporation (Ningxia State Grid)—high curtailment rate of wind and solar power violated Article 14(3) of China's Renewable Energy Law, which requires grid companies to purchase 'in full amount the on-grid electricity of the grid-connected power generated by projects using renewable resources which meet the grid connection technical standards in the coverage area of their grids'.[12] In the Ningxia case, the Yinchuan Intermediate People's Court concluded that

---

[11] Duong Quoc Hung, Md Rakibuzzaman Shah and N Mithulananthan, 'Technical Challenges, Security and Risk in Grid Integration of Renewable Energy' in Dilan Jayaweera (ed), *Smart Power Systems and Renewable Energy System Integration* (Springer International Publishing 2016).

[12] Renewable Energy Law of the People's Republic of China (2005/2009), art 14(3) (REL).

Ningxia State Grid had not violated the REL.[13] The plaintiff has appealed the decision, but the court of appeal has not yet announced its decision. In the Gansu case, Gansu State Grid and FON negotiated a mediation agreement that is final. Gansu State Grid committed to: (a) continuous investment to integrate renewable electricity into the grid before 2025; (b) provide an annual report during the execution of the aforementioned project to the court, with a corresponding copy furnished to FON; and (c) disclose information pertinent to the utilization of clean energy, subject to public oversight.[14]

2.2 Party Identity

Friends of Nature, registered with the Ministry of Civil Affairs in 1994, is the oldest environmental NGO in China. Like many early Chinese NGOs, FON had governmental ties.[15] One of its founding members, Liang Congjie, was a senior member of the People's Political Consultative Conference, a political advisory body without legislative power.[16] Despite initially relying on Liang's political influence, FON has developed institutional competence, maintaining a positive reputation with both the public and the government. It has become an iconic NGO that uses litigation to drive legislative change.[17] FON was the applicant in another landmark case, the Green Peafowl lawsuit, successfully halting the construction of a hydropower station due to inadequate environmental impact assessments, which failed to consider the destruction of China's largest green peafowl habitat.[18]

Gansu State Grid and Ningxia State Grid are wholly-owned provincial subsidiaries of the State Grid Corporation of China (State Grid), the world's largest power distribution and transmission company, managing 80% of China's

---

13 Ningxia case (n 1) [16].

14 Gansu case (n 1) Annex 2 (Mediation Agreement) [4]–[6].

15 Qiusha Ma, 'The Governance of NGOs in China since 1978: How Much Autonomy?' (2002) 31 *Nonprofit and Voluntary Sector Q* 305.

16 Kerry Brown, 'Liang Congjie Obituary' *The Guardian* (London, 15 November 2010) <www.theguardian.com/world/2010/nov/15/liang-congjie-obituary> accessed 7 June 2024.

17 Hao Zhuang, John A Zinda and James P Lassoie, '"Crouching Tiger, Hidden Power": A 25-Year Strategic Advocacy Voyage of an Environmental NGO in China' (2022) 31 *J of Environment & Development* 331.

18 See Mingzhe Zhu, 'Biodiversity Litigation in China: Confronting Degradation from All Sides in the Era of Ecological Civilization' in Guillaume Futhazar, Sandrine Maljean-Dubois and Jona Razzaque (eds), *Biodiversity Litigation* (OUP 2023) 134.

electricity grids.[19] Each provincial subsidiary of the State Grid is responsible for the investment, construction, maintenance, and operation of the grid within its province. Gansu State Grid manages the grid in Gansu Province and Ningxia State Grid oversees the grid in Ningxia Hui Autonomous Region. Since the electricity sector reform in 2002, which dismantled the monopoly of the gigantic State Power Corporation and separated power generation from grid transmission,[20] the State Grid now only manages transmission and does not control generation.

While State Grid itself has never been sued for GHG emissions, its provincial subsidiaries have faced climate litigation. For instance, a mining company sued Hunan State Grid and claimed that the grid company's infrastructure construction interfered with its prospecting activities. Hunan State Grid successfully argued that the network construction would support a public interest significant enough to outweigh the plaintiff's rights as it reduces CO_2 emissions by 44 billion tonnes per year.[21]

Supporting the FON on both the Gansu case and Ningxia case as a third party to the litigation is the Research and Service Center for Environmental and Resource Law at China University of Political Science and Law. According to Article 15 of the Civil Procedure Law, a third party can support the plaintiff.[22] Previously known as the Center for Legal Assistance to Pollution Victims, they are a prominent organization in China contributing to environmental litigation and law development. As 'one of the most successful environmental NGOs in China',[23] the center represents and provides legal aid to pollution victims, offers free capacity building programs for lawyers, promotes international research exchanges, and participates in the drafting of various legislations and regulations. Another third party supporting FON on the Gansu case is the Chinese Wind Energy Association of the China Renewable Energy Society, a

---

19 Bloomberg News, 'China's State Grid to Keep Spending High to Match Renewables' *Bloomberg.com* (15 January 2024) <www.bloomberg.com/news/articles/2024-01-15/china-s-state-grid-to-keep-spending-high-to-match-renewables> accessed 9 June 2024.

20 Hao Zhang, *In Pursuit of Carbon Neutrality: Energy Law and Regulation in China* (CUP 2023) 13.

21 *Qianyu Mining Co Ltd v Electricity Engineering Company of Hunan Province* [2016] Case No Jing 0102 Civ 1894 Court of Xicheng District Beijing Municipality, 承德县乾宇矿业有限责任公司与湖南省送变电工程公司等财产损害赔偿纠纷一审民事判决书，北京市西城区人民法院 (2016) 京0102 民初1894号 (28 Dec 2018).

22 Civil Procedure Law (1991/2017), art 15 (CPL).

23 See Tseming Yang and Adam Moser, 'Environmental Tort Litigation in China' (2011) 41 *Env't L Rep* 10895.

government-organized NGO established in 1981 for scientists and engineers specializing in wind power.

Several eminent scholars in environmental and civil law provided expert opinion in support of the defendants, Gansu State Grid and Ningxia State Grid. Although expert opinions have no statutory basis or legal impact under Chinese law, they are nonetheless often considered by judges.[24]

2.3 Summary of the Claims Made

FON's central claim in both cases was that the grid companies' high curtailment rates of solar and wind power have led to environmental damage. The most important provision upon which they made their case is Article 14 of the REL, known as the 'priority access and full purchase provision'. The plaintiffs argued that Article 14(3) of the REL imposes a statutory obligation on grid companies to purchase all electricity that is generated from renewable energy sources and provide necessary connection services. However, between January 1, 2015 and June 30, 2016, wind and solar curtailment reached 360 million kilowatt-hours (kWh) in Ningxia and 18.671 billion kWh in Gansu. FON further contended that this curtailed renewable energy was replaced by electricity generated from coal, leading to significant unnecessary air pollutant and carbon dioxide emissions from coal power plants, which contribute to global climate change and have substantial adverse environmental impacts.[25]

2.4 Challenges to Bringing a Claim in Court by an NGO

FON and environmental NGOs have standing to bring litigation against grid companies under the 2012 CPL amendments and the 2014 EPL amendments.[26]

[24] According to an empirical survey conducted by the High Court of Zhejiang Province, about 80% of the judges being interviewed admitted that they would read and carefully consider the experts' opinions. In some cases, they are even adopted in the decisions. See 中国法院网, 《专家法律意见书的是与非》 (8 May 2003) <www.chinacourt.org/article/detail/2003/05/id/55775.shtml> accessed 9 June 2024.

[25] Ningxia case (n 1) [2]; Gansu case (n 1) Annex 1 (Complaint).

[26] CPL, art 55; Environmental Protection Law of the People's Republic of China (1989/2014), art 58 (EPL). See Lei Xie and Lu Xu, 'Environmental Public Interest Litigation in China: Findings from 570 Court Cases Brought by NGOs, Public Prosecutors and Local Government' (2021) 34 *JEL* 53; Mengxing Lu, 'The Dilemma of NGOs in Vindicating Environmental Interests before the Courts: Normative and Empirical Observations from China' (2022) 10 *Chinese J of Comparative L* 222.

The CPL allows for statutory authorities and 'relevant social organizations' to file lawsuits regarding any 'environmental pollution' that threatens public interests.[27] The EPL amendment introduced 'ecological degradation' as an actionable cause, but a social organization seeking to qualify must be registered with a department of civil service at a level above the municipality and have specialized in environmental protection for at least five consecutive years.[28] Additionally, through judicial policy, the SPC established that environmental PIL can be filed to prevent acts that pose significant risks to environmental public interests before actual harm occurs.[29] Furthermore, standards of justiciability applicable to all civil adjudications must also be observed in PIL cases.[30]

There is little doubt that FON is a qualified organization who has standing to bring litigation against grid companies. The key issue, however, involves the connection between renewable power curtailment and environmental harm. The grid companies argued that they should not be held responsible for the coal burned, as State Grid does not operate the power stations.[31] The trial court in Gansu accepted this argument and ruled that FON's case was non-justiciable because the grid company does not burn coal. The High Court of Gansu, however, granted FON standing without explaining why grid companies could be held liable for emissions from thermal power plants beyond their control.[32] The High Court only referenced Article 119 of the CPL, which requires the plaintiff to identify (1) the defendant and (2) the specific claims, facts, and legal grounds. It appears that the Court took the position that a plaintiff needs to demonstrate only the existence of greenhouse gas emissions or, in this case, a high curtailment rate, to establish standing. It was not necessary at this stage to prove causation between the defendant's actions or inactions and the alleged environmental harm.

The grid companies questioned the effectiveness of legal action brought against them by asserting that such lawsuits would not resolve the curtailment rates issue for three main reasons. First, grid management itself does not produce excessive amounts of GHGs or pollutants. Second, the high curtailment

---

27 CPL, art 55.

28 EPL, art 58.

29 SPC Explanation on Several Questions Regarding the Application of Law in Environmental Public Interest Litigation 2015, 最高人民法院关于审理环境民事公益诉讼案件适用法律若干问题的解释（法释 [2015] 1号).

30 CPL, art 119.

31 Ningxia case (n 1) [16]; *The Friends of Nature Institute v Gansu State Grid* [2018] Decision No 679 of the High Court of Gansu Province，自然之友环境研究所诉国家电网甘肃公司案，甘肃省高级人民法院 [2018] 甘民终 679 号民事裁定书 (28 Dec. 2018).

32 ibid, *The Friends of Nature Institute v Gansu State Grid*.

rate is due to overproduction of renewable electricity in the western provinces, not inadequate purchases of renewable energy. Third, grid companies lack autonomy to make decisions about electricity production, transactions, and consumption; they must follow government instructions and plans. In an interview, the attorney for Ningxia Grid argued that 'the plaintiff cannot achieve its goal without engaging with the government agency responsible for planning'.[33] In response, FON's attorney stated, 'We hope the defendants will communicate to the government that it needs to do more to ensure the full purchase of renewable electricity'.[34]

FON is aware that the government agency that establishes developmental and microeconomic policy controls how much wind and solar power capacity can be built and how renewable electricity is integrated into the grid. Yet, FON chose to pursue an indirect approach by suing the grid companies, possibly because NGOs cannot directly sue the government. That said, the executive branch is not immune from climate litigation. Citizens affected by administrative decisions can file administrative lawsuits against government agencies.[35] Although there have been no administrative litigation cases specifically focused on climate governance so far,[36] some senior Chinese judges believe that citizens could challenge local or provincial governments' climate actions or inactions through administrative lawsuits.[37] In addition, prosecutorial offices can initiate PIL against an agency's actions or inactions, including in the area of environmental protection.[38] Prosecutors have an active role in urging government agencies to fulfill their statutory duties. Rather than through trial, prosecutors primarily exercise their prosecutorial power through the preliminary procedures where they advise government agencies on how to resolve issues.[39] China's climate governance could become more concrete and effec-

---

[33] The Paper, 'Ningxia Grid Said the High Curtailment Rates are Inevitable' (澎湃新闻, '"弃风弃光"被环保组织索赔3亿，宁夏电网：做不到全额收购') (April 11) <https://m.thepaper.cn/wifiKey_detail.jsp?contid=2069105&from=wifiKey#>.

[34] ibid.

[35] Administrative Litigation Law (1989/2017), art 12 (ALL).

[36] Air pollution cases that have climate co-benefits exist, of course. Yue Zhao, 'Potential Pathways for Climate Change Litigation in China: Empirical Analysis of 41 Public Interest Litigation Cases Involving Air Pollution' [2019] 6 *Journal of Shandong University* (Philosophy and Social Sciences) 26, 赵悦：《气候变化诉讼在中国的路径探究—基于41个大气污染公益诉讼案件的实证分析》山东大学学报（哲学社会科学版），2019年第6期，第26–35页.

[37] Personal communications, in file with the author.

[38] ALL, art 25(4).

[39] Mingzhe Zhu and Jun Kuang, 'Prosecutorial Preliminary Procedure of Public Interest Litigation in the Cooperative Approach to Environmental Governance'

tive if prosecutorial power would ensure the proper exercise of the executive branch's significant authority in climate actions.

PIL was the only option for NGOs to take any legal action until 2015 when a new legal mechanism called 'ecological and environmental damage compensation litigation' (EEDCL) was established.[40] However, it is unlikely that NGOs will be able to use this new channel any time soon. The policy document that introduced EEDCL authorizes government agencies to file civil lawsuits in cases of pollution or ecological degradation and encourages statutory authorities and qualified social organizations to initiate EEDCL.[41] The Civil Code later stipulated that if remediable ecological degradation occurs due to a violation of national provisions, but no personal harm is identifiable, the state or other legally specified organizations can demand timely remediation from the responsible party.[42] Although both the 'qualified organizations' mentioned in the policy document and the 'other organizations' in the Civil Code seem to imply that NGOs could use this new mechanism, the SPC has taken a different position. In a hearing before the Standing Committee of the Congress, Justice Yang Linping, Vice President of the SPC, explained that the EEDCL allows the government to sue ordinary citizens, while PIL involves disputes between private parties.[43] Therefore, it is realistic to assume that PIL remains the only viable option for NGOs seeking standing in climate litigation.

Recent cases have explored the possibility of challenging environmental impact assessments (EIAs) for insufficient consideration of climate factors.[44] The judiciary in the Green Peafowl case confirmed that NGOs have standing to challenge EIAs.[45] However, the courts may not yet be prepared to revoke project approvals

---

[2022] 22 *Journal of China University of Geosciences* (Social Sciences) 33, 朱明哲、匡俊：《环境合作治理中的行政公益诉讼诉前程序》中国地质大学学报（社会科学版）2022年第2期，第33–44页.

40 Wenjun Luo, 'Litigation for Ecological and Environmental Damage Compensation in China: EPIL with Typical Chinese Characteristics' in Xi Wang, Xiaobo Zhao and Noeleen McNamara (eds), *Environmental Public Interest Litigation in China* (Springer International Publishing 2023).

41 Pilot Plan for the Reform of the Compensation System for Ecological and Environmental Damage (2015), jointly issued by the General Office of the Central Committee of the Communist Party of China and the General Office of the State Council.

42 Civil Code of the People's Republic of China (2021), art 1234 (Civil Code).

43 <www.chinacourt.org/article/detail/2023/10/id/7588182.shtml>.

44 Xiangbai He, 'Mitigation and Adaptation through Environmental Impact Assessment Litigation: Rethinking the Prospect of Climate Change Litigation in China' (2021) 10 *TEL* 413.

45 Zhu, 'Biodiversity Litigation in China: Confronting Degradation from All Sides in the Era of Ecological Civilization' (n 18) 134.

solely due to inadequate climate impact assessments. Currently, the technical standards for EIAs only briefly address GHG emissions and climate effects without offering detailed guidance.[46] Asking judges to evaluate the adequacy of an EIA based on vague and general standards would require them to interpret technical standards and handle complex scientific expertise, which may be challenging.

Another theoretically possible but unlikely scenario is suing large emitters, such as major oil and gas state-owned enterprises (SOEs). The legal hurdles that are common to many jurisdictions, including establishing standing, proving harm, demonstrating causation, and showing redressability,[47] will also be challenges for Chinese claimants because Tort Law in China is similar in many respects to Tort Law in other jurisdictions.[48] Additionally, a court would have difficulty holding an emitter liable for climate damage if the company has not breached its statutory duties, highlighting the significance of Article 14 in the two cases reviewed in this chapter. While Chinese civil law recognizes the duty of care and social responsibility of enterprises, judges may be reluctant to sanction a business that has not violated any written laws.[49] A senior manager and lawyer from FON admitted that they cannot sue Chinese carbon majors or power plants for their GHG emissions because these emissions are legally permissible.[50] Although leveraging human rights claims can help litigants overcome legal and procedural challenges in some jurisdictions,[51] the 'rights turn' in China would require significant institutional and, possibly, cultural changes.[52]

2.5 The Tribunal

While the Ningxia case was handled by a general jurisdiction court, the Gansu case was tried by the Mineral Court of Gansu Province, which specializes in

46 Having said that, the EIA reports of new projects commonly contain an independent section on climate impacts.

47 Geetanjali Ganguly, Joana Setzer and Veerle Heyvaert, 'If at First You Don't Succeed: Suing Corporations for Climate Change' (2018) 38 *OJLS* 841.

48 Hongfei Xie, 'The Establishment and Obstacles of Tort Liability Resulting from Climate Change' [2022] 6 *Politics and Law* 2, 谢鸿飞：《气候变化侵权责任的成立及其障碍》政治与法律 2022年第6期，第2–17页.

49 ibid.

50 Personal correspondence, in file with the author.

51 Annalisa Savaresi and Joana Setzer, 'Rights-Based Litigation in the Climate Emergency: Mapping the Landscape and New Knowledge Frontiers' (2022) 13 *JHRE* 7.

52 Mingzhe Zhu, 'Human Rights Approach to Climate Change Litigation and Its Limits' in Wei Zhang and Huawen Liu (eds), *Chinese Yearbook of Human Rights*, vol 6 (Brill 2024).

environmental and natural resource cases.[53] The role of specialized environmental tribunals in China remains somewhat ambiguous. On the one hand, they follow the same procedural rules as other courts. On the other hand, some judges hope that their specialization and increased exposure to environmental issues will provide them with better scientific expertise and enable more comprehensive integration of environmental considerations into judicial decisions.[54] However, in the cases reviewed here, no significant differences were observed.

2.6 Evidence and Climate Science

The central issue in the Ningxia case and Gansu case was not climate science but rather the technical aspects of the power system. In the Ningxia case, there were differing interpretations among the plaintiff, the defendant, and the Court regarding the relationship between thermal power and renewable electricity. The plaintiff appeared to believe that any renewable power that was curtailed would be replaced by coal-generated electricity. In contrast, the grid company explained that electricity generation could sometimes exceed consumption, and renewable power might not replace coal power due to limitations in grid connectivity. Before the marketization reform in 2022, the National Development Reform Commission (NDRC) and the National Administration of Energy (NAE) projected annual electricity usage and instructed grid companies to communicate these expectations to electricity producers. Therefore, if renewable energy producers generated more electricity than anticipated, it would not necessarily lead to a reduction in coal power production or coal use. While the Court seemed to align with the plaintiff's view on electricity generation and consumption, it emphasized the critical role of coal power in maintaining grid security.

2.7 Reference to Transnational Climate Change Jurisprudence

None of the official documents in these cases reference foreign law or judicial decisions. Additionally, there is no evidence of any influence from other jurisdictions on the claimant's strategy. Although the claimant's legal officers and attorneys are knowledgeable about recent developments in climate litigation and participate in international workshops organized for academics and

53 An insightful account on the functioning of the environmental tribunals in China, see Rachel E Stern, 'The Political Logic of China's New Environmental Courts' (2014) 72 *China J* 53.

54 Personal correspondence, in file with the author.

practitioners, they may find that foreign examples are not directly applicable to their practice.

3. CRITICAL REFLECTIONS

3.1 Legal Innovations

FON's argument is innovative in two ways. Firstly, they frame GHG emissions as a matter of ecological degradation, moving beyond the conventional pollution-centered paradigm. Secondly, they attempt to activate energy law litigation as a proxy for climate change litigation.

3.1.1 From pollution to ecological degradation

For foreign scholars who try to understand climate law and policy in China, it is crucial to bear in mind that the Chinse government neither considers nor regulates GHGs as pollutants. This position is perhaps most obviously demonstrated in the oral submission of the Chinese delegation before the International Court of Justice in the hearing on the Obligations of States in Respect of Climate Change. The identification of GHG emissions as pollution, according to the representative of China, 'is not supported by scientific evidence' and 'fails to account for the unique nature of adverse effects from GHG emissions'.[55]

Whereas some jurisdictions regulate GHGs as pollutants,[56] Chinese law establishes different rules for GHGs and air pollutants. According to the Atmospheric Pollution Prevention and Control Law (APPCL), 'atmospheric pollution caused by coal, industries, transport, dust, and agriculture' shall be prevented and controlled, while containing GHGs is only regarded as a co-benefit of regulating atmospheric pollutants.[57] The text of this provision makes a clear distinction between atmospheric pollutants and GHGs. While the former are subject to prevention and control, the reduction in the latter is merely a co-benefit of air pollution control. Two climate change litigation strategies can help to illustrate this distinction. An NGO can challenge thermal

---

55 'Oral Submission of the People's Republic of China to the Request for an Advisory Opinion of International Court of Justice by the United Nations General Assembly', December 3, 2024.

56 The qualification of carbon dioxide as a pollutant was a crucial question in the landmark *Massachusetts v EPA* case. *Massachusetts v EPA* [2007] Supreme Court of the United States 549 US 497, 127 S Ct 1438, Supreme Court Reporter 1438; Richard J Lazarus, *The Rule of Five: Making Climate History at the Supreme Court* (Belknap Press 2021) 39.

57 APPCL (1987/2018), art 2(2).

power stations for their pollution, indirectly reducing GHG emissions as a co-benefit. Alternatively, an NGO might argue that the excessive GHG emissions of a thermal plant are pollution. While plaintiffs that apply the first strategy constantly win PIL cases, the NGO that recently applied the second strategy obtained only confirmation from the court that GHGs are not pollutants.[58]

Therefore, in China, a necessary transition in litigation concerning GHGs is to move beyond the pollution-centered paradigm. Indeed, the FON in the Ningxia and Gansu cases argued that GHGs, although not pollutants in themselves, lead to climate change, which is ecological degradation. They emphasized that 'global climate change caused by greenhouse gases from coal-based thermal power generation has significant negative impacts on the environment and ecology'.[59]

As explained prior, this innovation was made possible by the distinction between pollution and ecological degradation in Chinese law. Whereas the EPL allows NGOs to take action against pollution *and* ecological degradation, the Tort Law, prior to 2020, only held polluters liable for damage caused by pollution.[60] When FON sued in 2016, the discrepancy between the EPL and the Tort Law had apparently been addressed by an SPC judicial policy document. The policy has one clause that extends tort liability to ecological degradation not caused by pollution.[61] However, the rest of the document initially specified only rules regarding damages caused by pollution until its amendment in 2020 when these rules were made applicable to ecological degradation. The Civil Code, promulgated in 2020, eventually includes ecological degradation alongside environmental pollution as a source of civil liability.[62] However, the judiciary seems hesitant to differentiate the two terms, so the pollution-centered paradigm remains dominant in Chinese judicial practice.

In addition to other challenges in litigating climate cases via tort or PIL,[63] the lack of clear and coherent rules regarding ecological degradation remains another significant obstacle. In this context, successfully characterizing climate

---

58 *Beijing Grassland Alliance Environmental Protection Center v Xingyi Shangcheng Power Generation Company Limited* [2023] Court of Qianxinan Buyei and Miao Autonomous Prefecture, <climatecasechart.com/non-us-case/beijing-grassland-alliance-environmental-protection-center-v-xingyi-shangcheng-power-generation-company-limited/>.

59 Ningxia case (n 1) [4]; Gansu case (n 2) the plaintiff's arguments [16].

60 Tort Law of the People's Republic of China (abrogated in 2021), art 65.

61 SPC Explanation on Several Questions Regarding the Application of Law in Environmental Tort Liability Cases 2015, 最高人民法院关于审理环境侵权责任纠纷案件适用法律若干问题的解释（法释 [2015] 12号).

62 Civil Code, art 1229.

63 Xiangbai He (n 44).

change as ecological degradation could theoretically incentivize courts to consider whether existing rules and principles already offer solutions to phase out fossil fuels.

It is difficult to determine if FON has successfully established ecological degradation as a new ground for future climate PIL. The Court in the Ningxia case asserted that 'this case is a dispute of environmental pollution liability',[64] avoiding any discussion of the differences between pollution and degradation. The mediation agreement in the Gansu case does not mention pollution or degradation. However, in his report on environmental adjudication addressed to the legislative body, the Chief Justice of the SPC, Zhang Jun, mentioned that a court in Gansu Province 'guided the grid company that violated the provisions of the REL [. . .] to reach an agreement with the applicant to invest at least 913 million RMB in grid connection of renewable electricity, which will prevent air pollution and reduce GHG emissions'.[65] In addition to confirming that the company breached its duty, the Chief Justice further suggested that courts will consider GHG reductions as a positive outcome of environmental adjudication.

Other jurisdictions also witness some growth in 'polluter pays cases', in which plaintiffs 'seek monetary damages from defendants based on an alleged contribution to harmful climate change impacts'.[66] Judges' appreciation of climate impacts can vary greatly depending on the jurisdiction and the type of claim. In polluter pays cases, plaintiffs typically need to prove injury, causation, and sometimes redressability at the national or regional level, which can be challenging for domestic judges.[67] In contrast, in the cases in China reviewed here, the plaintiff only needs to claim that climate change generally has adverse effects on ecology to gain standing. Given China's ambitious climate goals and the political momentum of 'ecological civilization',[68] it is

---

64 Ningxia case (n 1) [16].

65 Zhang Jun, 'SPC's Working Report on Environment and Resources Adjudication before the Standing Committee of the National People's Congress' (张军：《最高人民法院关于人民法院环境资源审判工作情况的报告—2023年10月21日在第十四届全国人民代表大会常务委员会第六次会议上》) (21 October 2023) <www.npc.gov.cn/c2/c30834/202310/t20231021_432326.html> accessed 9 June 2024.

66 Joana Setzer and Catherine Higham, 'Global Trends in Climate Change Litigation: 2024 Snapshot' (Grantham Research Institute on Climate Change and the Environment 2024) Policy Report 4.

67 Mingzhe Zhu and Liyuan Fan, 'A Comparative Study of the Judicial Construction of Scientific Credibility in Climate Litigation' (2024) 33 *Rev of Eur, Comparative & International Environmental L* 250, 250.

68 Coraline Goron, 'Ecological Civilisation and the Political Limits of a Chinese Concept of Sustainability' (2018) 2018 *China Perspectives* 39.

likely that judges will not question the reality or undesirability of climate change. However, if required to find civil liability, they may conduct further assessments of the impacts of climate change and the legal requirements involved.[69] While some plaintiffs tend to center their arguments on climate liability, others might instead frame climate damage as a part of environmental damage.[70]

3.1.2 Activating energy law in climate change litigation

The judiciary's commitment to reducing GHG emissions provides context for FON's efforts to leverage the REL. This effort marks the second major innovation in these cases. In the absence of comprehensive climate legislation, judges in China have often infused climate considerations into statutory and contractual interpretations in various civil law, particularly contractual, disputes.[71] For instance, a court ruled that an energy conservation project outlined in the appendix of a disputed contract must be interpreted as a binding obligation, in the light of the national policy on information disclosure for new buildings.[72] In contrast, FON based its environmental claims on the energy law rather than contract doctrines. While there is a lack of binding legal rules in China's climate regime,[73] the Chinese legislative body has adopted several laws related to different energy domains[74] and voted on a framework energy law in 2024. Successfully litigating climate cases through energy law could address the issue of legal gaps and facilitate future judicial decision-making on climate litigation cases.

Litigating climate issues through energy law, however, highlights a significant systemic problem: accountability. The cases reviewed here render the

---

69 For the elements of purely ecological prejudice in Chinese civil law, see Mingzhe Zhu, 'The Ecologization of the Chinese Civil Code' (2020) 64 *Pravovedenie J* 511.

70 See Setzer and Higham (n 66) 32.

71 Zhu, 'The Rule of Climate Policy' (n 6).

72 *Mujie Jia v Shandong Hailiang Real Estate Co Ltd* [2018] Case No Lu 0104 civ 6255 Court of Huaiyin District Jinan Municipality, 贾木杰与山东海亮房地产开发有限公司商品房预售合同纠纷一审民事判决书，济南市槐荫区人民法院 (2018) 鲁0104 民初6255 号 (29 Jan. 2019).

73 Haomiao Du and Hao Zhang, 'Climate Neutrality in the EU and China: An Analysis of the Stringency of Targets and the Adaptiveness of the Relevant Legal Frameworks' (2022) 31 *Rev of Eur, Comparative & International Environmental Law* 495; Zhu, 'The Rule of Climate Policy' (n 6).

74 Energy Conservation Law of the People's Republic of China 1997, Circular Economy Promotion Law of the People's Republic of China 2009, Renewable Energy Law of the People's Republic of China 2006, Electric Power Law of the People's Republic of China 1995, and Law of the People's Republic of China on the Promotion of Clean Production 2002.

accountability issue visible without being able to solve it. Energy laws grant considerable power to central and local government agencies and impose numerous responsibilities on governments and large state-owned enterprises, but they lack effective legal mechanisms for accountability. For instance, Article 20 of the draft Energy Law states that the government should phase out fossil fuels and promote non-fossil fuels, yet it provides no specific provisions to ensure this happens.

Even in rare instances where accountability measures are prescribed, government and state-owned enterprises often find ways to circumvent them. Article 14(3) of the REL is a clear example. To ensure grid companies fulfill their duty to purchase renewable electricity, Article 29 stipulates that a grid company who fails to comply with these duties is liable for the economic losses of the renewable electricity generator caused by this failure. Furthermore, it states that the provincial government should impose a fine on the grid company if the failure continues.[75]

Article 14(3) also stipulates, however, that renewable energy generators must cooperate with power grid enterprises to protect grid security and stability.[76] This provision is motivated by the technical difficulties in maintaining power grid stability despite the variability and unpredictability of renewables. In 2016, the NDRC issued a policy document titled 'Measures for the Guaranteed Full Purchase of Renewable Electric Power'. This document also stipulates that priority access and full purchase of renewable electricity must not compromise grid stability.[77]

Even though prioritizing grid stability could be seen as a reasonable choice motivated by technical considerations, the implementation of this choice 'significantly weakened the regulatory support of renewable energy generation'.[78] To begin with, none of these provisions clearly defines the circumstances threatening grid security and stability. Therefore, grid companies and local development agencies have broad discretion to decide when integration of renewables threatens the equilibrium of grids.

The choice to prioritize grid stability also gave rise to a problematic planning and overseeing structure in policy implementation. The NDRC's policy requires the provincial branch of the NAE to work with local finance departments to devise a provincial implementation plan. Additionally, the provincial governments are tasked with overseeing grid companies to ensure they adhere to central government plans regarding the percentage of renewable electricity

75 REL, art 29.

76 Zhang (n 20) 39.

77 Measures for the Guaranteed Full Purchase of Renewable Electric Power 2024, art 3 (MGFPREP).

78 Zhang (n 20) 52.

in total electricity production.[79] However, grid companies, local governments, and state-owned enterprises have strong economic incentives to continue using thermal power that presents no technical challenges and whose price is kept low by the central government.

Despite the legislative mandate, the actual purchase of renewable electricity depends on government agencies' planning and judgment regarding grid access distribution. Provincial managing authorities can allocate the operating hours of electricity generation units and their dispatch sequence.[80] Although the policy measures used to govern provincial electricity generation and transmission have no legally binding effect, the courts often use them to determine the content of parties' rights and obligations. For example, if a renewable electricity supplier, rather than an NGO, sues a provincial grid company under Articles 4 and 19 of the REL, the Court may rule that the local regulations and plans specifying transaction modes and quotas are necessary. Consequently, it may find that the grid company has met its statutory duty as long as it purchases its government-mandated quantity of renewable electricity. Legally non-binding policy measures have practical value because non-binding government policies often influence courts' interpretations, especially in climate litigation.[81]

The Chief Justice's report evaluating the success of FON's creative attempt to activate energy law appears promising for climate litigation.[82] The report suggests that failing to fully purchase renewable energy constitutes a breach of duty. However, the Court's decision in the Ningxia case suggests that the accountability issue remains. The Court demonstrated considerable deference to industrial realities and the agency's macroeconomic policy judgments. It recognized that coal-fired electricity is crucial for maintaining grid stability due to the 'inherently unpredictable, intermittent, and volatile nature of wind and photovoltaic energy production'.[83] The Court also supported the defendant's argument that renewable energy production is mainly concentrated in the western part of the country, where local industrial and economic conditions did not create enough demand for this energy.[84] In China's western provinces, where giant solar and wind farms on thousands of acres are built but industrial energy consumption remains relatively low, mismatch between supply and demand is anticipated. Furthermore, the Court ruled that Article 14(3) of the REL should be interpreted in conjunction with the MGFPREP, which is

79 MGFPREP, arts 20, 21.
80 Zhang (n 20) 47–49.
81 Zhu, 'The Rule of Climate Policy' (n 6).
82 Zhang (n 66).
83 ibid.
84 Ningxia case (n 1) [15].

a non-binding policy document. It acknowledged Ningxia State Grid's critical role in ensuring national electricity security, stating that a full purchase of renewable energy is only required when grid connections are viable and secure.[85] In other words, when overload, frequency, and voltage anomality created by integration of renewables become too significant, the grid companies are not required to purchase renewable energy. Additionally, the Court pointed out that the quantity of renewable electricity that the grid company is required to purchase is determined by government planning rather than by electricity generation. The provincial Development and Reform Commission confirmed that the defendant had met the targets set by this commission.[86]

3.2 Access to Justice Challenges

As mentioned earlier, NGOs can only challenge emitters through PIL. Despite extensive scholarly attention,[87] the number of PIL cases initiated by NGOs remains low. In 2023, all courts in China heard 6,219 new environmental PIL cases, of which only 240 were filed by NGOs.[88] The majority were initiated by prosecutors. Although prosecutors are supposed to complement the work of NGOs in civil PIL litigation by bringing cases only where no environmental NGO is able to litigate,[89] in practice their role is predominant. This has raised concerns about the 'nationalization' of PIL.[90] These concerns have become more pronounced due to the requirement that courts must pause ongoing PIL procedures if the government files an EEDCL lawsuit regarding the same activities.[91]

The first instance court in the Ningxia case did not endorse the claimant's arguments but still ordered the defendant to pay half of the claimant's attorney fees (300,000 out of 600,000 yuan) and 20,000 yuan for other costs. The Court based its decision on the plaintiff's pursuit of public interests, its social

85 Ningxia case (n 1) [16].

86 ibid.

87 See, for instance, Hao Zhuang and Steven A Wolf, 'Environmental Public Interest Litigation: New Roles for Civil Society Organizations in Environmental Governance in China' (2021) 7 *Environmental Sociology* 393.

88 SPC, *White Paper on China's Environment and Resources Adjudication (2019)* (People's Court Press 2020) 6 (最高人民法院：《中国环境资源审判（2023）》，人民法院出版社2024年版，第6页。).

89 CPL, art 55(2); see Xie and Xu (n 26) 151.

90 Hangping Chen and Hanjun Zhou, 'Reflection on the Nationalization of Public Interest Litigation' (2019) 6 Northern Legal Science 70, (陈杭平、周晗隽：《公益诉讼"国家化"的反思》北方法学2019年第6期，第70–79页).

91 ibid 72.

responsibility, and the considerable efforts invested by the plaintiff's attorney in the investigation.[92] In the Ningxia case and Gansu case, gathering evidence was not excessively costly because FON relied primarily on publicly available information. In contrast, when forensic evaluations of environmental degradation or pollution are provided by qualified institutions, costs can exceed one million yuan. On average, each PIL lawsuit costs between 200,000 and 300,000 yuan, about one-third of an ordinary NGO's annual budget.[93] FON is perhaps better resourced than other NGOs; according to its website, they attracted an income of 5.7 million yuan in 2022. The financial burden of litigation on NGOs further constrains the scope of environmental PIL. In climate litigation in particular, the required legal and scientific expertise and the availability of evidence often increases the plaintiff's financial burden and deters further actions.

Admittedly, the wind and solar curtailment cases seem to confirm some scholars' speculation that NGOs could explore innovative opportunities through high-profile and strategic litigation.[94] However, under an authoritarian government that shows a high level of distrust toward civil society and the judiciary, the number of cases filed by NGOs remains low both in terms of their frequency and success rate. Therefore, portraying the synergy between NGOs and prosecutors as a 'duet' paints an overly optimistic picture, as a duet suggests the involvement of two equal and mutually complementary partners. Instead, expecting NGOs to make significant changes to climate governance through the courts is more akin to building and defending sandcastles in a storm.[95] The policy changes following the renewable curtailment cases, which we discuss in the next section, seem highly coincidental.

3.3 Impacts on the Law, Regulation, Policy, and Social Norms

The judiciary's reception of the curtailment cases has generally been positive. Meanwhile, the law and policy landscape in the power sector to reduce curtailment rates of renewable electricity has undergone significant changes since 2015. In its White Paper on environmental adjudication, the SPC highlights the Gansu case to demonstrate how the judiciary 'establishes the ideal of green

92 Ningxia case (n 1) [19].

93 Chen and Zhou (n 90) 74.

94 Ying Xia and Yueduan Wang, 'An Unlikely Duet: Public-Private Interaction in China's Environmental Public Interest Litigation' (2023) 12 *Transnational Environmental L* 396.

95 Daniel Carpenter-Gold, 'Castles Made of Sand: Public-Interest Litigation and China's New Environmental Protection Law' (2015) 39 *Harv Envtl L Rev* 241.

development'.[96] The SPC emphasizes that the Court contributed to the development of 'new forms of productivity' or 'new productive forces' because the grid company committed to investing in infrastructural assets for renewable energy connection.[97]

The government has sought to reduce the excessive curtailment rate of renewables since at least 2015. Zhang argues that the most effective tool to tackle renewable curtailment has been administrative mandates.[98] In 2018, the NDRC issued the Clean Energy Consumption Plan (2018–2020), which assigned provincial governments annual targets of renewable penetration based on local electricity demand, existing renewable energy capacity, and grid network transportation capacity. The Plan further instructed that smart grids should be extended to rural areas to improve grid efficiency.[99] Within the administrative regime, renewables consumption targets become key performance indicators for local bureaucrats and grid companies' managers. Provincial bureaus of energy, grid companies, electricity producers, and electricity consumers are jointly responsible for achieving the targeted renewable electricity consumption.[100] Provincial governments break down and assign the Plan's annual targets to responsible entities. Grid companies play a key role in managing renewable energy access to inter- and intra-provincial grid networks and often follow mandates to enforce these targets.[101] According to the NEA's first assessment in 2022, the target of increasing renewable energy consumption to 95% in 2020 was achieved.

However, the issue of coordination remains unresolved. In 2023, newly added renewable energy capacity was 292.78 GW, significantly surpassing the NAE's expectation of 160 GW. The power sector anticipates that the NDRC cannot ignore this overperformance of renewable energy capacity and the technical challenges of renewable integration.[102] The NDRC must reconsider the 95% consumption target that was determined upon a calculation that does not match reality.

Meanwhile, since 2015 China has also made significant strides in power sector reforms to introduce competition and to support renewable energy

---

96 SPC, *White Paper on China's Environment and Resources Adjudication (2019)* (People's Court Press 2020) 3.

97 ibid.

98 Zhang (n 20) 48.

99 Clean Energy Consumption Plan (2018–2020), s 16.

100 ibid, s 9.

101 Zhang (n 20) 48.

102 Yongping Zhang and others, 'The Current State of China's Electricity Market' *Dialogue Earth* (1 September 2023) <dialogue.earth/en/business/the-current-state-of-chinas-electricity-market/> accessed 28 July 2024.

integration. These reforms have introduced competition in the wholesale, distribution, and retail segments, promoting the normalization of medium- and long-term electricity trading and large-scale pilot projects for spot trading.[103] Nationwide electricity trade increased from 1,632 terawatt-hours (TWh) in 2017 to 5,254 TWh in 2022, with more than 80% of this trade occurring intra -provincially.[104] In January 2022, the NDRC and NAE issued guidelines aimed at establishing a unified national electricity market by 2030. These guidelines align with the country's carbon reduction and energy transition goals for 2030, including lowering its CO_2 intensity by over 65% from the 2005 level,[105] increasing the share of non-fossil fuels in primary energy consumption to 25%, and installing 1,200 GW wind and solar capacity.

In theory, marketization has the potential to solve the problem of high curtailment rates of renewables in China's west. One reason for wind and solar curtailment is that wind and solar farms in the western provinces produce more energy than local industrial and residential users can consume. If these provinces could trade their renewable energy with the industrially and economically developed provinces powered by coal, the renewables would not be wasted, and some coal might be spared. Therefore, growing inter-provincial electricity trading offers an opportunity to use resources more efficiently.

In practice, however, most trades are still intra-provincial. One institutional hurdle is that the pilot projects for power marketization were implemented only at the provincial level. As a result, each province has its own trading rules, technical standards, and pricing mechanisms, which increases the transaction costs of inter-provincial trading. Furthermore, inter-provincial trading involves the construction of ultra-high-voltage transmission networks, for which it takes an average of three years to get a license. In contrast, policies favoring renewables make it possible to build a large-scale centralized solar energy farm from scratch within two years.[106]

The Chinese government is implementing policies to address the curtailment problem, though the actual outcomes of these policies remain uncertain. However, none of these policies appear to be related to the cases reviewed in this contribution. Furthermore, these cases do not seem to have significantly impacted public climate awareness, as they were not reported in mainstream media. Even FON is not keen to publicize the cases because its lawyers are still trying to achieve a more favorable decision in the appeal process.

---

[103] ibid.

[104] ibid.

[105] CO_2 intensity refers to the CO_2 emissions per unit of gross domestic product (GDP). Focusing on carbon intensity has been China's position.

[106] Zhang and others (n 102).

4. SIGNIFICANCE BEYOND CHINA

4.1 Two Contextual Elements

The two case studies offer an opportunity to reflect on the role of the power sector in climate litigation beyond China. Before assessing the transnational significance of these cases, however, we need to discuss two contextual elements related to China's renewable electricity sector.

First, as Ningxia State Grid and Gansu State Grid highlighted, the high curtailment rates in Gansu and Ningxia are due to these provinces producing vast amounts of renewable electricity, a result of deliberate developmental policies. The wind and solar boom in western China began in the early 2000s as public and private investments, along with subsidies and other policy tools, reduced the costs of renewable energy components.[107] Measures such as cutting value-added tax for renewable projects, streamlining and decentralizing approval procedures, initiating pilot projects, and setting ambitious targets for renewable capacity in the NDRC's development plans contributed to western China, including Gansu and Ningxia, becoming the epicenter of the country's renewable power development. Installed wind capacity surged from 2 GW in 2006 to 186 GW in 2013, while solar capacity expanded from 2 GW in 2011 to 39 GW in 2015, with 49% and 44% of these capacities, respectively, located in western China.[108] However, the rapid growth in renewable energy led to uncoordinated overproduction.

It was only after 2015 that the NDRC and NEA adopted administrative mandates to increase renewable energy penetration.[109] For instance, each province in China must now meet the target for renewable energy integration. Other elements, such as improvements in inter-provincial electricity transmission, particularly through ultra-high-voltage transmission networks, and occasional coal shortages also contributed to the alleviation of renewable curtailment.[110] However, while there was a boom in renewables, the administrative mandates and the ultra-high-voltage grids were not in place. Grid companies and local governments had, therefore, neither policy incentives nor technological measures to promote renewable penetration.

---

107 Harlan provides a historical account of China's strategy of developing renewables in its west that is accessible to non-technical experts. See Harlan (n 10) 11.

108 ibid 12.

109 Zhang (n 20) 48.

110 Xian Zhang and others, 'What Is Driving the Remarkable Decline of Wind and Solar Power Curtailment in China? Evidence from China and Four Typical Provinces' (2021) 174 *Renewable Energy* 31.

Second, electricity sector governance in China is highly state controlled yet decentralized and fragmented.[111] Despite subsequent reforms, this sector remains predominantly owned, planned, financed, and operated by government or semi-government entities. However, managing capacities are not allocated to one 'independent, well resourced, and robust agency at the central government level',[112] but to several and at different levels.

As the leading state apparatus in developmental and microeconomic policy, the NDRC holds significant decision-making power over the electricity sector. It has the authority to set the bundled wholesale and retail electricity prices and transmission tariffs. Also, it approves significant infrastructure investments in the electricity sector.[113] Since 2016, the NDRC has also set the minimum operating hours for renewable energy equipment in each province.

The two major grid corporations in China—the State Grid and the China Southern Power Grid—are state-owned and managed by the State-owned Assets Supervision and Administration Commission (SASAC) of the State Council. At the central government level, the policy priorities of the two leading agencies contradict; the NDRC 'has kept electricity prices low and prevented them from contributing to the rise of the consumer price index', while the SASAC 'seeks to maximize the central SOEs' fiscal revenue and advance their market dominance'.[114]

The power of policymaking and planning is largely delegated to subnational governments whose primary concern is local economic development, not decarbonization. These subnational governments can resist large-scale decarbonization projects and plans for various reasons, such as securing electricity supply to local industries, differences in dispatch prices of coal and renewable energy, and the interests of local generators and grid companies.

The two major grid companies and the five largest power-generating corporations are also powerful actors in policymaking.[115] Their chairpersons hold positions equivalent to vice-ministers in the centralized Party-state system. While the direct oversight of SASAC, adherence to state policies, and Party leadership in senior management appointments impose strong political control on these companies,[116] their senior managers can leverage their social connec-

---

111 Zhang (n 20) 12.

112 ibid 14.

113 ibid 15.

114 ibid 16.

115 Zhang (n 20) 17–18.

116 Jiangyu Wang, 'The Political Logic of Corporate Governance in China's State-Owned Enterprises' (2014) 47 *Cornell Int'l LJ* 631; Xiankun Jin and others, 'Political Governance in China's State-Owned Enterprises' (2022) 15 *China Journal of Accounting Research* 100236.

tions and political influence, particularly as members of the National Peoples' Congress, to shape and influence regulatory policies.[117]

4.2 The Difficulty of Coordination

These contextual elements highlight both the significance and the challenges of efforts to coordinate the decarbonization of the energy sector. Transitioning from fossil fuels to renewable energy requires extensive infrastructure development and land use changes that involve numerous stakeholders, including various governmental agencies, large corporations, local communities, and both industrial and residential consumers. Each stakeholder has a unique set of rights, responsibilities, and roles and is embedded in an existing social and political network. In terms of climate change litigation, on the one hand, the involvement of multiple actors creates opportunities to advance decarbonization by invoking the responsibilities of state actors or the fundamental rights of citizens. The success of some strategic climate change litigation cases demonstrates already the potential for fostering systemic changes. On the other hand, energy transition projects could face delays or blockages if potential legal issues are not addressed during the planning stages. Moreover, powerful actors might evade responsibility by shifting blame to others, as demonstrated in the Ningxia case.

4.3 Incoherence in Law and Policy

Incoherence in law and policy is arguably a global issue. For example, Bogojević has shown that, despite national commitments to emission reductions, local planning authorities in the United Kingdom, which are not statutorily required to achieve climate neutrality within specific timelines, may continue to approve projects that result in increased emissions.[118] A similar tension exists in China and the power sector. While the law imposes statutory obligations on entities like grid companies, governmental authorities who oversee the energy sector, such as the NDRC, may limit these obligations through policy documents. Although Chinese administrative judges can refuse to apply administrative acts or policy documents that they believe violate statutory rules and can communicate their recommendations to the executive branch,[119] they rarely review industrial policies. As demonstrated in the Ningxia case, judges often defer

---

[117] Curtis Milhaupt and Wentong Zheng, 'Beyond Ownership: State Capitalism and the Chinese Firm' (2015) 103 *GeoLJ* 665.

[118] Sanja Bogojević, 'Squaring the Circle? Regional Airport Expansion, Climate Change and the Planning Regime' (2024) 87 *MLR* 967.

[119] ALL, arts 53, 64.

to the executive branch's explanations and interpret statutory rules in light of policy documents. Achieving decarbonization in the power sector will likely require a profound transformation of both the sector's governance structure and possibly the industrial structure itself.

Power sector management in China exemplifies an authoritarian policy model.[120] Authority is concentrated in a few governmental agencies and state actors who possess significant autonomy and powerful policy tools to make and enforce major decisions. Despite the profound economic, social, and environmental impacts of these decisions on stakeholders, public participation is entirely absent. While this undemocratic process might be theoretically justified by the expertise of the ruling elites, there is no assurance of rational decision-making, as evidenced by the uncoordinated expansion of renewable energy capacities in western China. Even if this disjointed investment in renewables and subsequent overproduction were viewed as policy failures, no legal accountability is imposed for these issues.

Given the need for structural transformations, evaluating the success of climate change litigation should focus more on the Ningxia and Gansu cases' catalytic effects rather than just wins or losses in courts. FON has not yet achieved its desired legal outcomes in court. The challenges it faces include not only legal and procedural obstacles but also the structure of China's power sector, which is marked by significant state involvement and the political role of state-owned enterprises. Nevertheless, by suing the grid companies, FON has brought attention to the issue of renewable energy curtailment, even though holding the government accountable directly is not feasible. As long as the lawsuits raise awareness and spark discussions about necessary changes, the claimant's efforts are not in vain.

4.4 Discrepancies Between Policy Ambitions and Material Realities

Last but not least, the renewable curtailment cases demonstrate the uneasy relationship between the ambition of promoting climate mitigation by law and policy and the sheer technical challenges that can be overcome only by large-scale infrastructure upgrades and intensive investment. Current literature on climate law primarily focuses on the analysis of legal provisions and requirements, the legal arguments and the rulings of leading cases, and, sometimes, the social dynamics behind these cases.[121] It largely ignores the material conditions necessary to achieve climate goals.

---

[120] See Bruce Gilley, 'Authoritarian Environmentalism and China's Response to Climate Change' (2012) 21 *Environmental Politics* 287.

[121] Benoit Mayer and Harro van Asselt, 'The Rise of International Climate Litigation' (2023) 32 *Rev of Eur, Comparative & International Environmental Law* 175, 183.

The technological complexity of energy systems and the economic calculation of energy transitions lie far beyond the craft of most lawyers. However, some basic technological knowledge and interdisciplinary analysis are needed in climate law. On the one hand, international organizations and central governments 'up there' can set ambitious climate goals and binding rules that can only be achieved by systemic changes of the entire energy systems and social organizations. On the other hand, local governments, energy enterprises, and other actors 'down here' who can claim better expertise in energy system matters can perpetuate the status quo on the basis of the unavailability of technologies and infrastructures necessary to phase out fossil fuels. What is missing is conversations on how to reorganize our societies and reimagine our energy networks with regard to the best available or prospective technologies, and how to use legal and policy instruments to make possible the desired changes.

The discrepancy between legal requirements and technological barriers is arguably more pronounced in Global South countries. Two broad observations are relevant here. First, economic and industrial development remains the policy priority of these countries and is seen as a means to improve the well-being of their peoples. Second, developmental projects, targets, plans, and policies in the Global South are still entrenched in unsustainable Global North developmental paradigms.[122] Therefore, particularly in the Global South, two seemingly contradictory imperatives are equally urgent in achieving industrial development and energy transitions. First, financial and technological assistance is crucial. Second, development paradigms that originated in the North but later globalized have to be challenged.

5. CONCLUSION

The cases reviewed in this chapter should be regarded as a respectable effort to litigate climate change within the unique legal and policy landscape of China. The plaintiff did not pursue strategies that are gaining visibility and allure in other legal systems, such as identifying GHGs as pollutants, framing climate change as a human rights issue, or advocating for the rights of nature. Instead, the arguments are rooted in the Chinese context, where development officials aim to manage climate issues through top-down industrial and economic directives. While the Chinese approach to climate governance is effective in delivering ambitious pledges, policy outputs, and promoting clean energy production, it often falls short in producing tangible positive outcomes and lacks public participation, transparency, and accountability. In this context, the

122 Ruth Gordon, 'Unsustainable Development' in Carmen G Gonzalez and others (eds), *International Environmental Law and the Global South* (CUP 2015).

plaintiff's efforts to hold powerful actors accountable and to challenge the gap between ambitious pledges and actual mitigation outcomes are particularly noteworthy.

Putting these cases in context highlights the challenges NGOs face in promoting climate governance through litigation in China. Environmental PIL remains, so far, the most practical legal avenue for action, but it has limitations. While establishing a legal stance may be straightforward, relevant legal rules often lack clear and enforceable responsibilities, making them difficult to leverage. Even when legal responsibilities are defined, as with the Renewable Energy Law, policy documents often enable governmental or influential industrial actors to circumvent the law.

The Ningxia and Gansu cases fall under the category of strategic litigation, where the plaintiff's goal was not merely to resolve a single dispute, but to raise awareness of the need for systemic changes to address the high curtailment of renewable energy and, more broadly, to decarbonize the power sector. The high curtailment issue in China's western regions stems from uncoordinated policy incentives issued by different governmental agencies and a lack of market mechanisms—challenges that extend far beyond the scope of any single lawsuit.

It is also worth noting that renewable curtailments might not be the most decisive factor in phasing out fossil fuels in China. In December 2021, China approved more coal power plants than in the entire previous year combined. This decision is a response to the sudden power cuts by many provincial governments in September 2021 to meet the central government's targets for reducing carbon intensity and energy consumption, which resulted in real power shortages. China's international commitments and climate goals are hindered by misaligned incentives among various government levels and the dominance of state-owned entities in an ill-functioning energy market that hinder effective implementation.[123] Indeed, China is not the only country where there is competition between different energy policy objectives, such as access to energy, energy security, and decarbonization.[124] However, in this country that uses industrial policy as the primary tool of climate mitigation, the outcome of such a competition is rather predictable; the makers and implementors of industrial policy tend to prioritize energy supply over decarbonization targets.

123 Henry Gao and Weihuan Zhou, 'Competition among Purposes: The Chinese Experience in the Governance of Climate Change and Energy Transition' (2023) 26 *J Intl Econ L* 803.

124 Oliver Hailes and Jorge E Viñuales, 'The Energy Transition at a Critical Juncture' (2023) 26 *J Intl Econ L* 627.

11. When corporate and financial law goes green: *Australian Securities and Investments Commission v Mercer Superannuation (Australia) Limited*

Rebekkah Markey-Towler

1. INTRODUCTION

Australian Securities and Investments Commission v Mercer Superannuation (Australia) Limited ("*ASIC v Mercer*") was the first "greenwashing" case brought by Australia's corporate regulator in the Federal Court of Australia.[1] It concerned false and misleading statements made by Mercer Superannuation (Australia) Limited ("Mercer") about the sustainable nature of seven of its "Sustainable Plus" investment options. There was no dispute over liability or penalties – rather, Mercer admitted to contravening s 12DB(1) of the *Australian Securities and Investments Commission Act 2001* and jointly proposed a declaration of contravention, pecuniary penalties and an adverse publicity order with the Australian Securities and Investments Commission. On 2 August 2024, the Court made the declaration of contravention, ordered Mercer to pay a $11.3 million penalty and issued an adverse publicity order.[2]

This chapter argues that *ASIC v Mercer* reflects the start of the "greening" of corporate and financial law in Australia. "Greening" refers to the practice of applying corporate and financial law provisions to sustainability issues, corporate and financial regulators integrating sustainability into their activities, and companies and financial entities seizing the opportunity to make "green" profits. This case also illustrates the limitations inherent in the current approach to the greening of corporate and financial law. It is unclear whether cases like *ASIC v Mercer* will contribute to the dramatic transformations needed, in law and beyond, to respond to the impacts and challenges of climate change.

1 [2024] FCA 850, BC202410429 [*Mercer*].
2 *Mercer* (n 1) [8].

The chapter proceeds as follows: section 2 examines the context, parties and forum of the *ASIC v Mercer* case and highlights their significance in the context of climate change litigation in Australia and beyond. Section 3 follows with a discussion of the claims, reasoning and evidence used in the case. Section 4 describes the possible impacts of the case on corporate and financial law provisions, regulator mandates and corporate and financial entity behaviour, as well as raising challenges and solutions.

2. CONTEXT

2.1 Climate Litigation Landscape

Australia is a hotspot for climate change litigation. The Australian and Pacific Climate Litigation Database adopts a broad definition of climate litigation that includes:

- disputes where climate change is a central issue;
- cases where climate change is a peripheral issue;
- lawsuits motivated by climate concerns; and
- cases with implications for mitigation or adaptation.[3]

As of October 2024, 450 cases meeting these criteria have been filed in state and federal courts in Australia.

Australia may be a jurisdiction leading climate litigation for several reasons. First, Australia is a carbon intensive economy.[4] In 2022, it was the highest emitter of greenhouse gases per capita of all Organization for Economic Co-operation and Development (OECD) member countries, due in large part to

[3] Melbourne Law School, 'Australian and Pacific Climate Litigation Database' (7 October 2024) <law.app.unimelb.edu.au/climate-change/index.php#overview> accessed 7 October 2024.

[4] Jacqueline Peel, 'Climate litigation is on the rise around the world and Australia is at the head of the pack' (*The Conversation*, 27 July 2023) <theconversation.com/climate-litigation-is-on-the-rise-around-the-world-and-australia-is-at-the-head-of-the-pack-210375> accessed 19 November 2024.

Australia's reliance on burning coal for energy production.[5] Coal is Australia's second-highest value export item, just behind ore and metal scrap.[6]

Second, until recently there had been limited climate change government regulatory action. In contrast, Australia's civil society has been active in the climate change space for many years and has shown a propensity for using innovative legal tools to hold governments and other actors accountable.[7] These factors have contributed to increased reliance on the courts to ensure that all actors, including government and the corporate and financial sectors, act to address climate change.[8]

Greenwashing litigation like *ASIC v Mercer* is on the rise worldwide. Cases that attempt to hold actors to account for misleading claims that their actions or products will address climate change are referred to as climate-washing litigation.[9] Climate-washing litigation cases have been filed in jurisdictions around the world, including in the United Kingdom and the United States.[10]

---

5 Ian Tiseo, 'Global per capita greenhouse gas emissions 2022, by select OECD country' (*Statista,* 19 April 2024) <https://www.statista.com/statistics/478783/leading-countries-based-on-per-capita-greenhouse-gas-emissions/>; Australian Government, 'Australian electricity generation – fuel mix' (*energy.gov.au*, 29 September 2023) <energy.gov.au/energy-data/australian-energy-statistics/data-charts/australian-electricity-generation-fuel-mix> accessed 19 November 2024.

6 Austrade, Annual Report 2022–23 (*Australian Government,* 2023) <https://www.austrade.gov.au/content/dam/austrade-assets/global/wip/austrade/documents-about-us/austrade-annual-report-2022-23.pdf> accessed 7 October 2024.

7 Jacqueline Peel, 'Climate litigation is on the rise around the world and Australia is at the head of the pack' (*The Conversation*, 27 July 2023) <theconversation.com/climate-litigation-is-on-the-rise-around-the-world-and-australia-is-at-the-head-of-the-pack-210375> accessed 19 November 2024.

8 Australia's financial sector, including superannuation funds, can also play a significant role in financing the green transition. As at 2024, Australians had $3.9 trillion invested in superannuation.

9 Lisa Benjamin & Joana Setzer, 'Climate-Washing Litigation: Legal Liability of Misleading Climate Communications' (January 2022) <cssn.org/wp-content/uploads/2022/01/CSSN-Research-Report-2022-1-Climate-Washing-Litigation-Legal-Liability-for-Misleading-Climate-Communications.pdf>.

10 Rachel Lidgate et al, 'Greenwashing dispute risks – International perspectives' (*Herbert Smith Freehills*, 5 December 2023) <herbertsmithfreehills.com/insights/2023-12/greenwashing-dispute-risks-international-perspectives> accessed 8 October 2024.

2.2 Plaintiff

Most climate change litigation has been initiated by non-governmental organisations (NGOs), individuals, or both. According to the Grantham Research Institute's *Global Trends in Climate Change Litigation 2024 Snapshot*, NGOs and individuals were the plaintiffs in 70% of global cases.

Strategic climate litigation where plaintiffs use litigation as a method to achieve social change beyond their own individual circumstances is a subset of climate change litigation.[11] Litigation can also be used as a regulatory tool to change laws, shape behaviour and increase public awareness.[12]

Australia has a long history of NGOs and concerned individuals bringing strategic climate litigation as plaintiffs. Many early cases in Australia sought to challenge government decision-making pursuant to environmental and planning legislation for projects like coal mines.[13] Recently, plaintiffs have started to bring cases against corporations and financial entities to hold them accountable for climate change and challenge their environmental representations, with the aim to drive energy transition and adaptation.[14]

In 2017, Guy and Kim Abrahams, longstanding shareholders in the Commonwealth Bank of Australia, brought a strategic climate litigation case against the bank. Represented by Environmental Justice Australia, the Abrahams argued that the Commonwealth Bank of Australia had breached its disclosure obligations under the *Corporations Act 2001* ("Corporations Act") by failing to disclose climate change risks in its annual reporting.[15]

ASIC v Mercer involved a different category of plaintiff. ASIC is Australia's corporate, markets and financial services regulator. It is mandated to "take whatever action it can take, and is necessary, in order to enforce and give

---

11 Amy Rose, Denitsa Mladenova & Vesselina Newman, 'The crucial role of strategic climate litigation' in Francesco Sindico, Kate McKenzie, Gaston A Medici-Colombo & Lennart Wegener (eds), *Research Handbook on Climate Change Litigation* (Edward Elgar Publishing 2024) 18, 20.

12 Jacqueline Peel & Hari Osofsky, *Climate Change Litigation: Regulatory Pathways to Cleaner Energy* (Cambridge University Press, 2015).

13 Jacqueline Peel, Rebekkah Markey-Towler & Thea Shields, 'Global perspectives on corporate climate legal tactics: Australia national report (*British Institute of International and Comparative Law*, 2024) 5 <biicl.org/documents/12234_global_perspectives_on_corporate_climate_legal_tactics_-_australia_national_report_v1.pdf> accessed 30 August 2024.

14 ibid.

15 'Guy Abrahams v Commonwealth Bank of Australia' (*Australian and Pacific Climate Change Litigation*, 2017) <law.app.unimelb.edu.au/climate-change/case.php?CaseID=571&browseChron=1> accessed 30 August 2024.

effect to the laws of the Commonwealth that confer functions and powers on it".[16] ASIC administers the *Corporations Act* and the *Australian Securities and Investments Commission Act 2001* ("ASIC Act"), among other legislation.

As part of its enforcement strategy, ASIC is vested with the power to commence civil proceedings and criminal investigations. Following an investigation or examination conducted pursuant to Part 3 of the ASIC Act, ASIC can commence proceedings to seek civil remedies from the courts, if it is in the public interest to do so. ASIC can also seek damages for fraud, negligence, breach of duty or other misconduct.[17]

ASIC v Mercer is an example of a case in which ASIC used its enforcement powers to address greenwashing. As defined in the case, greenwashing involves:

> making false or misleading environmental or sustainability claims in order to make a company or its business appear more environmentally friendly, sustainable or ethical, particularly in order to induce consumers to purchase its products or services or to invest in the company.[18]

Each year, ASIC sets out a list of enforcement priorities. In both 2023 and 2024, one of ASIC's enforcement priorities was "[m]isleading conduct in relation to sustainable finance including greenwashing".[19] ASIC filed *ASIC v Mercer* in the Federal Court of Australia on February 27, 2023. While this was the first greenwashing case brought by ASIC, it was not the first time a corporate and financial regulator had taken enforcement action to address greenwashing in Australia. The Australian Competition and Consumer Commission ("ACCC") had previously taken action against entities engaged in greenwashing pursuant to the Australian Consumer Law.[20] In *ACCC v Goodyear Tyres*, the tyre company made misleading representations about the environmental benefits of its new tyres. Goodyear Tyres provided an enforceable undertaking to the regulator that it would remove all material associated with the representations.[21] In

---

16 *Australian Securities and Investments Commission* Act, s 1(2)(g) [*ASIC Act*].

17 ibid s 50.

18 *Mercer* (n 2) [3].

19 'ASIC enforcement priorities' (Australian Securities & Investments Commission, 2023) <asic.gov.au/about-asic/asic-investigations-and-enforcement/asic-enforcement-priorities/> accessed 30 August 2024.

20 *Competition and Consumer Act 2010* sch 2.

21 'Australian Competition and Consumer Commission v Goodyear Tyres' (*Australian and Pacific Climate Change Litigation*, 2008) <law.app.unimelb.edu.au/climate-change/case.php?CaseID=618&browseChron=1> accessed 30 August 2024.

ACCC v DuluxGroup, the Federal Court of Australia found that DuluxGroup had made misleading or deceptive statements about the heat reducing potential of its paint products.[22] If the claims were true, the home might have been better adapted to the increased temperatures associated with climate change and reduce the health impacts on the inhabitants and reduce the need for use of cooling devices (thereby reducing carbon emissions).

The fact that a public regulator was the plaintiff in *ASIC v Mercer* means that some access to justice issues faced by NGOs and individuals in climate litigation in Australia, especially litigation against corporate and financial sector actors, may not be as significant. For example, in Australia, costs usually "follow the event", meaning that the unsuccessful party pays the costs of the successful party. This practice can be a prohibitive barrier for public interest groups or individuals bringing claims against well-resourced defendants.[23]

The cost of litigation is also a relevant concern to public regulators, as taxpayers will likely foot the bill for the costs of unsuccessful litigation. In addition, the budgets of regulators are limited, meaning that they will sometimes be incentivised to settle litigation, especially when facing defendants with deep pockets. However, regulators have a legal mandate to pursue enforcement action. Therefore, while it is important for regulators to act prudently when bringing litigation, the risks of adverse cost orders are a necessary part of regulators performing their essential function.

2.3 Defendant

Superannuation funds, like Mercer, have historically not been the primary defendant in climate change cases. A recent report from the Grantham Research Institute notes that the majority of climate cases have been filed against governments.[24] For example, in 2023 over 70% of all cases were filed against governments and only 26% involved corporate defendants.[25]

Nevertheless, there has been growing interest in litigation against corporations and other private entities, not limited to companies in the fossil fuel sector. Climate litigation has been brought against businesses in other sectors

---

22 *Australian Competition and Consumer Commission v DuluxGroup (Australia) Pty Limited* [2014] FCA 1158.

23 Peel, Markey-Towler and Shields (n 16) 31.

24 Joana Setzer and Catherine Higham, 'Global trends in climate change litigation: 2024 snapshot' (Grantham Research Institute on Climate Change and the Environment, London School of Economics and Political Science, 2024) 19 <https://www.lse.ac.uk/granthaminstitute/publication/global-trends-in-climate-change-litigation-2024-snapshot/> accessed 30 August 2024.

25 ibid.

such as airlines, food and beverages, e-commerce and financial services.[26] In particular, climate change litigation in the financial sector has been a growing area of interest.[27] For example, in 2018 Mr. McVeigh filed a case against his superannuation fund, REST, arguing that the entity had breached its obligations under the Corporations Act by failing to provide information about how it was going to manage climate-related financial risks. In addition, Mr. McVeigh argued that the superannuation fund trustee had breached its obligations under the *Superannuation Industry (Supervision) Act 1993* to perform its duties and exercise its powers in the best interests of beneficiaries and with care, skill and diligence.[28] The case was settled out of court when REST agreed to incorporate climate risks into its investments and adopt a "net zero by 2050" strategy.

ASIC v Mercer is another example of climate change litigation with a superannuation fund as the main defendant. Mercer is a trustee of the Mercer Super Trust, which holds significant assets for the benefits of its members. Between June 2021 and June 2022, the Mercer Super Trust held approximately 300,000 member accounts with total assets under management totalling around $29.8 billion. After Mercer Super Trust merged with BT Super in April 2023, it had approximately 850,000 members with assets of approximately $65 billion.[29] The Trust holds an Australian Financial Services Licence which authorises it to "provide financial product advice, deal in a financial product in relation to superannuation and life products, and provide a superannuation trustee service to retail and wholesale clients".[30]

Superannuation funds like Mercer have brought climate change litigation in part because of the significant impact these actions can have in addressing climate change. For example, superannuation (or pension funds) hold $55.7 trillion USD in assets worldwide. This means that "[t]heir investment choices will … affect the likelihood of establishing a low-carbon society".[31] Superannuation funds can use climate change as an opportunity to make "green" profits by offering sustainable investment options to attract customers.

---

26 ibid.

27 Javier Solana, 'Climate Litigation in Financial Markets: A Typology' (2020) 9(1) TEL 103.

28 'McVeigh v Retail Employees Superannuation Trust' (*Sabin Center for Climate Change Law*, 2017) <climatecasechart.com/non-us-case/mcveigh-v-retail-employees-superannuation-trust/> accessed 1 September 2024.

29 *Mercer* (n 2) [14].

30 ibid [13].

31 Esmeralda Colombo, 'From Bushfires to Misfires: Climate-related Financial Risk after *McVeigh v. Retail Employees Superannuation Trust*' (2022) 11(1) TEL 173, 176.

Superannuation funds protect the interests of their beneficiaries and can take steps to "assess, manage, and report on the risks of climate change to their assets".[32] More relevant to *ASIC v Mercer*, misleading and deceptive conduct relating to a superannuation fund's green activities can distort the information available to consumers and the market, causing harm.

2.4 Forum

Australia's court system is divided into federal and state or territory courts, with the High Court of Australia above both these jurisdictions. The High Court of Australia hears cases in its original jurisdiction and is the final court of appeal in Australia.[33] State and territory courts and tribunals deal with issues pursuant to state legislation. Federal courts hear disputes governed by federal laws. Federal laws include bankruptcy, native title, taxation, intellectual property and commercial regulation (including legislation relating to corporations and financial institutions).[34]

Most courts in Australia have been involved in climate change litigation, including the Federal Court of Australia. *ASIC v Mercer* was heard in the Federal Court. Historically, the Federal Court of Australia has been involved in climate litigation pursuant to provisions in Australia's federal *Environment Protection and Biodiversity Conservation Act 1999*. However, *ASIC v Mercer* represents a new type of case in Australia, where corporate and financial law provisions are applied to climate change disputes.[35] These provisions are discussed in section 4.

Cases in the Federal Court of Australia, including climate change cases, usually follow the same process as other cases in the Federal Court: a matter is filed in the Registry and allocated to a judge, various case management hearings and interlocutory disputes are held to prepare the matter for trial, a trial takes place, judgment is delivered, and parties have an opportunity to appeal the decision.[36]

---

[32] ibid 176.

[33] *Commonwealth of Australia Constitution Act 1900*, ss 71, 73, 75.

[34] ibid s 51.

[35] *ASIC v Mercer* is not the first case in which the Federal Court of Australia has heard a climate dispute involving provisions other than the *Environment Protection and Biodiversity Conservation Act 1999*. However, these types of disputes have historically been less common.

[36] JLB Allsop, 'Central practice note: National framework and case management (CPN-1)' (Federal Court of Australia, 2022) <fedcourt.gov.au/law-and-practice/practice-documents/practice-notes/cpn-1> accessed 1 September 2024.

ASIC v Mercer, however, followed a slightly different process as there was no dispute between the parties as to the merits of ASIC's allegations of greenwashing.[37] Instead, the parties jointly submitted proposed declarations of contravention, pecuniary penalties and an adverse publicity order. As such, the Court's primary role was to determine the appropriate penalty, after hearing the relevant facts of misleading and deceptive conduct.

3. THE DECISION

3.1 Causes of Action

ASIC v Mercer is an example of litigation to address greenwashing. As Horan J stated in the judgment, greenwashing refers to "false or misleading environmental or sustainability claims that aim to make a company or its business appear more environmentally friendly, sustainable or ethical, particularly in order to induce customers to purchase its products or services or to invest in the company".[38] In another greenwashing case brought by ASIC, *Australian Securities and Investments Commission v LGSS Pty Ltd*, O'Callaghan J, quoting an article by Das, Pearce and Henson, stated that greenwashing:

> pertains to the misleading and deceptive disclosures employed by financial institutions to entice environmentally conscious investors into purchasing their financial products that, in reality, fall short of meeting the expected Environmental, Social, and Governance (ESG) or green credentials. These ESG credentials encompass environmental compliance and measures to protect the environment, reduce greenhouse gas emissions, and manage natural resources; social compliance, which evaluates how a company treats its stakeholders; and governance compliance, focusing on appropriate governance practices such as executive transparency and accountability.[39]

Horan J noted that greenwashing is not regulated by a specific statutory regime in Australia. Rather, regulators such as ASIC rely on existing prohibitions in Australian law relating to misleading or deceptive conduct and false or misleading representations. Such provisions are included in subdivision D of Division 2 of Part 2 of the ASIC Act, Part 7.10 of the Corporations Act and Part 2-1 of the Australian Consumer Law.

---

37 *Australian Securities and Investments Commission Act 2001* ss 12DB, 12DF.

38 *Mercer* (n 2) [3].

39 [2024] FCA 587 [1]; Cecilia Anthony Das, Prafula Pearce & Tenielle Henson, 'What's Green and Ethical about Greenwashing in the Promotion of Financial Products' [2023] NZLJ 400, 401.

In this case, ASIC argued that Mercer had made false and misleading representations in contravention of s 12DB(1)(a) of the ASIC Act and that Mercer had engaged in conduct that was liable to mislead the public in relation to financial services in contravention of s 12DF(1) of the ASIC Act. These provisions provide as follows:

> *12DB False or misleading representations*
>
> 1) A person must not, in trade or commerce, in connection with the supply or possible supply of financial services, or in connection with the promotion by any means of the supply or use of financial services:
>
> a. make a false or misleading representation that services are of a particular standard, quality, value or grade; or
>
> . . .
>
> Note: Failure to comply with this subsection is an offence (see section 12GB).
>
> . . .
>
> *12DF Certain misleading conduct in relation to financial services*
>
> 1) A person must not, in trade or commerce, engage in conduct that is liable to mislead the public as to the nature, the characteristics, the suitability for their purpose or the quantity of any financial services.
> Note: Failure to comply with this subsection is an offence (see section 12GB).

3.2 Facts

As mentioned previously, *ASIC v Mercer* did not follow the path of a conventional dispute about liability. Instead, the case centred around determining an appropriate penalty and proceeded by way of the parties submitting a statement of agreed facts and admissions ("SAFA") pursuant to s 191 of the *Evidence Act 1995*. The allegations of misleading and deceptive conduct related to representations made by Mercer about its Sustainable Plus investment options during four periods of time.

In the first period, between 12 November 2021 and 25 January 2022, a video accessible on Mercer's website, on Vimeo and YouTube, included a statement that its Sustainable Plus options "exclude investments in certain sectors deemed not to be suitable" and that these options "will not invest in alcohol, gambling and carbon intensive fossil fuels like thermal coal" (the "Video Statement").[40]

[40] *Mercer* (n 2) [24].

During the second period, from 25 January 2022 to 4 April 2022, Mercer had a statement on its website that its Sustainable Plus options have "a higher proportion of sustainability-themed assets and exclude companies involved in alcohol production, carbon intensive fossil fuels, gambling and pornography" (the "First Website Statement").[41] Another page on its website included a similar statement (the "Second Website Statement").[42] During this time, the Video Statement remained on Vimeo and YouTube.

Between 4 April 2022 and 22 July 2022 (the third period), Mercer had a statement on its website that said that Sustainable Plus options "have a broader exclusion approach, for example, excluding companies involved in production of alcohol, carbon intensive fossil fuels like thermal coal, gambling and adult entertainment" (the "Third Website Statement").[43] The First Website Statement remained on its website and the Video Statement was still accessible on Vimeo, YouTube and its website.

During the fourth period from 22 July 2022 to 1 March 2023, Mercer continued to have the First Website Statement accessible on its website and continued to make representations in the Video Statement. During this same time period, Market Forces, an NGO concerned with climate issues, published articles stating that Mercer's Sustainable Plus investment options included investments in fossil fuel companies such as Whitehaven Coal and New Hope Corporation. Following these revelations, Mercer removed the words "carbon intensive fossil fuels like thermal coal" from the Third Website Statement (the "Amended Third Website Statement") but left the rest of the statement unchanged.[44]

Importantly, as Horan J noted, almost all of these representations were made without qualification. In particular, the statements in the Video Statement were made in "absolute terms", with no qualification about the level of exposure to unsuitable sectors.[45] In addition, the First, Second and Third Website Statements said that the investment options excluded companies involved in alcohol, gambling and carbon intensive fossil fuels. There was no qualification that the investment options could include minor exposure to these entities.[46] For example, some of the investment options had small investments in companies involved in fossil fuels, alcohol or gambling.[47]

41 ibid [28].
42 ibid.
43 ibid.
44 ibid [32]–[33].
45 ibid [26].
46 ibid [29].
47 ibid [45].

3.3 Reasoning

After setting out the facts, Horan J then turned to discussing the merits of the claim. In particular, Mercer admitted that all of the statements (the "Five Online Statements") constituted representations to the viewers that the Sustainable Plus options "excluded, and would continue to exclude, investments in companies involved in, or deriving profit from, the production or sale of alcohol, gambling, and (except in the case of the Amended Third Website Statement) the extraction or sale of carbon intensive fossil fuels".[48] The statements were made without qualification.

Horan J also noted that Mercer admitted to an inadequate decision-making process relating to the making of the representations. For example, there was no process for approving website communications, other than updates to the Product Disclosure Statements. There was also no process to monitor or require the use of its "Marketing Checklist" or "Legal & Compliance Checklist" which set out questions and procedures to ensure that the information was accurate and complete, and had received legal clearance. There was no evidence that these checklists had been followed.[49]

Contrary to the representations, the Sustainable Plus options did not exclude all investments in companies involved in alcohol, gambling or carbon intensive fossil fuels. Six of the seven Sustainable Plus options had some exposure "to companies involved in, or deriving profit from, the production or sale of alcohol, gambling, or the extraction or sale of carbon intensive fossil fuels" during the relevant periods.[50] For example, some of the options had shares in BHP Group Ltd and Origin Energy Limited, both fossil fuel companies.[51] Some of the options had investments in fossil fuel companies, companies involved in the sale or production of alcohol, and companies involved in gambling.[52]

As such, Horan J noted that Mercer had admitted that it made false and misleading representations about the standard, quality, value or grade of the Sustainable Plus options in trade or commerce in connection with the promotion and supply of financial services, in contravention of s 12DB(1). As Horan J said:

> Each of the Representations falsely conveyed that the Sustainable Plus Options did not include any investments in companies involved in, or deriving profit from, the production or sale of alcohol, gambling or the extraction or sale of carbon intensive

---

48 ibid [34].
49 ibid [36].
50 ibid [43].
51 ibid [44].
52 ibid [45].

> fossil fuels. Each of the Representations conveyed that the Sustainable Plus Options would not in the future include any such investments, even though Mercer did not have reasonable grounds to make such a representation.[53]

In addition, Mercer admitted that it had engaged in conduct in trade or commerce that was liable to mislead the public as to the nature and characteristics of the Sustainable Plus options, in contravention of s 12DF(1). As stated by Horan J:

> the Sustainable Plus Options actually included investments in companies involved in, or deriving profit from, the production or sale of alcohol, gambling and the extraction or sale of carbon intensive fossil fuels, and because [Mercer] did not have reasonable grounds to make the Representations.[54]

3.4 Remedies

ASIC and Mercer jointly sought orders from the Court for declarations of contravention of ss 12DB(1)(a) and 12DF(1) of the ASIC Act, pecuniary penalties of $11.3 million AUD, and an adverse publicity order that required Mercer to publish a notice about the contraventions and relief ordered by the Court.

3.4.1 Declarations

The Federal Court of Australia is empowered to declare that there has been a contravention of a civil penalty provision, such as ss 12DB and 12DF, pursuant to s 12GBA of the ASIC Act. The Court is required to make a declaration if it is satisfied that a person has contravened the provision.[55] The declaration must specify the court that made the declaration, the civil penalty provision that was contravened, the person who contravened the provision and the conduct that constituted the contravention.[56]

In the case, Horan J was satisfied that Mercer had contravened ss 12DB(1)(a) and 12DF(1) of the ASIC Act. His Honour was also satisfied that the declarations proposed by the parties accurately reflected Mercer's conduct that had contravened the provisions and provided "appropriate and adequate particulars" of how and why the conduct contravened the civil penalty provisions. As such, his Honour made declarations of contravention pursuant to s 12GBA.[57]

---

53 ibid [50].
54 ibid [51].
55 *ASIC Act* (n 19) s 12GBA(3).
56 ibid s 12GBA(4)(a)–(d).
57 *Mercer* (n 2) [69]–[71].

3.4.2 Pecuniary penalties

The Court is also empowered, pursuant to s 12GBB of the ASIC Act, to make a pecuniary penalty order against a person where a declaration of contravention has been made. While the parties jointly proposed pecuniary penalties of $11.3 million, the role of the Court was to determine the appropriate amount to be paid to the Commonwealth.[58]

The primary purpose of a civil (including a pecuniary) penalty is "to protect and promote the public interest in compliance with the law by deterring further contraventions of a like kind by the contravenor and by others".[59] In other words, the purpose is "deterrence, both specific and general, rather than punishment or retribution".[60]

The Court is required to consider all relevant matters in determining an appropriate pecuniary penalty, including:

> (a) the nature and extent of the contravention; and
> (b) the nature and extent of any loss or damage suffered because of the contravention; and
> (c) the circumstances in which the contravention took place; and
> (d) whether the person has previously been found by a court (including a court in a foreign country) to have engaged in any similar conduct; and
> (e) in the case of a contravention by the trustee of a registrable superannuation entity – the impact that the penalty under consideration would have on the beneficiaries of the entity.[61]

His Honour held that the contraventions were serious. In terms of general deterrence, he noted that there has been a "substantial increase" in demand for ESG investment products in recent years, creating an incentive for financial entities to offer such products. Greenwashing in relation to these products reduces consumers' ability to make informed choices, can result in consumers supporting entities they do not want to support and reduces consumer trust.[62] His Honour held that:

> it is vital that consumers in the financial services industry can have confidence in ESG claims made by providers of financial products and services. As is the case in many other industries, consumers may place great importance on ESG considerations when making investment decisions. Any misrepresentations in relation to ESG policies or practices associated with financial products or services, whether as

58 ibid s 12GBB.
59 c(n 2) [76].
60 ibid [76].
61 ibid s 12GBB(5).
62 ibid [100]–[102].

> an aspect of "greenwashing" practices or otherwise, undermines that confidence to the detriment of consumers and the industry generally.[63]

In relation to specific deterrence, his Honour noted that Mercer was a "well-resourced and sophisticated financial services provider" who ought to have had "adequate processes and procedures in place to prevent and/or identify and remove marketing material which was false or misleading".[64] Nevertheless, the cooperation Mercer offered reduced the need for specific deterrence.

There were other factors relevant to Horan J's decision. For example, the contraventions were not isolated. His Honour noted that they "occurred repeatedly over an extended period and involved multiple statements".[65] These representations were "absolute and unqualified".[66] The contraventions could be grouped into four periods of time, with an increasing number of contraventions in each successive period.[67] These contraventions "arose from failures by Mercer to implement adequate systems to ensure that ESG claims in relation to its superannuation products were accurate, and to monitor and enforce the application of any sustainability exclusions associated with such ESG claims".[68] Mercer did not undertake a comprehensive review of its ESG until after the proceedings had commenced.[69]

Horan J also considered mitigating factors. Mercer had cooperated with ASIC throughout the proceeding, admitted to the contraventions at the earliest possible opportunity, admitted matters contrary to its interests, disclosed relevant materials and agreed to substantial penalties.[70] Mercer also took remedial and corrective action to prevent future contraventions. For example, after the proceedings commenced, they removed the representations and published an information booklet that informed members that the Sustainable Plus options could be exposed to excluded companies.[71]

In light of those listed factors, his Honour considered that the total pecuniary penalty of $11.3 million for the period between 12 November 2021 and 1 March 2023 was appropriate. This comprised $1 million for the first period, $2 million for the second period, $3.3 million for the third period and $5

63 ibid [148].
64 ibid [104].
65 ibid [108].
66 ibid [109].
67 ibid [114], [117].
68 ibid [147].
69 ibid.
70 ibid [144].
71 ibid [139].

million for the fourth period.[72] The difference in penalties was justified by the increase in the number of contraventions over time and the failure to conduct a full review of its processes. The penalty did not exceed a total appropriate amount when the conduct was considered as a whole.[73]

3.4.3 Adverse publicity order

The Federal Court of Australia has the power to make an adverse publicity order pursuant to s 12GLB of the ASIC Act. Horan J noted that the purposes of an adverse publicity order are "to alert consumers to the fact that there has been contravening conduct, to protect the public interest by dispelling incorrect or false impressions created by such contravening conduct, and to support the primary orders by assisting in preventing repetition of the contravening conduct".[74] In contrast to a pecuniary penalty order, adverse publicity orders are punitive in nature.[75]

His Honour held that it was appropriate to make an adverse publicity order in the terms jointly submitted by the parties, requiring Mercer to display a notice on the sustainable investments page of its website for a period of six months. This was because the representations were made over an extended period of time and were likely viewed by many people who could not readily be identified.[76] Such an order was appropriate because it was a "relatively inexpensive" and "proportionate" way to ensure that consumers are provided with the corrected information. In addition, the order would bring the case to the public's attention, helping to enforce other court orders and prevent repeat conduct.[77]

There are questions about the effectiveness of such orders. The punitive effect of a fine is obvious: financial loss. However, there is little evidence that adverse publicity orders result in financial or reputational damage.[78]

4. IMPACTS

This section evaluates the impact of *ASIC v Mercer* on climate change mitigation and adaptation. Litigation is an important tool to tackle climate change as

---

72 ibid [154].
73 ibid [157]–[158].
74 ibid [163].
75 ibid.
76 ibid [164].
77 ibid [166].
78 Lloyd Freeburn, Ian Ramsay, 'An Analysis of Publicity Orders under the Australian Securities and Investments Commission Act 2001' (2022) 30 AJCCL 287, 302–304.

it provides an avenue to hold entities accountable for their actions. These cases send a signal to other actors that they too may need to change their behaviour as the world transitions towards a low carbon economy. Nevertheless, there are inherent limitations in *ASIC v Mercer* that may inhibit litigation's potential contribution to addressing the urgency of climate action.

4.1 Corporate and Financial Law

ASIC v Mercer is a significant case in Australia's climate litigation landscape. It involved an extension of Australia's corporate and financial law regime to address climate change issues. In particular, it was the first case filed by ASIC in the Federal Court of Australia that used Australian protections around misleading and deceptive conduct to address greenwashing. ASIC similarly brought successful greenwashing proceedings against Active Super and Vanguard pursuant to ss 12DB and 12 DF of the ASIC Act.[79]

Cases have been brought internationally pursuant to corporate and financial law provisions. For example, the Amsterdam District Court found that the Dutch airline company KLM had violated Dutch consumer law through misleading advertising campaigns about flying responsibly.[80] Greenpeace France and others argued that TotalEnergies and TotalEnergies Electricité et Gaz's advertising campaign misled consumers with claims of aiming for net zero by 2050, becoming a major player in energy transition and promoting the sustainability role of gas and biofuels.[81]

ASIC v Mercer might be seen as part of the "greening" of corporate and financial law in Australia. Enforcement action is being taken to ensure that there is trust in the market for green products. As ASIC Chair Joe Longo has said, "good quality, reliable information is essential to market integrity, and to confident and informed decision-making by investors … ASIC's role is to help shore up that trust, by finding the right balance between guidance, surveillance and enforcement".[82]

---

79 *Australian Securities and Investments Commission v LGSS Pty Ltd* [2024] FCA 587; *Australian Securities and Investments Commission v Vanguard Investments Australia Ltd* [2024] FCA 308.

80 'FossielVrij NL v KLM' (2022) <climatecasechart.com/non-us-case/fossielvrij-nl-v-klm/> accessed 19 November 2024.

81 'Greenpeace France and others v TotalEnergies SE and TotalEnergies Electricite et Gaz France' (2022) <climatecasechart.com/non-us-case/greenpeace-france-and-others-v-totalenergies-se-and-totalenergies-electricite-et-gaz-france/> accessed 19 November 2024.

82 Joe Longo, 'Greenwashing: A view from the regulator' (*RIAA Conference Australia*, 2 May 2024) <asic.gov.au/about-asic/news-centre/speeches/greenw

However, *ASIC v Mercer* could also be seen as a simple application of existing laws relating to misleading and deceptive conduct. Horan J explicitly acknowledged that Australia does not have a specific statutory framework to address greenwashing. Rather, these cases apply existing provisions in the Corporations Act, Australian Consumer Law and the ASIC Act. In recognition of this, Horan J applied principles that typically apply to misleading and deceptive conduct cases to the circumstances in *ASIC v Mercer*.

These principles are, first, to precisely identify the contravening conduct or representation. The second step is to consider whether the conduct was engaged, or representation made, in trade or commerce. The third step is to consider the meaning conveyed by the conduct or the representation to the relevant audience. The final step is to determine whether the representation was false or misleading or whether the conduct was liable to mislead the public, as required under ss 12DB(1)(a) and s 12DF(1) of the ASIC Act.[83]

In *ASIC v Mercer*, the Court did not expand existing legal doctrines relating to misleading and deceptive conduct. Arguably, the only difference between this case and previous cases brought pursuant to misleading and deceptive conduct provisions was that the subject matter related to green issues. ASIC shares the view that a greenwashing enforcement action is merely an application of the current legal regime. As stated by ASIC Chair Joe Longo:

> ASIC's greenwashing interventions are founded on enforcing what are long-standing and well-established legal obligations – those that prohibit misleading and deceptive conduct. The idea that it is not acceptable to engage in conduct liable to mislead the public is not new – and the courts are no stranger to considering that concept.[84]

The use of existing doctrine to address climate disputes could mean that the real harm of greenwashing is not adequately analysed or remedied. *ASIC v Mercer* focused primarily on the potential harm to consumer confidence and choice and to competitive markets. As Horan J stated, consumers place "great importance on ESG considerations" when making investment decisions but greenwashing "undermines that confidence to the detriment of consumers and the industry generally".[85] Greenwashing reduces the ability to make informed decisions, misleads customers into supporting industries they may not want to support and reduces consumer trust.[86] Greenwashing also distorts the infor-

---

ashing-a-view-from-the-regulator/> accessed 1 September 2024.

[83] *Mercer* (n 2) [57]. See also *Self Care IP Holdings v Allergan Australia* [2023] HCA 8.

[84] Longo (n 87).

[85] *Mercer* (n 2) [148].

[86] ibid [102].

mation an investor needs to make informed decisions and "undermines the efforts of businesses that are pursuing ESG goals accurately and fairly".[87]

The case also considered the possibility of financial harm caused by Mercer's actions. The parties jointly submitted that it was not possible to identify any financial harms suffered by consumers due to several unknown factors. In particular, it was not known who selected the Sustainable Plus options on the basis of the representations, the alternative choices they would have made if they had known the truth, and the difference in performance of an alternative superannuation fund. Horan J did recognise, however, that consumers were prevented from ensuring their money was not invested in industries they did not want to support, potentially distorting confidence in their choices.[88]

The focus on harm to consumer confidence and choice, while consistent with the existing provisions in corporate and financial law, meant that there was no consideration in the judgment of the possible significant climate harms caused by greenwashing. For example, misrepresentations can falsely convey that industries are taking steps to mitigate and adapt to the impacts of climate change or mean that finances that could have been channelled into "green" companies and projects are instead directed to "brown" finance. There was no consideration or discussion in the decision of these potential impacts.

For these reasons, it is arguable that existing corporate and financial laws may not be sufficient to mitigate harms caused by greenwashing or to spur the action needed to achieve climate goals such as keeping global temperature increases below 1.5 or 2 degrees Celsius. This does not, however, necessarily mean that there is a need to develop new statutory provisions to prohibit greenwashing – it could indicate that there should be further consideration of how corporate and financial law like misleading and deceptive conduct might be "greened".

Sarah Light has argued, for example, that existing corporate, securities, anti-trust and bankruptcy laws can contribute to environmental goals if these laws integrate an environmental priority principle.[89] This principle "would require that the environment be valued as a first-order concern on par with efficiency and the related market-based considerations that underlie conventional accounts of the law of the corporation" and when there is conflict "long-term environmental values and goals" ought to prevail.[90]

Applying this view to *ASIC v Mercer*, an environmental priority principle could serve as a preventative measure to prevent greenwashing by expanding

---

[87] ibid [4].

[88] ibid [133].

[89] Sarah E Light, 'The Law of the Corporation as Environmental Law' (2019) 71 SLR 137, 200.

[90] ibid 205.

environmental interpretation of existing superannuation fund trustees' duties to act with care, skill and diligence and to perform their duties and exercise their powers in the best interests of beneficiaries.[91] This expanded environmental interpretation could require superannuation fund trustees to prioritise investments that promote the long-term interests of their beneficiaries, including living on a planet kept within safe global warming levels. If an environmental priority principle had applied in *ASIC v Mercer*, it is possible that Mercer would have taken care to prevent greenwashing in the first place. The Court might also have analysed the impact of this environmental priority principle on the interpretation of the existing law in a greenwashing context, offering guidance for future conduct.

Another method to reform existing corporate and financial law (specifically the misleading and deceptive conduct provisions) would be to require consideration of harm beyond that posed to consumers and the financial market, which would include consideration of the actual or potential environmental harm caused by the misrepresentations. This would mean that consumer and financial harms are not the only important factors in determining an appropriate penalty. In *ASIC v Mercer*, Horan J could have reflected on the possibility that the money consumers were investing in these Sustainable Plus options was diverted from actual green projects and companies.

These two changes – an environmental priority principle and incorporating climate harms into the determination of a pecuniary penalty – could help green corporate and financial law. Factors relevant to climate change would no longer be limited to the factual circumstances of the case (i.e. the actions of the superannuation fund in marketing its products as sustainable). Rather, climate change considerations would become an important part of judicial and financial decision-making. Trustees would be required to consider the long-term health of the planet and beneficiaries when offering products. Climate change considerations would also be relevant in determining the appropriate penalty applied in greenwashing cases.

4.2 Corporate and Financial Regulators

ASIC v Mercer also reflects the significant impact climate change has had on the "greening" of corporate and financial regulators in Australia. Under the ASIC Act, ASIC has the function of "monitoring and promoting market integrity and consumer protection in relation to the Australian financial system".[92] Within this mandate, ASIC is addressing greenwashing. For example, as of

---

[91] *Superannuation Industry (Supervision) Act 1993* s 52A(2)(b).

[92] ASIC Act (n 19) s 12A(2).

2024, they had commenced three civil penalty proceedings (including *ASIC v Mercer*), issued 19 infringement notices and had 60 corrective disclosure outcomes.[93]

This trend of corporate and financial regulators addressing greenwashing is also visible in the United States.[94] The US Securities and Exchange Commission has brought enforcement actions and administrative proceedings against companies for greenwashing, including in *SEC v Vale S.A.*,[95] *BNY Mellon Investment Adviser, Inc.*,[96] *Goldman Sachs*[97] and *DWS Investment Management Americas, Inc.*[98]

However, while cases such as *ASIC v Mercer* arguably represent an important extension of corporate and financial regulator mandates in relation to sustainability issues, these cases cannot be divorced from concerns about the role of ASIC in addressing climate change. For example, ASIC's remit spans an extraordinary number of companies, markets, financial services and professionals. It regulates over 95,000 entities of varying sizes and complexities. Its remit is "one of the widest of any corporate regulator in the world".[99] The

---

[93] Australian Securities & Investments Commission, 'ASIC's interventions on greenwashing misconduct: 2023–2024' (*Report 791*, August 2024) 4 <download .asic.gov.au/media/lbygvudn/rep791-published-23-august-2024.pdf> accessed 1 September 2024.

[94] Barbara Ballan & Jason J Czarnezki, 'Disclosure, Greenwashing and the Future of ESG Litigation' (2024) 81 Washington & Lee LR 545, 589.

[95] *SEC v Vale S.A.* No. 22-cv-02405 (E.D.N.Y. Apr. 28, 2022).

[96] Order Instituting Administrative and Cease-And-Desist Proceedings, Making Findings, and Imposing Remedial Sanctions and a Cease-And-Desist Order, BNY Mellon Inv. Adviser, Inc., Investment Advisors Act Release No. 6032, Investment Company Act Release No. 34591 (23 May 2022).

[97] Order Instituting Administrative and Cease-and-Desist Proceedings, Making Findings, and Imposing Remedial Sanctions and a Cease-and-Desist Order, Goldman Sachs Asset Mgmt., L.P., Investment Advisers Act Release No. 6189 (22 Nov. 2022).

[98] Order Instituting Administrative and Cease-and-Desist Proceedings, Making Findings, and Imposing Remedial Sanctions and a Cease-and-Desist Order, DWS Inv. Mgmt. Ams., Inc., Investment Advisers Act Release No. 6432 (25 Sept. 2023).

[99] The Senate Economics References Committee, 'Australian Securities and Investments Commission investigation and enforcement' (*Commonwealth of Australia*, 2024) <parlinfo.aph.gov.au/parlInfo/download/committees/reportsen /RB000117/toc_pdf/AustralianSecuritiesandInvestmentsCommissioninvestigat ionandenforcement.pdf> accessed 1 September 2024.

regulator already faces challenges in fulfilling its mandate to investigate and enforce the law, without considering greenwashing interventions.[100]

Pending regulatory developments in Australia will further expand ASIC's remit in terms of sustainability issues. For example, Australia's mandatory climate change disclosure regime will take effect in 2025.[101] Entities will be required to disclose information about their climate risks and opportunities, including their governance, strategy and risk management metrics and targets. ASIC will have considerable oversight of these disclosures, including for the purposes of enforcement action. There has been little discussion of the resource implications and the impact that increased involvement in sustainability issues will have on the regulator's capacity.

If these capacity constraints are not addressed, ASIC may not be able to contribute to addressing the harms caused by corporate and financial sector greenwashing and to contribute to the achievement of sustainability goals more generally. While a recent Senate Inquiry about ASIC's capacity recommended that the regulator be split into separate company regulatory and financial conduct authorities, it provided no detail as to what this might look like and the complex issues this raises.[102] More immediate reforms could include ASIC continuing to prioritise greenwashing as a thematic issue and ensuring that an adequate portion of its budget is dedicated to addressing this issue.

There remain questions about ASIC's role in addressing greenwashing and climate change more generally. For example, ASIC does not have a specific regulatory mandate, nor a statement of expectations (a directive from the Australian Government as to how the regulator ought to act in fulfilling these new sustainability activities). There is also no publicly available information specific to how ASIC prioritises enforcement action to address greenwashing.[103]

---

[100] ibid; for example, in a previous case, the ACCC settled litigation against the tobacco industry because it would have used most of their annual litigation budget.

[101] *Treasury Laws Amendment (Financial Market Infrastructure and Other Measures) Bill 2024*.

[102] See additional senator comments in the following report: The Senate Economics References Committee (n 105) 164–165.

[103] Instead, ASIC lists some of the general factors that it considers in deciding which enforcement actions to pursue. These include the nature and seriousness of the suspected misconduct, conduct of the person or entity after the suspected misconduct, strength of the evidence, expected public benefit, likelihood of improvement and deterrence, and mitigating factors: Australian Securities & Investments Commission, 'ASIC's approach to enforcement' (Information Sheet 151, August 2023) <asic.gov.au/about-asic/asic-investigations-and-enforcement/asic-s-approach-to-enforcement/#How_we_decide> accessed 1 September 2024.

While it is important that ASIC has made greenwashing a priority, more clarity is needed about the objectives ASIC is trying to achieve through its involvement in climate change issues.[104] Addressing climate change may require more than application of its existing regulatory mandate. ASIC might also play a role in steering corporations and financial sector entities to support green outcomes. Instead of merely undertaking greenwashing enforcement action, ASIC could be given a clear statement of expectations from the government to support the achievement of good green outcomes. ASIC could also issue regulatory guidance to corporations and financial sector entities about what supports green transition and what might constitute greenwashing.

4.3 Corporate and Financial Actors

Finally, *ASIC v Mercer* is significant because it indicates the possible role that financial institutions (and superannuation funds in particular) can play in addressing climate change. Horan J referred to the substantial increase in demand in recent years for ESG products. Mercer offered several Sustainable Plus options to its members as part of this development.

However, as seen through the remedies sought in *ASIC v* Mercer, greenwashing litigation does not necessarily incentivise superannuation funds to contribute to climate change goals. The purpose of pecuniary penalties is deterrence, to discourage entities from engaging in misleading and deceptive conduct. In other words, the main object of civil penalty proceedings is to disincentivise bad behaviour, rather than encourage good behaviour. Moreover, the purpose of adverse publicity orders is specifically punitive, to punish misleading and deceptive conduct. This does not necessarily encourage entities to "do the right thing" by punishing the entity for engaging in greenwashing. The case does not provide corporations and financial sector entities with guidance about actions that would support the achievement of good green outcomes, instead primarily deterring other superannuation trustees from engaging in greenwashing.

The limited potential of these remedies to incentivise good climate action by superannuation funds is arguably reflected in the remedial action undertaken by Mercer to address its misleading and deceptive conduct. After the commencement of proceedings, Mercer removed the false and misleading representations from its website, but did not undertake divestment from investments in unsustainable industries such as fossil fuels. Instead, Mercer published a booklet informing members that the Sustainable Plus options

---

104 These are issues relating to the mitigation of greenhouse gas emissions, in particular, as well as adapting to the impacts of climate change.

could be exposed to excluded companies.[105] The fund did not undertake to further divest from these entities involved in the excluded areas nor seek to engage with these companies to encourage them to support the transition to net zero emissions. This demonstrates the limitations of enforcement actions to address greenwashing unless combined with broader incentives and strategies for industries to contribute to green goals.

It would be incorrect to conclude that greenwashing enforcement action is not a necessary part of the regulator toolkit to address climate change. It is important that consumers and businesses have confidence and trust when participating in financial markets. Nevertheless, litigation alone is unlikely to incentivise superannuation funds to contribute to achieving climate change goals. Superannuation funds are not providing the finances needed to address sustainability goals. For example, institutional investors (which includes superannuation funds) contributed less than 1% of all direct private climate change finance globally in 2021/2022.[106]

Therefore, further legal reform may be necessary to incentivise entities to play a positive role in addressing climate challenges such as mitigation of emissions or providing finances for adaptation activities. For example, mandatory disclosure provisions might require entities, including superannuation funds, to disclose their climate risks and opportunities, including providing information about their investments in green transition. However, information mechanisms alone are unlikely to spur sufficient, radical transformation of superannuation fund investment choices. Instead, the government could set targets for superannuation funds to provide finances that contribute to mitigation and adaption goals. In *ASIC v Mercer*, this may have provided Mercer greater incentive to refrain from engaging in greenwashing activities.

5. CONCLUSION

This chapter has analysed the significance of *ASIC v Mercer* as the first greenwashing case brought by ASIC to the Federal Court of Australia. This case is significant for its potential "greening" of corporate and financial laws. It also illustrates the limitations of litigation as a tool to achieve the green transformations required in law and beyond to address climate change.

In *ASIC v Mercer*, concerns about climate change did not expand the application of existing legal doctrines or the regulator's mandate; corporate and financial law provisions were applied in a conventional sense. Regulators like

---

105 *Mercer* (n 2) [139].

106 Barbara Buchner et al, 'Global Landscape of Climate Finance 2023' (*Climate Policy Initiative*, November 2023) <https://www.climatepolicyinitiative.org/publication/global-landscape-of-climate-finance-2023/> accessed 1 September 2024.

ASIC continue to face challenges in fulfilling their mandates: they have unaddressed capacity constraints and lack a clear mandate regarding their role in addressing climate concerns. Litigation does not necessarily incentivise businesses to contribute to climate change goals, instead focusing on punishing greenwashing behaviour rather than providing an incentive and model for good green investments.

These challenges suggest that litigation is an important, but limited, regulatory tool. It must be accompanied by substantive law reform that aims to provide the changes needed to keep global warming to safe levels. This chapter advocates for the adoption of an environmental priority principle and for making climate harm a consideration in sentencing regimes. It also advocates for clear mandates for corporate regulators to address climate change and sufficient resourcing to do so. Finally, enforcement action is not enough. It needs to be coupled with disclosure and target setting requirements to ensure entities contribute to addressing long-term sustainability challenges.

12. Threatened birds meet creative judges: India's climate litigation takes flight in *MK Ranjitsinh and Others v Union of India and Others*

Parul Kumar

1. INTRODUCTION

In *MK Ranjitsinh and Others v Union of India and Others* ('*MK Ranjitsinh*') the Indian Supreme Court made history by articulating a constitutionally guaranteed fundamental right to be free from the adverse effects of climate change.[1] This is the first time that the Indian Supreme Court has extensively discussed climate change in the context of India's international obligations and constitutional rights guaranteed under the Indian Constitution. The new constitutional climate right emerged in a somewhat unexpected manner as it was pronounced in a case focused on the conservation of endangered birds where such a right was not argued. This chapter examines the specific elements of this judgment and assesses its significance for climate litigation in India.

Section 2 of the chapter provides a detailed explanation of the case, including the arguments raised by the petitioners and respondents and the trajectory of the Supreme Court's reasoning in arriving at its conclusion. Section 3 highlights the role of the Indian Constitution in facilitating this landmark climate litigation judgment. Section 4 delves into the Court's creative yet staunchly anthropocentric approach in arriving at its conclusions. Section 5 engages with the tension between biodiversity protection and renewable energy objectives, which lies at the heart of this case. Lastly, section 6 deals with the future implications of this important judgment.

[1] *MK Ranjitsinh and Others v Union of India and Others* [2024] INSC 280 (*MK Ranjitsinh*) <https:// api.sci.gov.in/supremecourt/2019/20754/20754_2019_1_25_51677_Judgement_21-Mar-2024.pdf>.

2. CASE SUMMARY

In a judgment dated March 21, 2024 (made available on April 8, 2024), in the case of *MK Ranjitsinh*, the three-judge bench of the Indian Supreme Court established a new constitutional right against the adverse effects of climate change.[2] This right was pronounced in a public interest litigation case primarily focused on the conservation of certain endangered birds, particularly the great Indian bustard. Although such a right was not directly argued by the parties, its recognition emerged from broader discussions related to India's international obligations regarding climate change mitigation and the importance of promoting renewable energy to this end. The emergence of the right, therefore, requires a deeper understanding of the legal proceedings that paved the way for it, albeit not in the most predictable or obvious manner.

2.1 The Public Interest Litigation for the Conservation of Endangered Birds and the Supreme Court's Order in 2021

In 2019, conservationists and interested citizens led by MK Ranjitsinh approached the Supreme Court in a public interest litigation (filed as a writ petition under Article 32 of the Constitution) 'to seek urgent directions to save two species that are on the brink of extinction, the Great Indian Bustard and the Lesser Florican'.[3] At the time of filing of the petition in 2019, only about 100–150 great Indian bustards and 700 lesser floricans reportedly survived worldwide, with the majority living in India.[4]

The petitioners identified a variety of factors that threatened the survival of these endangered birds, including the depletion of grasslands forming their habitat diverted for agriculture, grazing, and other redevelopment purposes, and the development of mines, saltpans, and human habitat in and around their natural habitats.[5] Other factors include collision with infrastructure in their flight path, particularly with power lines and wind turbines. The petitioners sought the preservation of these endangered birds through 'the formulation

---

2 The bench was composed of Chief Justice DY Chandrachud, Justice JB Pardiwala, and Justice Manoj Misra.

3 Public interest litigations are filed as a writ petition under the Constitution of India 1950, art 32; *MK Ranjitsinh & Others v Union of India, through the Secretary, Ministry of Environment & Others*, Writ Petition (Civil) No 838 of 2019 before the Supreme Court of India (2019 Writ Petition) [1].

4 ibid [B].

5 ibid [C].

and implementation of an Emergency Response Plan that incorporates adequate in situ and ex situ conservation measures'.[6]

The Supreme Court, cognisant of these threats, directed in an order of April 19, 2021, that in the priority and potential areas for the great Indian bustard (spanning approximately 99,000 square kilometres in India's western states of Gujarat and Rajasthan),[7] overhead cables would be converted to underground power lines, wherever feasible, and future low voltage power lines would be laid underground (the '2021 Order').[8]

In November 2021 certain respondents to the petition – the Ministry of Environment, Forest and Climate Change (MOEFCC), the Ministry of Power, and the Ministry of New and Renewable Energy (MNRE) (the 'Respondent Ministries') – filed an interim application seeking to modify the court order.[9] The Respondent Ministries contended that this order 'has vast adverse implications for the power sector in India and energy transition away from fossil fuels' and also highlighted that the MNRE was not heard before the order was passed.[10] The thrust of the application was based on the technical, economic, and practical infeasibility of converting overhead power lines into underground power lines.[11] Additionally, the Respondent Ministries made their argument based on India's international commitment to climate change mitigation and the important role of the states of Gujarat and Rajasthan (where the endangered birds are predominantly found) in meeting India's renewable energy targets and Nationally Determined Contributions under the Paris Agreement due to their solar and wind energy potential.[12]

2.2 The Crafting of a Constitutional Climate Right in the Supreme Court's Judgment of March 21, 2024

Through its judgment of March 21, 2024, the Supreme Court declared that the Constitution of India guarantees a right against the adverse effects of climate

6 ibid [D].

7 See *MK Ranjitsinh* (n 1) [5], [51].

8 *MK Ranjitsinh & Others v Union of India and Others*, Order of 19 April 2021 in Interim Application No 85618 of 2020 in Writ Petition (Civil) No 838 of 2019 (2021 Order) [14], [18] <https://api.sci.gov.in/supremecourt/2019/20754/20754_2019_31_1502_27629_Judgement_19-Apr-2021.pdf>.

9 *MK Ranjitsinh & Others v Union of India and Others*, Order of 19 April 2021 in Interim Application No 149293 of 2021 in Writ Petition (Civil) No 838 of 2019 (Respondents' Interim Application for Modifying the 2021 Order).

10 ibid [2].

11 ibid [2], [12]–[14], [41].

12 ibid [2], [48]–[52].

change, derived from the constitutionally guaranteed fundamental rights to life (Article 21) and equality (Article 14).[13] The Court also relied on the Directive Principle of State Policy (DPSP) in Article 48A of the Constitution that mandates that '[t]he State shall endeavour to protect and improve the environment and to safeguard the forests and wild life of the country' and the Fundamental Duty in Article 51A(g) that casts a duty on citizens of India to protect and improve the natural environment.[14] In contrast to the Fundamental Rights in Part III of the Constitution that are enforceable against the state, the DPSP in Part IV and the Fundamental Duties in Part IV-A of the Constitution are prescriptive but not enforceable. The Supreme Court in its judgment recognises this and links the DPSP and Fundamental Duties to enforceable fundamental rights found in Articles 21 and 14 of the Constitution:

> Although these are not justiciable provisions of the Constitution, they are indications that the Constitution recognises the importance of the natural world. The importance of the environment, as indicated by these provisions, becomes a right in other parts of the Constitution. Article 21 recognises the right to life and personal liberty while Article 14 indicates that all persons shall have equality before law and the equal protection of laws. These articles are important sources of the right to a clean environment and the right against the adverse effects of climate change.[15]

By recognising the fundamental rights under Articles 21 and 14 as the source of the right against the adverse effects of climate change, the Court makes this new constitutional climate right enforceable against the state.

In arriving at its conclusions, the Court was persuaded by India's international commitment to climate change mitigation[16] and India's jurisprudence on the right to a healthy environment.[17] It concluded that the right against the adverse effects of climate change and the right to a clean environment are 'two sides of the same coin'.[18] The Court also referred to other climate litigation

---

[13] The judgment was made available on 8 April 2024; Constitution of India 1950, art 21: 'Protection of Life and Personal Liberty. No person shall be deprived of his life or personal liberty except according to procedure established by law' and art 14: 'Equality before Law. The State shall not deny to any person equality before the law or the equal protection of the laws within the territory of India'; *MK Ranjitsinh* (n 1) [24], [25].

[14] *MK Ranjitsinh* (n 1) [20].

[15] ibid [20].

[16] ibid [11]–[17]. The Supreme Court relied on *MC Mehta v Kamal Nath*, (2000) 6 SCC 213; *Virender Gaur v State of Haryana*, (1995) 2 SCC 577.

[17] ibid [20]–[24].

[18] ibid [24].

cases worldwide,[19] though without specifically highlighting their bearing on the *MK Ranjitsinh* case.[20]

The Court devotes considerable attention in the judgment to the importance of solar power as a source of renewable energy in the Indian context.[21] It notes, in particular, its potential to significantly reduce reliance on fossil fuels and curb GHG emissions and to enable India's rural population to meet their energy needs in an environmentally sustainable manner, contributing to India's sustainable growth agenda and energy security.[22] It further highlights the role of solar energy in reducing air pollution caused by fossil fuels and its advantage in not straining India's scarce groundwater supplies.[23] It concludes that '[t]he extensive use of solar power plants is a crucial step towards cleaner, cheaper, and sustainable energy'.[24] Implicit in the Court's engagement with the topic, therefore, is the view that solar energy is crucial for helping fulfil India's constitutional climate right.

2.3 The Recall of the 2021 Order and the Court's Justification

Following the discussion on renewable energy, the Court seeks to balance the concerns about the endangered birds with India's renewable energy needs (which it connects to the right against the adverse effects of climate change). The Court finds that although the great Indian bustard is a gravely endangered species, 'there is no basis to impose a general prohibition in regard to the installation of transmission lines for the distribution of solar power in an area about 99,000 square kilometres'.[25] The Court expresses the view that the reliefs sought by the petitioners in this public interest litigation may hinder the fulfilment of India's climate obligations and interfere with India's international climate commitments and the right to be free from the adverse effects of climate change.[26] This is further buttressed with reference to India's pursuit of

---

19 ibid [44]–[50].

20 However, one commentator views the very recognition of climate litigation as a mode of addressing climate change as a welcome contribution of the judgment. See Ritwick Dutta, 'How a Critically Endangered Bird Helped Further the Cause of Climate Justice in India' (*The Wire*, 4 December 2024) <https://thewire.in/environment/how-a-critically-endangered-bird-helped-further-the-cause-of-climate-justice-in-india>.

21 *MK Ranjitsinh* (n 1) [36]–[43].

22 ibid [36]–[37].

23 ibid [42].

24 ibid [42].

25 ibid [52].

26 ibid [58].

sustainable development, a pathway reflecting 'the complex interplay between environmental conservation, social equity, economic prosperity and climate change',[27] which requires balancing the conservation of the great Indian bustard with 'the conservation of the environment as a whole'.[28]

The Court, critical of the implications of a blanket prohibition on overhead transmission lines, recalls the 2021 Order and delegates the task of formulating appropriate measures to preserve the great Indian bustard in a manner compatible with sustainable development to an expert committee.[29] This is justified on the grounds that the conversion of overhead power transmission lines into underground lines 'is a matter of environmental policy', and the expert committee can facilitate evidence-based conservation efforts 'by leveraging scientific expertise and engaging stakeholders in meaningful consultations'.[30]

The judgment concludes with the Court appointing the expert committee and outlining its tasks, and setting a deadline of July 31, 2024 for the submission of a report.[31] The case was then listed before the Supreme Court on August 12, 2024, and an extension was sought for the filing of the report.[32] The Court was informed at the time of a hearing on November 7, 2024, that while the Expert Committee had completed its task in relation to the state of Rajasthan, it still needed to visit the state of Gujarat and required an additional six weeks to complete the report, and at a subsequent hearing on May 13, 2025, the report pertaining to the state of Gujarat had been filed.[33] At the time of writing this chapter, the case had just been disposed of through the Supreme Court's judgment on 19 December, 2025. The details of this judgment are found in the appendix to this chapter.

---

[27] ibid [59].

[28] ibid [60].

[29] ibid [61].

[30] ibid [61].

[31] ibid [64]–[67], [69]–[70].

[32] *MK Ranjitsinh and Others v Union of India and Others*, Order of August 12, 2024 in Writ Petition (Civil) No 838/2019 <https://api.sci.gov.in/supremecourt/2019/20754/20754_2019_1_40_54681_Order_12-Aug-2024.pdf>.

[33] *MK Ranjitsinh and Others v Union of India and Others*, Order of 7 November 2024 in Writ Petition (Civil) No 838/2019 <https://api.sci.gov.in/supremecourt/2019/20754/20754_2019_1_2_56972_Order_07-Nov-2024.pdf>; and *MK Ranjitsinh and Others v Union of India and Others*, Order of 13 May 2025 in Writ Petition (Civil) No 838/2019 <https://api.sci.gov.in/supremecourt/2019/20754/20754_2019_4_44_61744_Order_13-May-2025.pdf>.

3. THE INDIAN CONSTITUTION AS THE FACILITATOR FOR INDIA'S CLIMATE LITIGATION

The Indian Supreme Court adopted a constitutional law route to declare a new right against the adverse effects of climate change. Global South jurisdictions have, in general, displayed a proclivity to rely on established socio-economic rights jurisprudence to chart their pathway into the territory of climate litigation.[34] A recent study of climate litigation in Global South jurisdictions found that 62.5% of cases relied on a rights-based approach.[35] Heightened interest in climate litigation, particularly in the Global South, has sparked closer engagement with the role of constitutions and constitutional law in facilitating climate litigation.[36] It is, therefore, important to assess the manner in which the Indian Constitution facilitated the emergence of a new constitutional climate right in India.

3.1 The Versatility of the Indian Constitution

The climate litigation judgment in *MK Ranjitsinh* is an excellent example to showcase how the Indian Constitution lends itself to both litigants and judges, procedurally and substantively.

Public interest litigation (PIL) petitions are entertained by the Supreme Court of India and High Courts in their writ jurisdiction under Articles 32 and 226 of the Constitution (provisions to enforce constitutionally guaranteed fundamental rights) to address grievances with a public character.[37] Discussing the emergence of this unique form of litigation in India, Divan notes: 'PIL now occupies an unrivalled space in Indian constitutional law that bridges the citizen, the court, and the State'.[38] Some of the noteworthy features of the facilitation of PIL by the Supreme Court have included judicial innovations such as

---

[34] César Rodríguez-Garavito, 'Human Rights: The Global South's Route to Climate Litigation' (2020) 114 AJIL Unbound, 40–41.

[35] Jolene Lin and Jacqueline Peel, *Litigating Climate Change in the Global South* (1st edn, OUP 2024) 63.

[36] See Parul Kumar, 'Of Comparative Constitutionalism and Definitional Inclusivity: Rights-Based Climate Litigation in India and the European Union' (2024) 17:3 CCLR 136; Navraj Singh Ghaleigh, Joana Setzer and Asanga Welikala, 'The Complexities of Comparative Climate Constitutionalism' (2022) 34:3 JEL 517.

[37] Shyam Divan, 'Public Interest Litigation' in Sujit Choudhry, Madhav Khosla and Pratap Bhanu Mehta (eds), *The Oxford Handbook of the Indian Constitution* (OUP 2016) 664.

[38] ibid 663.

the relaxation of rules of standing (*locus standi*) and allowing letters to the court to be treated as writ petitions.[39] Therefore, in addition to the substantive rights-based provisions of the Indian Constitution, the *MK Ranjitsinh* litigation has also been facilitated by procedural leniency afforded by the Constitution and its judicial interpretation. The petitioners benefited from this leniency by being permitted to bring a public interest litigation on endangered birds as concerned citizens, unencumbered by strict rules of standing and causality.

The petitioners argue that '[t]he Constitution of India mandates the preservation and protection of wildlife, which include avifauna such as the Great Indian Bustard and the Lesser Florican', citing the DPSP in Article 48-A and the Fundamental Duty under Article 51A(g).[40] The petitioners also rely on the principle of inter-generational equity, which they characterise as 'a facet of both Articles 14 and 21 of the Constitution', contending that the present generation's heritage includes the great Indian bustard and the lesser florican, which future generations should not be deprived of.[41]

Interestingly, the three judge bench delivering the judgment in *MK Ranjitsinh* relies on these same substantive constitutional provisions – the fundamental rights to equality and life (guaranteed by Articles 14 and 21, respectively), the DPSP in Article 48-A, and the fundamental duty in Article 51A(g) – as the basis for pronouncing the right against the adverse effects of climate change as an extension of the recognised right to a healthy environment.[42] Further, the new right against the adverse effects of climate change is purported to 'prioritize environmental protection and sustainable development, thereby addressing the root causes of climate change and safeguarding the well-being of present and future generations'.[43] Interestingly, the judges invoke inter-generational equity, like the petitioners, but ultimately reach a different outcome than that desired by the petitioners. This highlights on the one hand the versatility of India's constitutional provisions offering avenues to invoke diverse interests but on the other hand, a great degree of flexibility that can leave tremendous scope for wide-ranging judicial interpretations with differing outcomes.

This flexibility also means, in turn, that the judgment does not reveal a clear trajectory for addressing climate change in the future. The Court does not provide adequate guidance on how this right can be claimed and by whom, which could lead to contradictory outcomes. Dubash and Ghosh highlight this with a hypothetical example: the right could be used in the future to seek the acceleration of the closure of a coal power plant but equally, by an emission-intensive

39 ibid 668–675.

40 2019 Writ Petition (n 3) [35(A)].

41 ibid [35(Q)].

42 *MK Ranjitsinh* (n 1) [20]–[24].

43 ibid [35].

thermal power plant relying on coal-based power to claim that it aids people's adaptation to increasingly hot summers (an adverse effect of climate change).[44]

In fact, the Indian Supreme Court's broad judicial discretion in applying the right to a clean environment (from which the new right against the adverse effects of climate change is derived), allowing as it does for inconsistent outcomes, has come under similar criticism. Rajamani observes that '[i]nstead of defining the environmental right in question, and therefore drawing in its boundaries, the Court appears to have moulded, and in some cases expanded, the right to fit the case before it'.[45] This could also be the pitfall of the extension of the right to environment to the adverse effects of climate change.

3.2 How MK Ranjitsinh Contributes to India's Right to Environment Jurisprudence

Over the past four decades, India has developed a body of jurisprudence around the right to environment (and its variants, such as the right to a healthy environment/clean/hygienic/pollution-free environment).[46] This right has been derived primarily from the fundamental right to life under Article 21 of the Constitution[47] but has also been connected to the DPSP in Article 48A and the Fundamental Duty in Article 51A(g).[48] In this respect, the constitutional

---

[44] Navroz Dubash and Shibani Ghosh, 'Towards Operationalising a New Climate Right for India: Unpacking the Ranjitsinh Judgment' (*The India Forum*, 19 September 2024) <www.theindiaforum.in/climate-change/toward-operationalising-new-climate-right-india>.

[45] Lavanya Rajamani, 'The Right to Environmental Protection in India: Many a Slip between the Cup and the Lip?' (2007) 16:3 Rev Eur Comp & Int'l Envtl L 274, 279.

[46] For a detailed study of the emergence of the right to environment, see Lovleen Bhullar, 'The Judiciary and the Right to Environment in India: Past, present and future' in Shibani Ghosh (ed), *Indian Environmental Law: Key Concepts and Principles* (Orient Blackswan 2019).

[47] See e.g., *Virender Gaur and Others v State of Haryana and Others*, (1995) 2 SCC 577 [7]; *AP Pollution Control Board II v Prof MV Nayudu (Retd) and Others*, (2001) 2 SCC 62 [7]; *Hinch Lal Tiwari v Kamala Devi and Others*, (2001) 6 SCC 496 [13].

[48] See e.g., *Intellectuals Forum, Tirupathi v State of Andhra Pradesh*, (2006) 3 SCC 549 [82], see also [86]; *Indian Council for Enviro-Legal Action v Union of India*, (1996) 3 SCC 212 [49]; *Virender Gaur and Others v State of Haryana and Others*, (1995) 2 SCC 577 [7]; *Jitendra Singh v Ministry of Environment and Others*, (2020) 20 SCC 581 [19]; *MC Mehta v Union of India and Others*, (2002) 4 SCC 356 [1].

right against the adverse effects of climate change mirrors past judgments that invoke these constitutional provisions.

It is clear from the wording of the judgment that the right to environment is foundational for the Court's pronouncements on climate change. The Court's invocation of familiar and relatable vocabulary found in Indian case law on the right to environment might be indicative of a deliberate attempt to maintain a continuum between the established right to environment and a new right against the adverse effects of climate change. One example of this is the Court's emphasis, in the judgment, on the role of solar energy in mitigating air pollution and its effects, with air pollution being unequivocally connected to the use of fossil fuels.[49] Through reference to air pollution, the Court finds an easy link to past Supreme Court judgments that frame the right to environment as the right to pollution-free air and water (one of which it also cites).[50] This approach creates greater legitimacy for the new constitutional climate right and signals this to future judges who may be called upon to apply this right through familiar terminology.

Unlike many of the examples of climate litigation from the Global North, which rely on an emission reduction-based framing, the use of a pollution-based framing is more instinctual in a Global South jurisdiction like India for reasons of legal strategy and direct experience with effects of environmental degradation and drivers of climate change.[51] Litigants in the Global South may prefer to avoid a direct climate change framing and instead choose tried and tested arguments from broader environmental litigation.[52] Furthermore, because of their greater vulnerability to the impacts of climate change than in Global North jurisdictions, they might also be naturally predisposed to refer to the immediate effects of climate change rather than to abstract GHG budgets of the future, which are a key feature of climate litigation in the Global North.

The recognition of the human rights dimension of the impacts of climate change in *MK Ranjitsinh* is an important contribution to the established jurisprudence on the right to environment in India. The Supreme Court emphasises India's international obligations towards climate change mitigation, connecting them to the constitutional right to environment, which was an unprecedented and much-needed advancement in the environmental discourse in India. The Court does this by engaging with India's signing of the Paris Agreement of 2015 (including a specific reference to human rights in the Preamble of the

49 *MK Ranjitsinh* (n 1) [36], [37], [42].

50 The Supreme Court in *MK Ranjitsinh* cites *Virender Gaur and Others v State of Haryana and Others*, (1995) 2 SCC 577.

51 Kumar (n 36) 145–146.

52 Jacqueline Peel and Jolene Lin, 'Transnational Climate Litigation: The Contribution of the Global South' (2019) 113:4 AJIL 679, 685.

Paris Agreement)[53] and submission of its Nationally Determined Contribution (NDC).[54] Such reliance on the Paris Agreement is consistent with climate litigation in Global North jurisdictions.[55]

In this case the Supreme Court also recognises the disproportionate adverse impacts of climate change, highlighting the greater vulnerability of certain groups due to intersectionality of various factors, such as geographical location (for example, island and forest dwellers being more at risk of losing their homes and culture) and cultural identity (for example, indigenous communities depending on activities such as fishing and hunting susceptible to losing their source of sustenance).[56] The Court recognises that this could lead to the violation of the fundamental right to equality under Article 14 of the Indian Constitution and connects this fundamental right to the 'right to be free from the adverse effects of climate change'.[57] This adds an additional layer of constitutional protection in the context of climate change, since the right to environment has thus far been linked only to the fundamental right to life and personal liberty under Article 21. This display by the Supreme Court of India of a progressive stance on climate change, by connecting the well-established right to environment to the right against the adverse effects of climate change and, in turn, amplifying the importance of this right for particularly vulnerable communities using Article 14, is a welcome development. In the specific context of the facts of this case, however, these observations are used in a convoluted manner. The Court states that poor energy access adversely affects women, girls, vulnerable, and indigenous communities to a greater degree and implies that this violates their right against the adverse effects of climate change.[58] This is ultimately used to justify the lower priority accorded to the endangered birds sought to be protected in the case. This is symptomatic of the staunchly anthropocentric motivation of the pronouncement of the new constitutional climate right, which is discussed in the next section.

---

53 *MK Ranjitsinh* (n 1) [30].

54 ibid [14], [15].

55 Some prominent examples include *Verein KlimaSeniorinnen Schweiz and Others v Switzerland* App No 53600/20 (ECHR, 9 April 2024) (*Verein KlimaSeniorinnen Schweiz*); Civ Brussels (4e ch), 17 June 2021, 2015/4585/A, C.1.8 (unofficial translation); Bundesverfassungsgericht [BVerfG] [Federal Constitutional Court] Mar 24, 2001, 157 Entscheidungen des Bundesverfassungsgerichts [BVERFGE] 30 (Ger); HR 20 December 2019, ECLI:NL:HR:2019:2006 (state of the Netherlands) (Ministry of Economic Affairs and Climate Policy/Stichting Urgenda) (Neth).

56 ibid [26].

57 ibid [25], [27].

58 ibid [35], [36].

4. DEFYING CLIMATE LITIGATION CONVENTIONALISM: JUDICIAL CREATIVITY AND MISPLACED ANTHROPOCENTRICITY TO CHART INDIA'S CLIMATE LITIGATION PATH

Many prominent climate litigation databases use a narrow definition of climate litigation where a prerequisite for classifying a case as climate litigation is the centrality of a climate change-related issue of law or fact. For example, the Sabin Center, which maintains a database of climate litigation cases in the United States of America (USA) and globally, uses two criteria for defining a climate litigation case: first, the cases must be brought before judicial bodies; and second, 'climate change law, policy, or science must be a material issue of law or fact in the case'.[59] A narrow definition of climate litigation is shaped by Global North experiences[60] and tends to exclude cases emerging from the Global South jurisdictions, where many disputes brought to courts may be linked to climate change issues indirectly.[61]

A broader definition of climate litigation includes not just litigation that makes climate change a central issue ('core' cases), but also litigation that raises climate change as a peripheral issue and litigation that is motivated by climate change or has implications for climate change ('periphery' cases).[62] In recent literature, scholars from the Global South have called for broad-based conceptualisation of climate litigation that extends to environmental jurisprudence and diverse experiences of Global South jurisdictions, including rights-based litigation and socio-political vulnerability to climate change.[63] Indeed, *MK Ranjitsinh* is an excellent example for highlighting the limitations of the conventional climate litigation definitions: in India's most prominent climate litigation judgment to date, climate change was not a central issue raised by the parties to the litigation.

---

59 'Climate Change Litigation Databases: About' (*Sabin Center for Climate Change Law Climate Change Litigation Databases*) <https://climatecasechart.com/about/>.

60 Lin and Peel (n 35) 24.

61 ibid 25.

62 Jacqueline Peel and Hari M Osofsky, *Climate Change Litigation: Regulatory Pathways to Cleaner Energy* (1st edn, CUP 2015) 8.

63 Kumar (n 36); Birsha Ohdedar, 'Climate Adaptation, Vulnerability and Rights-Based Litigation: Broadening the Scope of Climate Litigation Using Political Ecology' (2022) 13:1 JHRE 137; Tracy-Lynn Field, 'Towards a Risk-Thematic Approach for African Climate Litigation' in Kim Bouwer et al (eds), *Climate Litigation and Justice in Africa* (Bristol University Press 2024).

In the petition filed by the conservationists led by MK Ranjitsinh in 2019, climate change was not raised as an issue. In fact, the petition highlighted the decidedly anthropogenic reasons for the sharp decline in the endangered birds' population. These included the birds' collision with power infrastructure and the destruction of the birds' natural habitat due to anthropocentric causes, including the depletion and diversion of grasslands.[64] Based solely on the issue brought to court by the petitioners and their very specific intentions, this case can be clearly classified as a conservation or biodiversity protection case. It would certainly not be viewed as a climate litigation case, as commonly understood in the Global North, especially in the context of more traditional definitions of climate litigation.

In contrast, the respondents to the petition, who filed an interim application to modify the 2021 Order, brought climate change and the need to take action against it into issue in *MK Ranjitsinh*.[65] The respondents highlighted the infeasibility of converting the power lines to underground lines, contending that the 2021 Order 'has vast implications for the power sector in India and energy transition away from fossil fuels'.[66] They argued that the energy transition is vital for climate change mitigation through emission reduction and to meet India's international commitments in the Paris Agreement.[67] Notably, the respondents neither referred to the Indian Constitution nor raised any rights-based arguments in this context.

Undoubtedly, one of the contributions of the judgment is to lay down constitutional foundations for human rights protection in the context of climate change in India – uncharted territory in India's environmental and constitutional law discourse until *MK Ranjitsinh*.[68] For this reason, this case falls squarely within the realm of climate litigation. Furthermore, it reflects a unique situation where the petitioners neither focus on climate change nor make a human rights argument in the face of climate change; the first reference to climate change was made by the aggrieved respondents (but without reference to constitutional rights). This reference is then amplified in a constitutional context by an opportunistic judiciary seemingly keen to enter this new territory.

---

[64] 2019 Writ Petition (n 3) [4].

[65] Respondents' Interim Application for Modifying the 2021 Order (n 9).

[66] ibid [2].

[67] ibid [2]; see also ibid [10], [48].

[68] Abhayraj Naik and Parul Kumar, 'Endangered Birds, Renewable Energy, and India's New Constitutional Climate Right' (2025) 37 Journal of Environmental Law 159; Parul Kumar and Abhayraj Naik, 'India's New Constitutional Climate Right' (*Verfassungsblog*, 25 April 2024) <https://verfassungsblog.de/indias-new-constitutional-climate-right/> accessed 28 May 2024.

Conspicuous in this articulation of a climate right – argued by neither the petitioners nor the respondents but proactively pronounced by a creative judiciary – is the anthropocentric focus of this constitutionally guaranteed human right. This is a particularly ironic feature of the judgment in *MK Ranjitsinh*, since the petitioners set out to protect endangered birds from the anthropogenic actions threatening the birds' existence. The Supreme Court of India rejects the corrective course to protect the birds identified in the 2021 Order by recalling that Order and reframes the case to shift the focus to human beings needing protection from climate change and carves out a right for them to take recourse to in the future.

The anthropocentricity lying at the heart of this judgment is, however, unsurprising. This approach has characterised India's right to environment jurisprudence (focusing on the fundamental right to life under Article 21 of the Indian Constitution),[69] and now forms the bedrock for the new constitutional right against the adverse effects of climate change. The Supreme Court's invocation of a more ecocentric interpretation of the right to environment has been the exception rather than the norm. One example is the Supreme Court's 2006 judgment in *TN Godavarman Thirumulpad v Union of India and Others*, which focused on the protection of forests and forest land.[70] Notably, the Supreme Court, in its 2021 Order to promote the conservation of the great Indian bustard and lesser florican, emphasised the duty of the state and central government to preserve the endangered bird species.[71] It referred to a 2012 Supreme Court judgment dealing with the conservation of an endangered wild buffalo species, which relied on Article 48A and Article 51-A(g), to make a case for favouring an ecocentric approach over an anthropocentric approach to achieve environmental justice.[72] However, the 2024 judgment reverts to the more mainstream anthropocentric approach commonplace in India's environmental jurisprudence.

5. THE CLASH BETWEEN BIODIVERSITY AND RENEWABLE ENERGY

The privileging of renewable energy infrastructure over biodiversity concerns is the dominant facet of anthropocentricity in a case where, in fact, the petitioners had ecocentric motivations. In a country with very ambitious goals to

---

69 Bhullar (n 46) 43–44.

70 *TN Godavarman Thirumulpad v Union of India and Others*, (2006) 1 SCC 1 [1]–[3], [77].

71 2021 Order (n 8) [5].

72 *TN Godavarman Thirumulpad v Union of India and Others*, (2012) 3 SCC 277 [17].

expand renewable energy capacity (to the tune of 500GW by 2030, through the addition of 50GW each year),[73] this rapid augmentation has meant laying claim to scarce land and water resources.[74]

India's commitment to transition away from fossil fuels has been framed in the context of 'environmental preservation' and a 'plethora of socio-economic benefits', including enhanced energy security and curbing air pollution.[75] But the Court displays a narrow understanding of 'environmental preservation' by not giving adequate importance to preserving endangered birds alongside India's energy transition. This is further evident in the Court's oversimplification in a subsequent part of the judgment: '[b]y recognizing the right to a healthy environment and the right to be free from the adverse effects of climate change, states are compelled to prioritize environmental protection and sustainable development, thereby addressing the root causes of climate change and safeguarding the well-being of present and future generations'.[76] A healthy environment, environmental protection, and sustainable development are effectively only viewed through the narrow prism of renewable energy and not the preservation of biodiversity. This mirrors the conflict between sustainable development and environmental concerns that is often seen in the right to environment jurisprudence in India.[77]

In the specific context of the feasibility of moving power lines underground, the Court fails to consider that examples of laying power lines underground in densely populated cities in India already exist and that this is consistent with government policies.[78] On a broader level, through highlighting the importance and urgency of meeting India's international commitments to climate change mitigation, the Court apparently carves out a zone of impunity for renewable energy projects. By emphasising the indispensability of shifting away from fossil fuels to renewable energy as a 'fundamental necessity for environmental

---

73 'Government Declares Plan to Add 50 GW of Renewable Energy Capacity Annually for Next 5 Years to Achieve the Target of 500 GW by 2030' (*Government of India Press Information Bureau*, 4 May 2023) <https://pib.gov.in/PressReleaseIframePage.aspx?PRID=1913789>.

74 Kundan Pandey, 'As Renewable Sector Grows, Need for Regulating It Comes to the Fore' (*Mongabay*, 20 April 2023) <https://india.mongabay.com/2023/04/as-renewable-sector-grows-need-for-regulating-it-comes-to-the-fore/>.

75 *MK Ranjitsinh* (n 1) [16], [17].

76 ibid [35].

77 See eg, Prashant Bhushan, 'Sacrificing Human Rights and Environmental Rights at the Altar of Development' (2009) 41:2 Geo Wash Intl L Rev 389.

78 Debadityo Sinha, 'Right Against Climate Change and the Need to Go Beyond "Greenwashing"' (*NDTV*, 4 November 2024) <www.ndtv.com/opinion/right-against-climate-change-and-the-need-to-go-beyond-greenwashing-5420179>.

preservation',[79] the Court implicitly seems to conclude that any environmental concerns arising from such projects would belong to a no-go zone, as far as judicial scrutiny is concerned. The Court rather obtusely chooses not to engage with the environmental costs of renewable energy projects, even though this issue was at the heart of the petition filed in *MK Ranjitsinh*. The Supreme Court does not consider whether renewable energy companies also have a duty to protect the environment in which they operate.

This is, in fact, not inconsistent with the treatment of the issue by policy-makers in India. Solar power projects are exempt from the process of environmental impact assessment in India.[80] This has meant that aggrieved residents in areas where wind and solar farms are being established, for example on rural land traditionally used as commons for cattle grazing, are deprived of adequate opportunities to be heard.[81] Some civil society actors in India have rightfully made a case for *ex ante* regulations to assess and address the environmental impact of large-scale renewable energy projects, to be treated on par with the proponents of major infrastructure projects.[82]

The Supreme Court has set a historical precedent and established an important right against the adverse effects of climate change. However, the judgment simultaneously raises questions about whether future judges who apply this right to new cases may be influenced by this absolutist anthropocentrism, support for big businesses, and disregard for concerns linked to biodiversity and the preservation of nature and wildlife. Although India's international commitments to climate change mitigation find extensive discussion in the judgment, its obligations under international treaties focused on biodiversity, such as the Convention on Biological Diversity and the UN Convention on the Conservation of Migratory Species (which includes the great Indian bustard within its scope) were, in contrast, completely ignored.[83] Climate change poses serious threats to biodiversity, and the preservation of endangered spe-

---

[79] *MK Ranjitsinh* (n 1) [17].

[80] Ministry of Environment, Forest and Climate Change (Government of India), *Office Memorandum No J-11013/41/2006-IA.II(I)* (dated 7 July, 2017) <https://environmentclearance.nic.in/writereaddata/public_display/circulars/2YME8DSI_OM%20Solar%20Parks%20dated%207th%20July%202017.pdf>.

[81] Joydeep Gupta, 'Climate Judgement Reveals Nuanced Interplay between Biodiversity and Renewable Energy' (*Mongabay*, 30 April 2024) <https://india.mongabay.com/2024/04/commentary-climate-judgement-reveals-nuanced-interplay-between-biodiversity-and-renewable-energy/>.

[82] Kundan Pandey, 'As Renewable Sector Grows, Need for Regulating It Comes to the Fore' (*Mongabay*, 20 April 2023) <https://india.mongabay.com/2023/04/as-renewable-sector-grows-need-for-regulating-it-comes-to-the-fore/>.

[83] Sinha (n 78).

cies requires greater attention and not neglect. The judgment's championing of renewable energy using a very narrow and uncritical lens is, therefore, short-sighted and lamentable.

6. FAST-FORWARDING TO THE FUTURE: WHERE CAN *MK RANJITSINH* TAKE US?

6.1 Future Climate Litigation in India: Whose Climate Right?

The Supreme Court portrays renewable energy as a silver bullet to address climate change, one with significant socio-economic co-benefits and as the basis for meeting India's international climate obligations. It uses this as the basis to recall the 2021 Order that directed the conversion of overhead power transmission lines to underground lines. However, it ultimately promotes a 'business-friendly green growth agenda' to serve the interests of large-scale energy companies, ignoring the fact that renewable energy projects come at a cost to the land and livelihood of rural communities in India and the energy produced has predominantly benefited urban and industrial users.[84]

A major critique of the Indian Supreme Court's deliberations on the environment, including through a proactive role that often enters the terrain of policymaking, has been that it serves to benefit elite litigants, rather than the most vulnerable sections of society.[85] It has been contended that the judiciary in cases concerning the right to environment tends to protect middle-class interests, 'unlikely to be moved by or on behalf of the poor on "urban poverty" or "livelihood" issues for the outcomes are predictable and unfavourable'.[86]

It has been argued that the Supreme Court of India in many PIL cases has favoured large-scale development projects over environmental causes and has remained insensitive to the rights of poor and disadvantaged members of society.[87] For example, in the capital city of Delhi, privileged urban stakeholders in India are able to utilise PIL in the Supreme Court to project and successfully

---

84 Birsha Ohdedar, 'India's New "Climate Right": Boon or Bane for Climate Justice?' (*Socio-Legal Review*, 10 February 2024) <www.sociolegalreview.com/post/india-s-new-climate-right-boon-or-bane-for-climate-justice>.

85 Prashant Bhushan, 'Supreme Court and PIL: Changing Perspectives under Liberalisation' (2004) 39:18 Economic and Political Weekly 1770, 1772; Geetanjoy Sahu, *Environmental Jurisprudence and the Supreme Court: Litigation, Interpretation, Implementation* (Orient Blackswan 2014).

86 Lavanya Rajamani, 'Public Interest Environmental Litigation in India: Exploring Issues of Access, Participation, Equity, Effectiveness and Sustainability' (2007) 19:3 JEL 293, 302

87 Bhushan (n 77) 1772, 1770.

advance elite causes as matters of public interest to the detriment of poorer groups in society – a phenomenon labelled 'bourgeoisie environmentalism'.[88]

One of the potential pitfalls of the significant climate litigation ruling in *MK Ranjitsinh* is a continuation of this trend, despite its rhetorical espousal of the need to protect vulnerable communities disproportionately affected by the impacts of climate change.

6.2 Climate Legislation and Positive Obligations?

India's climate policy framework comprises various policy documents, many of which predate the Paris Agreement, resulting in a rather disjointed and inadequate framework.[89] The Court in *MK Ranjitsinh* takes note of the fact that India does not, at present, have overarching climate legislation.[90] By noting India's lack of climate legislation and the diverse instruments that form India's climate policy, 'the ruling amplifies the absence of a robust national climate legislation in India'.[91]

In contrast to several rights-based climate litigation judgments in Global North jurisdictions (particularly in Europe) that challenge the efficacy of overarching frameworks of climate policy and climate legislation (thereby being named 'framework litigation' as a substream of climate litigation),[92] there is no comparable overarching legal framework to take to court in India. In contrast to the 'macro' approach taken by framework litigation in certain Global

---

88 Amita Baviskar, 'Cows, Cars and Cycle-Rickshaws: Bourgeois Environmentalists and the Battle for Delhi's Streets' in Amita Baviskar, Raka Ray (eds), *Elite and Everyman: The Cultural Politics of the Indian Middle Class* (Routledge/Taylor & Francis Group 2011) 404, 405.

89 Parul Kumar and Abhayraj Naik, 'India's Domestic Climate Policy Is Fragmented and Lacks Clarity' (2019) 54(7) Economic and Political Weekly <www.epw.in/engage/article/indias-domestic-climate-policy-fragmented-lacks-clarity>.

90 *MK Ranjitsinh* (n 1) [19].

91 Shawahiq Siddiqui, 'SC's Landmark Verdict on Climate Change Sows Seeds, but Stops Short of Delivery on Climate Justice' (*The New Indian Express*, 20 April 2024) <www.newindianexpress.com/xplore/2024/Apr/20/scs-landmark-verdict-on-climate-change-sows-seeds-but-stops-short-of-delivery-on-climate-justice> accessed 2 July 2024.

92 Catherine Higham, Joana Setzer and Emily Bradeen, 'Challenging Government Responses to Climate Change through Framework Litigation' (September 2022) Grantham Research Institute on Climate Change and the Environment and Centre for Climate Change Economics and Policy <www.lse.ac.uk/granthaminstitute/publication/challenging-government-responses-to-climate-change-through-framework-litigation/> accessed 2 July 2024.

North jurisdictions, Indian environmental jurisprudence has typically utilised a 'micro' approach, focusing instead on specific causes and projects,[93] with the judgment in *MK Ranjitsinh* being one such example. Although momentous in its own right for paving the way for future cases to seek the enforcement of a constitutional climate right against the state, a judgment like *MK Ranjitsinh* cannot offer a substitute for climate legislation in India.[94]

India's new constitutional right against the adverse effects of climate change is derived from fundamental rights (the right to life and personal liberty and the right to equality under Articles 21 and 14 of the Constitution, respectively). This right is, therefore, enforceable against the state. The Indian constitutional jurisprudence contains many examples of substantive rights being derived from the right to life under Article 21, which are enforceable against the state. Apart from the right to environment discussed in this chapter, these include, for example, the right to shelter, the right to health, the right to water, the right to food, and the right to privacy.[95] The exact nature and degree of enforcement of these rights against the state has varied in different contexts, a fuller analysis of which is beyond the scope of this chapter.

However, the extent to which these rights have been framed as positive obligations of the state has been a function of judicial interpretation and statutory codification. One example is the right to education. This right was first read into Article 21 of the Constitution by the Supreme Court, but then codified by the legislature as a new fundamental right in the Constitution,[96] in the form of Article 21-A, to provide 'free and compulsory education to all children of the age of six to fourteen years'.[97] This has been characterised as a shift from Article 21 protecting against the state proactively depriving persons of education to the imposition of a positive obligation on the state to provide for education.[98] It has also been argued that the bare text of Article 21 points towards

---

93 Kumar (n 36) 145, 146

94 Kumar and Naik (n 68).

95 For a comprehensive discussion on these rights, see e.g., Anup Surendranath, 'Life and Personal Liberty' in Sujit Choudhry, Madhav Khosla and Pratap Bhanu Mehta (eds), *The Oxford Handbook of the Indian Constitution* (1st edn, OUP 2017); Vrinda Bhandari et al, 'An Analysis of Puttaswamy: The Supreme Court's Privacy Verdict' (*The Leap Blog*, 20 September 2017) <https://blog.theleapjournal.org/2017/09/#gsc.tab=0>; Bhullar (n 46).

96 The evolution of the right to education has been analysed by Gautam Bhatia, 'Directive Principles of State Policy' in Sujit Choudhry, Madhav Khosla and Pratap Bhanu Mehta (eds), *The Oxford Handbook of the Indian Constitution* (1st edn, OUP 2017) and Surendranath (n 95) 768.

97 Constitution of India, 1950, art 21-A.

98 Surendranath (n 95) 768.

a negative right ('[n]o person shall be deprived of his life or personal liberty except according to procedure established by law') but one that has been judicially interpreted to give rise to positive socio-economic rights enforceable against the state.[99] This has been facilitated, in particular, through the incremental infusion of the normative content of the (non-enforceable) DPSP.[100]

As with other expansions of the right to life to bring specific thematic rights within its fold, the contours of the right against the adverse effects of climate change can be expected to be defined in future cases. The extent to which this would be framed as a positive right will also be defined by future litigation and the context of the cases brought before courts. The codification of climate-related obligations of the state in a statutory framework would arguably aid the chances of enforcing this right as a positive obligation of the state.

The judgment in *MK Ranjitsinh* was first made available by the Supreme Court on April 8, 2024. One day later, on April 9, 2024, the European Court of Human Rights delivered a historic judgment in the case of *Verein KlimaSeniorinnen Schweiz and Others v Switzerland*.[101] The Court ruled that Switzerland has a positive obligation to tackle the effects of climate change under Article 8 of the European Convention of Human Rights (ECHR),[102] and in failing to comply with this positive obligation, had violated Article 8 of the ECHR.[103] The Court observed:

> In line with the international commitments undertaken by the member States, most notably under the UNFCCC and the Paris Agreement, and the cogent scientific evidence provided, in particular, by the IPCC (see paragraphs 104–120 above), the Contracting States [to the European Convention of Human Rights] need to put in place the necessary regulations and measures aimed at preventing an increase in GHG concentrations in the Earth's atmosphere and a rise in global average

[99] Nikhil Pratap, 'Conflicting Fundamental Rights Under the Indian Constitution: Analyzing the Supreme Court's Doctrinal Gap' (LLM thesis, Columbia Law School 2022) <https://scholarship.law.columbia.edu/llm_essays_theses/7> 12, 13.

[100] ibid 13; Bhatia (n 96) 660–661.

[101] *Verein KlimaSeniorinnen Schweiz* (n 55).

[102] Convention for the Protection of Human Rights and Fundamental Freedoms (European Convention on Human Rights, as amended) (ECHR), art 8: 'Right to respect for private and family life. 1. Everyone has the right to respect for his private and family life, his home and his correspondence. 2. There shall be no interference by a public authority with the exercise of this right except such as is in accordance with the law and is necessary in a democratic society in the interests of national security, public safety or the economic well-being of the country, for the prevention of disorder or crime, for the protection of health or morals, or for the protection of the rights and freedoms of others'.

[103] See *Verein KlimaSeniorinnen Schweiz* (n 55) [544]–[554], [573].

> temperature beyond levels capable of producing serious and irreversible adverse effects on human rights, notably the right to private and family life and home under Article 8 of the Convention.[104]

In contrast, the Indian Supreme Court in *MK Ranjitsinh* creates a right against/to be free from the adverse effects of climate change but stops short of casting a positive duty on the state in this regard. Theoretically, such a positive obligation could stem from India's international obligations regarding climate change mitigation and adaptation, but domestic legislation further concretising these obligations would undoubtedly strengthen the basis for claiming or enforcing such positive duties. In *MK Ranjitsinh*, the Court notes the absence of a climate legislation framework in India[105] but does not make any prescriptive remarks about the need for one. In contrast to the case in *Verein KlimaSeniorinnen Schweiz*, where the positive obligation to tackle the effects of climate change was directly in issue, *MK Ranjitsinh* crafts a constitutional climate right for human beings that was not connected to the cause brought before it – the preservation of endangered birds. This is another reason why the right pronounced by it could not, in this instance, be connected to positive obligations upon the state in the direct context of the case.

7. CONCLUSION

The creation of a constitutional climate right by the Supreme Court of India in *MK Ranjitsinh* is a landmark moment for climate litigation in India. This chapter has provided an analysis of the contributions and shortcomings of this judgment.

On the one hand, the contribution of the Indian Supreme Court proactively recognising both India's international obligations towards climate change mitigation as well as recognising the human rights dimension of the climate crisis, and 'constitutionalising' it, is undoubtedly very valuable. Even though the Supreme Court has developed rights-based environmental jurisprudence over the past four decades, this is the first time that climate change finds such direct and extensive engagement in India's judicial discourse. For a country that will be significantly impacted by the adverse effects of climate change due to factors such as its geographical location and large population, the articulation of a constitutionally guaranteed fundamental right against the adverse effects of climate change is a significant climate litigation precedent.

On the other hand, the judgment is problematic for its blinkered treatment of the multi-faceted crisis that is climate change. In its single-minded

---

104 ibid [546].

105 *MK Ranjitsinh* (n 1) [19].

championing of renewable energy as the solution to the climate crisis, the Supreme Court is uncritical of the impacts of large-scale renewable energy projects and loses sight of the importance of preserving the endangered bird species that brought this case to court in the first place. Although the Court lays down the fundamental right against the adverse effects of climate change as an extension of the well-recognised constitutional right to environment in India, it does not provide any guidance about its future application and the exact role of the state (against which fundamental rights can be enforced) in guaranteeing this right. This raises questions about who will benefit from this constitutional climate right in the future. Will this right remain the exclusive preserve of big businesses or will it be used as a tool to protect those identified by the Court as most vulnerable to the effects of climate change due to their geographical and socio-economic position? This will remain a question to be assessed based on the future invocation and application of this right by Indian courts. The Supreme Court's recognition of the right to environment has been applied in a multitude of ways, depending on the circumstances of each case. This has also garnered criticism for allowing for too much amplitude for the right to be moulded to fit the specific circumstances of any case, particularly with 'sustainable development' easily used to support both ecocentric and anthropocentric causes. Since the right against the adverse effects of climate change is an extension of the right to environment, future judgments pertaining to this right could also exhibit this trend.

Following this judgment, legal scholars and commentators have embarked upon a rich and vibrant discussion on what this judgment could mean for India's climate change mitigation discourse. In particular, this judgment has opened up conversations on the meaning of climate justice and the need for climate legislation in India. This chapter has tried to engage with some of the literature that has emerged as direct reactions to this judgment. Undoubtedly, an enduring contribution of the judgment has been to spark interest to engage more deeply with the effects of climate change in India and its judicial redress and move the discussion forward in a more progressive direction.

APPENDIX

On 19 December 2025, the Supreme Court of India delivered a judgment with a final set of directions on the conservation of the Great Indian Bustard. In this judgment, the Supreme Court examined the contents of the two reports filed by the expert committee (one each for the states of Gujarat and Rajasthan) and the arguments of various parties concerning those expert committee reports. This appendix focuses on the key findings of the judgment.The Supreme Court considered the recommendations of the expert committee and the arguments of parties to arrive at its final directions, which are found in Paragraph 76 of the judgment. The most important directions are summarised below.

1. The priority areas in the states of Rajasthan and Gujarat, revised as recommended by the expert committee, are to include an additional 850 square kilometres in Rajasthan (for a total of 14,013 square kilometres) and an additional 240 square kilometres in Gujarat (for a total of 740 square kilometres). The Supreme Court also accepted the expert committee's recommendations on measures to monitor the Great Indian Bustard in these revised priority areas, and directed that future renewable energy projects and new overhead power lines within these areas are to be restricted.
2. The Supreme Court accepted the expert committee's recommendations for in-situ and ex-situ conservation of the Great Indian Bustard. These include the restoration, conservation, and consolidation of grassland ecosystems in India; the establishment and improvement of enclosures; improved predator management; jump-starting fertile eggs; GPS tagging of birds; weed removal and grass plantation; food and water management, and community engagement.
3. The Supreme Court accepted the expert committee's recommendation regarding long-term studies on the effects of climate change on the Great Indian Bustard and directed that such studies 'must be conducted'.
4. Power corridors are designated for the routing of overhead power lines. The aim of these power corridors is to restrict the areas within which power lines can be installed, thus reducing the risk to the Great Indian Bustard.
5. Separate measures for different types of power lines are to be adopted depending on location, including rerouting of some power lines, the use of insulated cables, and conversion to underground lines, to mitigate potential impacts on the Great Indian Bustard and its habitat.[106]

106 *MK Ranjitsinh and Others v. Union of India and Others* [2025] INSC 1472 [76].

It is not uncommon for the Supreme Court to appoint expert committees to seek their considered view on a complex issue and to arrive at its final directions based on the committees findings. Undoubtedly, the nature of judicial intervention in such cases can be seen as incursions into the realm of policy-making and raises questions about the separation of powers, a more comprehensive discussion of which is beyond the scope of this appendix.

The Supreme Court's observations on Corporate Sustainable Responsibility (CSR) are another notable feature of the judgment, found at Paragraphs 35–40. Referring to the mandatory CSR provisions in India's Companies Act, 2013, the Supreme Court noted that environmental responsibility is an inherent feature of 'social responsibility' within the CSR concept. The Supreme Court also emphasised the fundamental duty to protect the environment in Article 51A(g) of the Constitution, and drew a link between this constitutional provision and the Polluter Pays Principle. The Supreme Court found that in the context of activities such as mining, power generation, and infrastructure, which threaten the habitat of endangered species, corporate responsibility necessarily includes conservation efforts. The Supreme Court concluded with this observation in paragraph 40:

> *[t]he non-renewable power generators operating in the priority as well as non-priority areas in Rajasthan and Gujarat must always remember that they share the environment with the Godawan [the local name for the Great Indian Bustard], the Great Indian Bustard and must undertake their activities as if they are guests in its abode.*[107]

[107] ibid [40].

Index

Printed and bound by CPI Group (UK) Ltd, Croydon, CR0 4YY

17/06/2026

14902649-0004